The Inorganic Radiochemistry of Heavy Elements

Ivo Zvára

The Inorganic Radiochemistry of Heavy Elements

Methods for Studying Gaseous Compounds

Ivo Zvára
Joint Institute for Nuclear Research
Dubna
Russian Federation

ISBN 978-1-4020-6601-6   e-ISBN 978-1-4020-6602-3

Library of Congress Control Number: 2007940367

All Rights Reserved
© 2008 Springer Science + Business Media B.V.
No part of this work may be reproduced, stored in a retrieval system, or transmitted in any form or by any means, electronic, mechanical, photocopying, microfilming, recording or otherwise, without written permission from the Publisher, with the exception of any material supplied specifically for the purpose of being entered and executed on a computer system, for exclusive use by the purchaser of the work.

Printed on acid-free paper

9 8 7 6 5 4 3 2 1

springer.com

To my wife, Tamara

Contents

Preface xi

Symbols and Abbreviations xiii

Introduction xix
        Chapter Synopsis xxi
        Terms xxiii
        Chemical Character of the Transactinoid Elements xxvi
    References xxvii

1 Experimental Developments in Gas-Phase Radiochemistry 1
  1.1 Early Gas-Solid Chromatography Studies 1
  1.2 Techniques for Isolation of Short-lived Accelerator
      Produced Nuclides 4
      1.2.1 Off-line Simulation with Recoiling Fission Products 4
      1.2.2 On-line Experiments with Spontaneously Fissioning
           Nuclides 5
  1.3 Techniques for α-active Nuclides: Corrosive Reagents 9
      1.3.1 Relative Merit of Isothermal- and Thermochromatography . . 12
  1.4 Techniques for α-active Nuclides: Non-corrosive Reagents 14
      1.4.1 Thermochromatography of Hassium Tetroxide 14
      1.4.2 Chemical Identification of Metallic Element 112 16
  1.5 Prospects for Future of Radiochemical Studies of Heavy Elements . 18
      1.5.1 Classes of Compounds 18
      1.5.2 Groups of Related Elements 23
  References 30

2 Physicochemical Fundamentals 35
  2.1 Molecular Kinetics 36
      2.1.1 Concentration and Speed of Gaseous Molecules 36
      2.1.2 Number of Collisions with Wall 37

		2.1.3	Collisions in Gas and Rate of Chemical Interactions	38
		2.1.4	Diffusion in Gases	40
		2.1.5	Elementary Adsorption–Desorption Event	42
		2.1.6	Integrals Containing Boltzmann Factor	42
	2.2	Diffusional Deposition of Particles in Channels		44
		2.2.1	Diffusion Coefficients of Aerosols	44
		2.2.2	Deposition from Laminar Flow	45
		2.2.3	Diffusional Deposition — Engineering Approach	48
	References			51
3	**Production of Transactinoid Elements, Synthesis and Transportation of Compounds**			53
	3.1	Production of the Elements by Heavy Ion Accelerators		54
		3.1.1	Recoil Separation from Targets	56
		3.1.2	Thermalizing Recoils	56
	3.2	Rapid Synthesis of Volatile Compounds		60
		3.2.1	Experimental Findings on Kinetics	62
		3.2.2	Thermochemistry and Kinetics — Chlorination in Gas	65
		3.2.3	Synthesis of (Oxy)chlorides of Group 4 and 6 Elements	67
		3.2.4	Chlorination in the Adsorbed State	70
		3.2.5	Chemistry on Hot Aerosol Filters	72
	3.3	Scavenging of Gaseous Chemically Active and Radioactive Impurities		73
		3.3.1	Removing Water and Oxygen	73
		3.3.2	Chemical Filter After the Target Chamber	74
		3.3.3	Diffusional Deposition of Nonvolatile Species in Gas Ducts	75
		3.3.4	Deposition of Heat	78
	3.4	Transportation of Molecular Entities by Aerosol Stream		79
		3.4.1	Optimal Parameters of Aerosol	80
		3.4.2	Peculiarities in Aerosol Transportation of Short-lived Activities	82
	References			84
4	**Gas–Solid Isothermal and Thermochromatography**			87
	4.1	Characteristics of Methods		87
	4.2	Theory		89
		4.2.1	Ideal Isothermal Chromatography	89
		4.2.2	Ideal Thermochromatography	91
		4.2.3	Shapes of Chromatographic Peaks	93
	4.3	Mathematical Modeling of Gas–Solid Chromatography		100
		4.3.1	Monte Carlo Simulation of Individual Molecular Histories	101
		4.3.2	Calculational Procedure	104
		4.3.3	Sample Results of Simulations	106
	4.4	Vacuum Thermochromatography		112
		4.4.1	Retention Time	112

	4.4.2	Description by Random Flights 114
	4.4.3	Monte Carlo Simulation 116
References 117

5 Evaluation and Interpretation of the Experimental Data 119

- 5.1 Adsorption Enthalpy on Homogeneous Surface 120
  - 5.1.1 Thermodynamic Approach 121
  - 5.1.2 Experimental Values from Second Law 126
  - 5.1.3 Quasi Third Law Approach – Entropy from Statistical Mechanics 128
- 5.2 Adsorption Enthalpy from Thermochromatographic Experiments .. 135
  - 5.2.1 Basic Equations 136
  - 5.2.2 Third Law-based Results for Halides 137
- 5.3 Real Structure of Column Surfaces 139
  - 5.3.1 Geometrical and Chemical Structure of Fused Silica Surface 141
  - 5.3.2 Silanols and Siloxanes on Silica Surface 148
  - 5.3.3 Modification of Silica Surface by Haloginating Reagents ... 155
  - 5.3.4 Morphology of Metal Surfaces 157
  - 5.3.5 Modification of Metal Surfaces......................... 158
- 5.4 Lateral Migration of Adsorbate 159
  - 5.4.1 Surface Diffusion 159
  - 5.4.2 Surface Diffusion and Entropy of Adsorbate.............. 162
- 5.5 Evaluation of Adsorption Enthalpies on Real Surfaces 165
  - 5.5.1 Thermodynamic Parameters of Adsorption on Heterogeneous Surface 167
  - 5.5.2 Adsorption Entropy on Heterogeneous Surfaces with Surface Diffusion 169
- 5.6 Revised Approach to Interpretation of the Data on Transactinoid Halides 171
  - 5.6.1 Microscopic Picture of the Modified Silica Surface 171
  - 5.6.2 Rationale for the Correlation of Adsorption and Sublimation Energies 172
  - 5.6.3 Required New Experimental Data 177
  - 5.6.4 Real Picture of Adsorption and Monte Carlo Simulations ... 180
- 5.7 Non-trivial Mechanisms in Gas-Solid Chromatography 180
  - 5.7.1 Dissociative Adsorption – Associative Desorption 181
  - 5.7.2 Associative Adsorption – Dissociative Desorption 183
  - 5.7.3 Substitutive Adsorption – Substitutive Desorption 183
  - 5.7.4 Physical Adsorption – Substitutive Desorption 184
  - 5.7.5 Existence of Yet Unknown Compounds.................. 187
- References 187

6 Validity and Accuracy of Single Atom Studies 191
    6.1   Validity of Single Atom Chemistry 191
            6.1.1   Monte Carlo Simulation of Single Atom Experiments 192
            6.1.2   Theoretical Kinetic Limits 194
            6.1.3   Equivalent to Law of Mass Action 194
            6.1.4   More Considerations 195
    6.2   Analysis of Poor-Statistics Data 196
            6.2.1   Bayesian Approach to Statistical Treatment 197
            6.2.2   Half-life from Fraction of Decay Curve 202
            6.2.3   Adsorption Enthalpy from IC Experiment 204
            6.2.4   Adsorption Enthalpy from TC Experiment 208
            6.2.5   Adsorption Enthalpy from Corrupted
                     Thermochromatogram 209
            6.2.6   Conclusions 211
    References 212

Author Index 215

Subject Index 219

Preface

Throughout my life's work in science I have been greatly influenced by the standing problem of synthesis and studies of the heaviest chemical elements. In 1960 I joined the then-young Laboratory of Nuclear Reactions of the Joint Institute for Nuclear Research at Dubna. It was headed by G. N. Flerov who, with K. A. Petrzhak, discovered the spontaneous fission of uranium. The laboratory was equipped with a powerful cyclotron which could accelerate boron and heavier ions to energy of some 10 MeV per nucleon. A most ambitious goal was to discover new chemical elements. The first "planned" new nuclide, $^{260}104$, was expected to be produced by the bombardment of ^{242}Pu with ^{22}Ne. Estimates of its half-life were very uncertain, spanning many orders of magnitude. Necessarily, the initial emphasis was on physical methods of identification of the atomic and mass numbers because, in general, the physical techniques are effective down to very short lifetimes. On the other hand, element 104 was also of great interest for chemists. It was expected to be the first "transactinoid," resembling in its properties hafnium, the first "translanthanoid." As such it would strongly differ in chemical properties from all the lighter transuranium elements. This might facilitate and accelerate its chemical identification, which is an independent reliable method for the assignment of the atomic number and could eventually strengthen the primary physical evidence. The chemical identification of element 104 was the first task I got involved in. It was soon recognized that, with the availability of only one short-lived atom at a time, the processing of the accelerator bombardment products must be continuous and allow immediate chemical transformation of the new atom, once created. The goal was to achieve this, as well as the subsequent chemical isolation of the new molecules, in less than a second, which was the optimistic higher limit of $t_{1/2}$. Also required was highly efficient detection of the decay events of element 104 because the expected production rate was, by orders of magnitude, smaller than for any previous element. The more unusual was the combination of all these musts. The existing exclusively batchwise isolation techniques for hafnium and most other metallic elements took at least minutes to accomplish.

Our team did not see prospects of achieving the goal by simply upgrading the existing methods. In those times An. N. Nesmeyanov, head of the Chair of

Radiochemistry at the Moscow University, consulted the Flerov's laboratory in Dubna on radiochemical problems. He pointed to the expected considerable volatility of higher halides of the transactinoid, compared with that of similar compounds of actinoids, as a possible basis of fast separations. When seeking an experimental method which would make the most of the dissimilar volatility, I benefited from the experience and ideas I gained as a student of Professor Nesmeyanov. In his laboratory I separated various volatile brominated methanes to solve a problem in "hot atom chemistry." After a few years our small group of chemists did come with an efficient technique capable of isolating hafnium as tetrachloride in tenths of a second. The method combined the principles of hot atom chemistry and gas-solid chromatography. We successfully applied it to element 104 and subsequent transactinoids. A generation later, around 1990, other world laboratories involved in transactinoid studies also started experiments with gaseous compounds. Fortunately, all the transactinoid elements up to $Z = 118$ must either be volatile in elemental state or form some characteristic volatile compound(s), so that the gas phase techniques are a universal research tool in radiochemistry of the transactinoid elements.

The aim of this book is to outline and analyze some fundamental aspects of the work performed at Dubna and elsewhere, and to discuss prospects for the future.

My sincere thanks go to my colleagues: V. Z. Belov, Yu. T. Chuburkov, V. P. Domanov, B. Eichler, S. Hübener, M. R. Shalaevskii, L. K. Tarasov, A. B. Yakushev, B. L. Zhuikov, T. S. Zvarova – my wife, and others. Together we pioneered and conducted transactinoid studies as well as tried to analyze the fundamental aspects of what we were doing – the gas phase radiochemistry of metallic elements. We were a small group of chemists embedded in a large physical laboratory. Hence, it was of decisive importance for us that the late Prof. G. N. Flerov put much emphasis on the role of chemical identification of new elements. He actually initiated, and then invariably supported, radiochemical studies in the Dubna laboratory.

Symbols and Abbreviations

Symbols and abbreviations which are defined and then used only within a page or a short section are not listed.

Symbols

A	standard state molar area of adsorbed tracer, cm^2 mol^{-1}
a_m	dimension of molecular potential box, cm
a_z	column surface area per unit length, cm^2 cm^{-1}
C_a	surface concentration of adsorption sites, cm^{-2}
c_a	surface concentration of tracer molecular entity, cm^{-2}
c_g	gas phase concentration of tracer molecular entity ($\equiv n_1$), cm^{-3}
c_p	number concentration of aerosol particulates, cm^{-3}
c_C	Cunningham slip correction factor
D	shorthand for $D_{1,2}$, if obvious, cm^2 s^{-1}
$D_{1,2}$	coefficient of mutual diffusion in gas, cm^2 s^{-1}
D_a	coefficient of surface diffusion, cm^2 s^{-1}
D_{Bn}	Brownian diffusion coefficient (aerosols), cm^2 s^{-1}
D_{Kn}	effective diffusion coefficient in evacuated tubes, cm^2 s^{-1}
d_c	diameter of cylindrical channel or equivalent, cm
d_m	diameter of a molecule, cm
$d_{m,1}, d_{m,2}$	like above, for tracer and bulk gas, respectively, cm
E_{act}	molar activation energy, J mol^{-1}
E_b	molar energy of surface diffusion barrier, J mol^{-1}
E_d	molar desorption energy, J mol^{-1}
E_d^{min}	minimal molar desorption energy in spectrum, J mol^{-1}
E_d^{max}	maximal molar desorption energy in spectrum, J mol^{-1}
E_d^{het}	effective molar desorption energy for heterogeneous surface, J mol^{-1}

$Ei(x)$	integral exponential function of x
$F_p(z)$	penetration of irreversibly deposited particles through channel of length z
$F_p(z_\psi)$	penetration as function of channel reduced length $z\psi$
$F_p^\circ(z_\psi)$, $F_p^\square(z_\psi)$	particular penetration functions for circular and rectangular channels
$F_{pos}(\Theta\|k)$	Bayes posterior distribution
$F_{pri}(\Theta)$	Bayes prior distribution
g	temperature gradient of the linear column temperature profile, K cm^{-1}
h	Planck constant, 6.626×10^{-27} erg s
I	geometrical mean of the molecule rotational moments of inertia, g cm^2
K	thermodynamic adsorption equilibrium constant
K_c^{lc}	concentration equilibrium constant – localized adsorption
K_c^{mb}	concentration equilibrium constant – mobile adsorption
K_p^{lc}	partial pressure equilibrium constant – localized adsorption
K_p^{mb}	partial pressure equilibrium constant – mobile adsorption
k	data in likelihood formula
k_a	distribution coefficient in adsorption ($= c_a/c_g$), cm
k_a^{lc}	distribution coefficient in localized adsorption, cm^3
k_B	Boltzmann constant, 1.381×10^{-16} erg $\cdot$ K^{-1}
k_m	mass transfer coefficient, cm s^{-1}
$k_{r1,2}$	rate constant of chemical reaction involving tracer
L_N	Loschmidt number 2.687×10^{19} cm^{-3}
$L(k\|\Theta)$	likelihood function
l_c	length of open, constant geometry channel or column, cm
M	molar mass, g mol^{-1}
m	mass of molecule (or m_1, if evident), g
m_1	mass of molecule of tracer, g
m_2	mass of molecule of bulk gas, g
$m_{1,2}$	reduced mass of two colliding molecular entities, g
$\tilde{m}_n$	the n'th moment of pdf
$\bar{m}_n$	the n'th central moment of pdf
N_A	Avogadro number, 6.022×10^{23} mol^{-1}
N_a	mean number of displacements and adsorptions in simulated sequence
n_p	number concentration of aerosol particles, cm^{-3}
n	number concentration of molecules, cm^{-3}
n_1	number concentration of molecules of tracer ($n_1 \equiv c_g$), cm^{-3}
n_2	number concentration of molecules of bulk gas, cm^{-3}
$n^{(2)}$	number concentration of two-dimensional gas, cm^{-2}
n_a	number of molecules striking unit surface per unit time, cm^{-2} s^{-1}
n_{rot}	number of rotational degrees of a molecule

Symbols and Abbreviations

P	steric factor in binary reaction
p	pressure of three-dimensional gas, dyn cm^{-2}
$p^{(2)}$	pressure of two-dimensional gas, dyn cm^{-1}
p_0	arbitrary reference pressure, dyn cm^{-1} or bar
Q	volume flow rate, cm^3 s^{-1}
Q_0	flow rate at arbitrary reference p_0 and T_0, cm^3 s^{-1}
q_{int}	partition function of a free molecule
q_{el}, q_{rot}, q_{vib}	electronic, rotational, and vibrational components of q_{int}
$q_{tr3}, q_{tr2}, q_{tr1}$	translational partition functions per molecule in boxes of appropriate dimensions
R	universal gas constant, 8.315×10^7 erg mol^{-1}K^{-1} or 8.315 J mol^{-1} K^{-1}
$\mathfrak{R}$	velocity of peak relative to mean gas velocity w
r_λ^{IC}	fraction of radioactive nuclei surviving at exit of IC column
r_p	radius of aerosol particulate, cm
S	molar entropy, J mol^{-1} K^{-1}
S_{cf}	molar configurational entropy of localized adsorbate, J mol^{-1} K^{-1}
S_{int}	molar entropy of internal degrees of freedom, J mol^{-1} K^{-1}
S_{tr2}, S_{tr3}	molar translational entropy of two- and three-dimensional gases, J mol^{-1} K^{-1}
S_a^{lc}	molar entropy of localized adsorbate including S_{int}, J mol^{-1} K^{-1}
S_a^{mb}	molar entropy of mobile adsorbate including S_{int}, J mol^{-1} K^{-1}
S_{vib}^{mb}	molar vibrational entropy of mobile adsorbate, J mol^{-1} K^{-1}
S_{vib}^{lc}	molar vibrational entropy of localized adsorbate, J mol^{-1} K^{-1}
T	thermodynamic temperature, K
T_0	arbitrary reference temperature, K
T_A	temperature at peak of TC zone, K
T_A^{id}	temperature of ideal TC zone, K
T_S	temperature of TC column hot end, K
T_c	temperature of IC column, K
T_{mp}	melting point of a solid, K
T_z	temperature as function of z in TC column, K
t	running time, s
$t_{1/2}$	half-life of radioactive decay, s
t_f^*	random processing time available for particular molecule, s
t_g^{IC}	gas hold-up time in IC column (transport channel), s
t_g^{TC}	gas hold-up time in TC column, s
t_R^{IC}	net experiment duration (or retention) time in IC, s
t_R^{TC}	net experiment duration in TC, s
t_R^{VTC}	duration of VTC experiment, s
t_λ	mean lifetime towards radioactive decay, s
u_m	mean speed of gaseous molecules, cm s^{-1}

$u_m^{(2)}$	mean speed of two dimensional gas molecules, cm s^{-1}
$u_{1,2}$	mean relative speed of tracer and carrier gas molecules, cm s^{-1}
V	standard state molar volume, cm^3 mol^{-1}
v_z	free volume per unit length of column, cm^3 cm^{-1}
w	linear velocity of carrier gas, cm s^{-1}
w_0	linear velocity of gas at arbitrary reference p_0 and T_0, cm s^{-1}
w_{ps}	velocity of gravitational settling of particles, cm s^{-1}
x, y	general (or ad hoc) quantity or variable
Z	molar partition function
Z_{tr}	translational molar partition function
$Z_{1,2}$	collision rate of tracer molecule in carrier gas, s^{-1}
Z_z	mean number of collisions of molecule with surface when passing a volume (column of length z)
z	linear coordinate along column or channel, cm
z_ψ	reduced coordinate ($\equiv \psi z$)
z_A	mean coordinate of adsorption zone (first moment of pdf), cm
z_S	coordinate of the sample injection in TC column, cm
z_D^2	mean square diffusional displacement in time t, cm^2
$\bar{z}_G$	parameter of exponentially modified Gaussian (Eq. 4.21), cm
α	confidence level in traditional or Bayesian statistics, %
α_i, α_j	numerical coefficient in equations of Sect. 2.2.2
β_i, β_j	numerical coefficient in equations of Sect. 2.2.2
γ	parameter of exponential column temperature profile, K cm^{-1}
γ_T	heating rate in temperature programmed chromatography, K s^{-1}
$\Delta_{ads}H$	molar enthalpy change in adsorption, J mol^{-1}
$\Delta_{des}H$	molar enthalpy change in desorption, J mol^{-1}
$\Delta_r H$	molar enthalpy change in reaction, J mol^{-1}
$\Delta_r G$	standard Gibbs free energy change in chemical reaction, J mol^{-1}
$\Delta_{ads}S$	molar entropy change in adsorption, J · mol^{-1}K^{-1}
$\Delta_{des}S$	molar entropy change in adsorption, J · mol^{-1}K^{-1}
$\delta(x)$	Dirac's delta function
ε_b	diffusion barrier between adsorption sites, erg
ε_d	desorption energy per molecule, erg
ε_z	kurtosis excess of a peaking pdf or experimental profile
ς	mean molecular flight length in VTC, cm
ς_z	mean projection of molecular flight length on z axis in VTC, cm
η	parameter of exponential distribution of a length, cm
η_i	length parameters in equations of Sect. 2.2.2, cm
$\eta_i^\circ, \eta_i^\square$	parameters η_i for circular and rectangular channels
η_e	length parameter of exponentially modified Gaussian in Eq. 4.21, cm
η_0	mean deposition length in developed laminar or turbulent flow, cm
Θ	parameter in Bayesian statistics
θ^*, φ^*	random polar coordinates

Symbols and Abbreviations

θ	fractional coverage of surface
θ^o	coverage of surface in the standard state
Λ_m	de Broglie wavelength of molecule, cm
λ	constant of radioactive decay, s^{-1}
λ_m	mean free path length of a molecule, cm
$\lambda_{m,1}$	mean free path of a tracer molecule, cm
μ_2	dynamic viscosity of carrier gas, g cm^{-1} s^{-1}
μ_f	dynamic viscosity of fluid, g cm^{-1} s^{-1}
ν_0	frequency of vibrations of adsorbate normal to the surface ($\equiv 1/\tau_0$), s^{-1}
ν_x, ν_y	frequencies of vibrations of adsorbate in the surface plane, s^{-1}
ν_{xy}	geometric mean of ν_x and ν_y, s^{-1}
ν	geometric mean of ν_0, ν_x, and ν_y, s^{-1}
ξ^*	random number variable uniformly distributed in $[0,1)$
ρ	density of solid or liquid, g cm^{-3}
ρ_1, ρ_2	densities of condensed tracer or carrier gas, g cm^{-3}
ρ_p	density of particulate matter, g cm^{-3}
$\rho(y^*)$	general notation for pdf of random values with the mean y
$\rho(z)$	normalized peaking curve, in particular, chromatographic zone profile, cm^{-1}
$\rho^{IC}(z), \rho^{IC}(T_z)$	normalized profile of adsorption zone in IC, cm^{-1} or K^{-1}
$\rho^{TC}(z), \rho^{TC}(T_z)$	normalized profile of adsorption zone in thermochromatography, cm^{-1} or K^{-1}
Σ_z	skewness of peaking pdf or of experimental profile
σ_y^2	variance of pdf of y
σ_z^2	variance of chromatographic peak, cm^2
σ_G^2	variance of Gaussian peak, cm^2
σ_D^2	variance of IC peak due to longitudinal diffusion, cm^2
$\sigma_\Re^2$	variance of IC peak due to laminar flow patterns and retention, cm^2
τ_0	elementary adsorption sojourn time ($\equiv 1/\nu_0$), s
τ_a	mean adsorption sojourn time (mobile adsorption), s
τ_d	mean desorption time of localized molecule, s
τ_{Ns}	mean residence time in random series of adsorptions, s
τ_r	average time of chemical interaction of a molekule, s
$\phi_p(z_\psi)$	density distribution of irreversible deposit
$\phi_p^\circ(z_\psi), \phi_p^\square(z_\psi)$	density distribution of irreversible deposit in channels of particular cross sections
$\phi_p^\circ(z), \phi_p^\square(z)$	density distribution of irreversible deposit in channels of particular cross sections, cm^{-1}
$\psi, \psi^\circ, \psi^\square$	reduced distance parameters, proportional to D/Q, cm^{-1}
$\omega_{1,2}$	collision diameter $\omega_{1,2} \equiv (d_{m,1}, +d_{m,2})/2$, cm

Abbreviations

An	any actinoid
BI	Bayesian interval
CI	confidence interval
EOB	end of bombardment
EVR	evaporation residue
FWHM	full width at half maximum
IC	isothermal chromatography, -phic
MC	Monte Carlo
i.d.	internal diameter
LET	linear energy transfer
Ln	any lanthanoid
pdf	probability density function
pµA	particle microampere
PTFE	polytetrafluoroethylene
s.f.	spontaneous fission, spontaneously fissioning
SHE	superheavy element
SSTD	solid-state track detectors
STP	standard temperature and pressure: 273.15 K; 1 bar ($= 10^6$ dyn cm^{-2})
TAE	transactinoid element
TC	thermochromatography, -phic
VTC	vacuum thermochromatography

Subscripts

$\mid_1$	related to tracer
$\mid_2$	related to carrier gas
$\mid_d$	related to desorption
$\mid_m$	per molecule value of quantity

Superscripts

$\mid^\circ$	related to the standard state specified by Eq. 5.39
$\mid^*$	random value of the quantity with the symbol as the mean value
$\mid^{het}$	related to heterogeneous surface
$\mid^{lc}$	related to localized adsorption model
$\mid^{mb}$	related to mobile adsorption model
$\mid^{opt}$	optimized value of particular quantity

Introduction

Abstract Volatility of elements and compounds has been occasionally used for radiochemical separations since the studies of Mme. Curie. Steady progress in nuclear sciences and technologies posed new problems in separation of mixtures of radionuclides. In the 1960s radiochemists paid attention to numerous volatile halides and oxyhalides of metallic elements. These compounds then helped in discovery and studies of the first, short-lived transactinoid elements. Needs of this particular field continue to motivate and stimulate further development. In this book, the author will attempt a coherent look on the physicochemical background of all steps involved in the experiments. The emphasis will be on gas-solid chromatography, which provides the required data on volatile compounds. So far, conventional interpretation of the adsorption data in terms of thermodynamic characteristics has ignored heterogeneity of real column surfaces and their modification by the employed gaseous reagents. It necessitates a major revision of some conclusions from past studies. The Introduction, besides others, specifies some terms used in the book, especially in view of the problems of single atom chemistry. A schematic illustrates the place of the heaviest elements in the Periodic Table and the classes of volatile species, which are used for their chemical identification and further characterization.

A radiochemical experiment involves two principal steps. The first is obtaining the desired compound. Second, generally, is isolation of the compound. It may mean the isolation from a bulk matrix, separation of a complex mixture of radionuclides, or both. The separations are behind both applied and fundamental radiochemistry studies, which is the major emphasis in this book.

Already in the early days of radiochemistry some radionuclides were isolated from matrices and their mixtures were separated, making use of different volatility of various elements and compounds. Well-known is the role of the extreme volatility of radon in the discovery of "emanations" by Dorn and Rutherford (see a detailed story in Reference [1]). In her logbooks Mme. Curie noted purification of polonium by sublimation, when collecting deposits obtained at different temperatures [2]. After the discovery of nuclear fission, the volatile species — Kr and Xe in the elemental state, As and Sb as gaseous AsH_3 and SbH_3, as well as Ru in the state of RuO_4 — were often used for isolating these elements from mixed fission products. It was noted that the separations can be made quickly. Such early works were covered in review papers and reports [3–5].

In the beginning of the 1960s, Merinis and Boussieres [6] faced the task of separating products of spallation reactions. They pioneered the gas-phase separations of transition metals that are capable of forming highly to moderately volatile higher (oxy)halides. Successful employment of such an abundant class of compounds, albeit easily hydrolyzable, was an important achievement. These authors also introduced a novel variant of gas-solid chromatography — the thermochromatography, which is carried out in columns with downstream negative temperature gradient. The gaseous species of different adsorbability get practically irreversibly deposited in different temperature ranges along the column, thus yielding a preparative chromatogram. The latter possibility is an advantageous feature for application in radiochemistry: the isothermal elution gas chromatography, even though providing better resolution, requires much more time and rather sophisticated techniques for collecting the separated fractions, and for subsequent radiation measurement. Soon after, various authors occasionally employed thermochromatography to separate some fission products, both metals and nonmetals, as chlorides. All these experiments dealt with mixtures of relatively long-lived activities and were done by batch processing. Fast performance or high chemical yield was not of major concern.

In the mid-1960s, making use of volatile compounds and of gas chromatographic techniques, the present author with co-workers at JINR, Dubna first solved a new, most difficult radiochemical problem — "on-line" chemical identification and further studies of transactinoids, the elements with atomic numbers larger than 103. The latter can be produced only at accelerators of "heavy ions" (particles heavier than He) by the bombardment of mostly radioactive actinoid targets. They are obtained as single and very short-lived atoms — on the "one-atom-at-a-time" scale. The solution was found by combining the following principles:

- The use of "thin" solid targets for prompt extraction of the bombardment products owing to their recoil energy
- Thermalizing the recoils in a gas flowing through the "target chamber" to enable their fast interaction with gaseous reagents
- Feeding the necessary gaseous reagents into the target chamber or next to its exit to realize in situ "chemical volatilization"
- The use of the beneficial properties of gases for rapid transportation of the volatile species to a separation equipment and radiation detectors
- Isolation of the new element by a gas-solid chromatography technique

The JINR group had been alone in the field of gas-phase chemistry of transactinoid elements until the 1990s. From the very beginning they felt it necessary to provide at least limited physicochemical background for the initially empirical conclusions. Otherwise, one could not be sure that the regularities disclosed in the test runs would also be valid for yet unknown elements. One of the goals of this book is to survey, generalize and discuss the present knowledge, as well as ignorance, about the fundamentals of the gas-phase techniques, which are used in inorganic

Introduction

radiochemistry of metallic elements. The time sequence of the physicochemical problems which attracted major interest was determined mostly by the progress in transactinoid experiments. This, in turn, was greatly influenced by the changing chemical character of the transactinoids with higher atomic number. For these reasons the book starts with a chapter, which quotes some important experimental studies.

On the other hand, one cannot be satisfied with successful separations or with understanding the comparative data on the adsorption behavior of new elements and their lighter congeners. A much more exciting, as well as difficult, goal is to translate the data (e.g., adsorption enthalpies) on the behavior of a few molecules into the characteristics of bulk volatility of the compounds. To that end, it is necessary to go far beyond empirical facts and apply, whenever possible, some theoretical concepts.

Chapter Synopsis

The author attempts to present a coherent, and in some respects novel, view of the processes behind the experimental methods of gas-phase radiochemistry. It requires considering the microscopic picture of the behavior and history of the atoms and molecules; revealing the situations when the rate of chemical interactions may be more important than thermodynamics; taking into account the heterogeneous structure of real surfaces and lateral diffusion of adsorbates; and paying due attention to chemical modification of the surfaces by the employed gaseous reagents. As a rule, the microscopic picture of the adsorption–desorption events is as follows: upon striking the surface, the particle adsorbs on a particular site; it migrates laterally, visiting a number of other adsorption centers and, as a result, it desorbs from a different site, which is characterized by dissimilar adsorption energy. In literature, when evaluating the experiments, this complicated picture has been mostly oversimplified through too crude and sometimes even unsounded assumptions. The latter cases call for some criticism.

Chapter 1 presents some details of the experimental developments which were important steps forward. It also briefly describes a few typical concrete studies of the transactinoid elements. The emphasis is on novel approaches, techniques and classes of the utilized compounds, as well as on posing and solving problems. Consistently, the presentation tends to comply, if possible, with both the historical and logical sequences of the events. Closing sections of the chapter consider various classes of inorganic compounds, other than the widespread chlorides, as to their prospects for future studies.

Chapter 2 calls some fundamental laws and formulae of physical chemistry that are relevant to the content of the book. It deals with diffusion and reactions in gases, as well as with adsorption upon collisions with surfaces, emphasizing the molecular level. It provides some formulae for estimating molecular properties of uncommon compounds. Formulae describing irreversible diffusional deposition of molecular entities and aerosols from flowing gas on the walls of channels are presented quite extensively.

Chapter 3 is devoted to step-by-step discussion of the history of a newborn radioactive nucleus, which is included in a molecule and then fed into a chromatographic column. It considers the production reaction, recoil-enhanced separation from the target, the conditions of thermalizing the recoils, halogenation and the transportation by gas or aerosol flow. Some non-trivial peculiarities of the delivery by aerosol flow are mentioned.

Chapter 4 starts with some basic equations, which relate the molecular-kinetic picture of gas-solid chromatography and the experimental data. Next come some common mathematical properties of the chromatographic peak profiles. The existing attempts to find analytical formulae for the shapes of TC peaks are subject to analysis. A mathematical model of migration of molecules down the column and its Monte Carlo realization are discussed. The zone position and profile in vacuum thermochromatography are treated as chromatographic, diffusional and simulation problems.

Chapter 5 deals with the evaluation of desorption enthalpies from the experimental data using Second Law-like approaches, and through calculating adsorption entropy for the mobile model of adsorption on an ideal surface from first principles. The data obtained so far have pointed to a similarity between adsorption enthalpies of gaseous compounds (on silicas) and desublimation enthalpies of their bulk quantities. The search for the rationale of the unexpected finding drew attention to the actual surface structures of the adsorbents in use – their heterogeneity and modification by the employed gaseous reagents, which is extensively reviewed for the purpose. Because such parameters strongly affect the real picture of the adsorption state and the choice of its thermodynamic model, the approach to evaluating the results had to be greatly revised. It showed that the former prescriptions probably underestimated the actual values. Briefly considered are the eventual non-trivial mechanisms involved in the chromatographic processing, other than simple adsorption–desorption events of unaltered molecules.

Chapter 6 quotes some principal problems of the validity of the conclusions drawn from the experiments with only a few atoms. The last sections witness the advantages of using the Bayesian approach for the treatment of the data with poor statistics, in particular, in radiochemical experiments. When combined with Monte Carlo simulations of the experiment, the approach can easily account for peculiar situations like unsteady physicochemical conditions and performance of the radiation detectors during long runs on accelerator beam. It is demonstrated by concrete examples.

The material of this book refers to various fields of chemistry, to some mathematics and to numerous experimental works. Therefore, it was not the aim to present an exhaustive list of the appropriate references. Preferably cited are the works that seemed stimulating, representative and thorough. A few interesting older studies concerning the methods were published only in Russian and are not easily accessible; here are presented some details of them. Also included are some yet unpublished results obtained by the author and his co-workers, which contribute to the analysis of the published data and of the conventional concepts.

This book is mainly about the foundations of the experimental methods, and about the evaluation and interpretation of the raw experimental data. For this reason, in particular, only a fraction of the concrete results obtained in the one-atom-at-a-time studies of the very heaviest elements is mentioned. The available data on the concrete properties of the transactinoid elements and superheavy nuclides, as well as the appropriate theories, have been reviewed in several papers [7, 8], multiauthor books [9, 10] and proceedings of dedicated conferences [11–13]. Preparation of an experiment with TAE includes estimation of the adsorption properties of the expected TAE compound. It is based on the estimations of thermochemical properties. The problem is not directly related to the content of this book, however important and interesting it may be. A technology of such estimates is presented in [14]; see also references therein. These empirical estimations may be inaccurate if the anticipated relativistic effects in the chemical properties of TAEs are strong. A recent review of the long-standing theoretical studies on this subject has been presented by Pershina [15].

Terms

The content of this book refers to several different fields of chemistry and physics. It seems necessary to specify the meaning of some terms used throughout this book for the purpose.

Trace, tracer and derived terms here consistently imply that the concentration of the element or compound is so low that it makes the collisions between two identical tracer entities in the course of a real experiment (i.e., limited in duration) completely improbable. In practice, it means sub-microgram quantities of the element, that is, the range from commercial "carrier-free" radionuclides down to single atoms. The behavior of elements must not depend on the available quantity in this range. Such definition of the tracer scale is somewhat truncated from the higher side – usually, the term extends until concentrations at which infrequent encounters can occur. On the other hand, from the lower side, here the definition also comprises single atoms. This slight shift of the conventional scale will help us to avoid repeating statements about the limitations of some conclusions.

Chemical volatilization is the production of volatile compounds or the elemental state of the radionuclides by treating the original recoil atoms or other samples with appropriate, mostly gaseous, reagents, usually at elevated temperatures.

Volatility is essentially a property of bulk amounts of the compound and may by quantitatively characterized by the energy or enthalpy of sublimation (or vaporization) as well as by the equilibrium vapor pressure. At the tracer level, we deal with the interaction between adsorbate and foreign surface. Therefore, the adsorption behavior of different compounds in the particular chemical system may or may not change similar to their bulk volatility. Hence, this term is to be used carefully – only when such parallelism is empirically established and, even better, if its base is understood. Usually, such correlation takes place in the absence of specific chemical mechanisms of adsorption interaction. *Microvolatility* might better serve the purpose, but is not in common use.

Adsorbability is strictly used to characterize a trace adsorbate as to its strength (energy) of interaction with surfaces because the term is derived from the word adsorbable. In the literature the term mostly does have this straightforward meaning, though occasionally it serves to characterize the adsorptive power of substrates.

Standard state has two easily distinguishable meanings in the book. When the conventional thermodynamic standard state is meant (10^5 Pa = 1 bar, ideal gas), it is not denoted here by any superscript ($\Delta_r G$, $\Delta_{vap} H$, $\Delta_{sub} S$, etc.) In the meantime, the thermodynamic consideration of the adsorption phenomena requires accepting less common standard states; these will be specified in Sect. 5.1.1 and denoted by the superscript "°", like in $S°$, $\Delta_{des} S°$ and $K°$.

Gas-solid isothermal chromatography (IC) of volatile species is realized in columns with steady, longitudinally constant temperature. The analytes are detected at the column exit by a variety of techniques.

Thermochromatography (TC) is a variant of the gas-solid chromatography characterized by a steady, monotonically decreasing temperature profile along the column. Due to it, the migration velocity of the analyte gradually slows down. It offers a convenient way to realize preparative separations by choosing such temperature range that the elements under study come to practical stop, each in a particular temperature range. Concrete techniques for measurement of these inner chromatograms differ; experiments with short-lived nuclides need very specific techniques of detecting, which are described in appropriate sections of the book. Outside the radiochemical community, the term thermochromatography also occurs in a dissimilar meaning: continuous analysis of the gaseous products of the thermal destruction of polymers, the composition of which changes with higher temperature.

Carrier gas here usually consists of a major component, chemically inert towards tracer(s) and of minor (still macroscopic) quantities of reactive gases or vapors (see below).

Reagent and *carrier* are the chemically active minor components of the carrier gas, while its major component is inert. In the first place, they both serve to react with the tracer and produce the required chemical compound, but they also play other roles discussed in Chapters 3 and 5. *Reagent* is preferred when emphasizing the gaseous state of the agent. Occasionally, a reagent like chlorine is the major component of the carrier gas. *Carrier* is usually a vapor rather than gas and is closer in reactivity and volatility to the tracer compounds under study. One can employ *isotopic carrier* when experimenting with radioisotopes of a common element, but this, of course, changes the original tracer status of the activity.

Group numbers follow the IUPAC recommended [16] system of 18 groups based on the long variant of the Periodic Table; see Table I.1 and Fig. I.1 below. This convention removes the former ambiguity of having A and B in the original eight-group table [17].

Lanthanoids and *actinoids* are well sounded replacements for the historical "lanthanides and actinides." They are recommended by IUPAC because *–oid* emphasizes

Table I.1 Place of the Transactinoid/Superheavy Elements in the Mendeleev Periodic Table of the Elements. Emphasized are the **calculated** and just **expected** ground state electronic configurations

Group 3	4	5	6	7	8	9	10
Y	Zr	Nb	Mo	Tc	Ru	Rh	Pd
$4d5s^2$	$4d^25s^2$	$4d^45s$	$4d^55s$	$4d^75s$	$4d^85s^2$	$4d^55s^2$	$4d^{10}$
Lu	Hf	Ta	W	Re	Os	Ir	Pt
$5d6s^2$	$5d^26s^2$	$5d^36s^2$	$5d^46s^2$	$5d^66s^2$	$5d^56s^2$	$5d^76s^2$	$5d^96s$
Lr 103	**Rf 104**	**Db 105**	**Sg 106**	**Bh 107**	**Hs 108**	**Mt 109**	**Ds 110**
$7s^2p$	104	105	$6d^47s^2$	$6d^57s^2$	$6d^67s^2$	$6d^77s^2$	$6d^87s^2$
	$6d7s^2p$	$6d^37s^2$					

Transactinoids ———————————————————————→

11	12	13	14	15	16	17	18
Ag	Cd	In	Sn	Sb	Te	J	Xe
$4d^{10}5s$	$4d^{10}5s^2$	$5s^2p$	$5s^2p^2$	$5s^2p^3$	$5s^2p^4$	$5s^2p^5$	$5s^2p^6$
Au	Hg	Tl	Pb	Bi	Po	At	Rn
$5d^{10}6s$	$5d^{10}6s^2$	$6s^2p$	$6s^2p^2$	$6s^2p^3$	$6s^2p^4$	$6s^2p^5$	$6s^2p^6$
Rg 111	112	113	114	115	116	117	118
$6d^97s^2$	$6d^{10}7s^2$	$7s^2p$	$7s^2p^2$	$7s^2p^3$	$7s^2p^4$	$7s^2p^7$	$7s^2p^6$

Superheavies - →

similar chemical character of elements (cf. metalloid), while –ide is systematically used for binary compounds of a particular element or group (cf. oxides, pnictides, etc.). We use the symbol Ln in the sense of "any lanthanoid" and An — for any actinoid.

Transuranium, transplutonium, (and so forth) elements are essentially all elements with atomic numbers higher than that specified by the name; no upper limit is implied. In the meantime, for what are essentially translawrencium elements, the radiochemical community uses the improper term *transactinoid elements* (TAEs). Nevertheless, we will stick to the latter because it emphasizes that, after a long row of elements of the actinoid character, one again deals with *d*-elements, each of which must have distinct individual chemistry.

Superheavy elements (SHEs) is not a well-defined term because it takes into account nuclear physical characteristics and methods of production, rather than chemical properties or position in the Periodic Table. A better-defined, though still vague, term is *superheavy nuclides* — they have the atomic numbers around 112 or higher and are produced by the bombardment of actinoid targets with neutron rich projectiles; the latter now range from ^{48}Ca, a "doubly-magic" nuclide, to ^{58}Fe. In principle, the transactinoid elements with $Z \geq 112$ can be also produced by bombarding the doubly magic target ^{208}Pb and adjacent nuclides with ions much heavier than ^{48}Ca. However, the *superheavy nuclides* obtained with actinoid targets are the heaviest attainable isotopes of the corresponding transactinoid elements. In the chart of nuclides, they are positioned on the slopes of the anticipated "island of stability" towards spontaneous fission. The center of the island is to be a new doubly magic nucleus with Z around 120; theories yet give different concrete values for Z and A. The *superheavy nuclides* might be better called "superheavy isotopes of TAEs." As such, these nuclides, generally, possess much longer total half-lives than the lighter isotopes of the same element obtained by the lighter targets. It makes their chemical studies feasible. To date, the known superheavy nuclides extend to $Z = 118$.

Names and symbols of transactinoid elements used herein are those recommended by IUPAC. The procedure of naming suggests that the discovery of a new element is established by a joint IUPAC–IUPAP Working Group. Then the discoverers are invited to propose a name and a symbol to the IUPAC Inorganic Chemistry Division. Their proposal is one of several criteria, which are taken into account by the IUPAC when choosing a "suitable" name; see Ref. [18]. The concrete recommendations for elements numbers 101 to 109 are published in Ref. [19], for element 110 in [20] and for element 111 in [21]. Before the late 1990s, some groups called element 104 "kurchatovium, Ku" and element 105 "nielsbohrium, Ns" or "hahnium, Ha." These names are used in older publications.

Chemical Character of the Transactinoid Elements

The expected or experimentally assigned positions of transactinoids in the Mendeleev Periodic Table are shown in Table I.1. The ground state electronic structures were obtained mostly by sophisticated relativistic calculations [10]; their quantitative characteristics are being refined. Results of different authors may vary

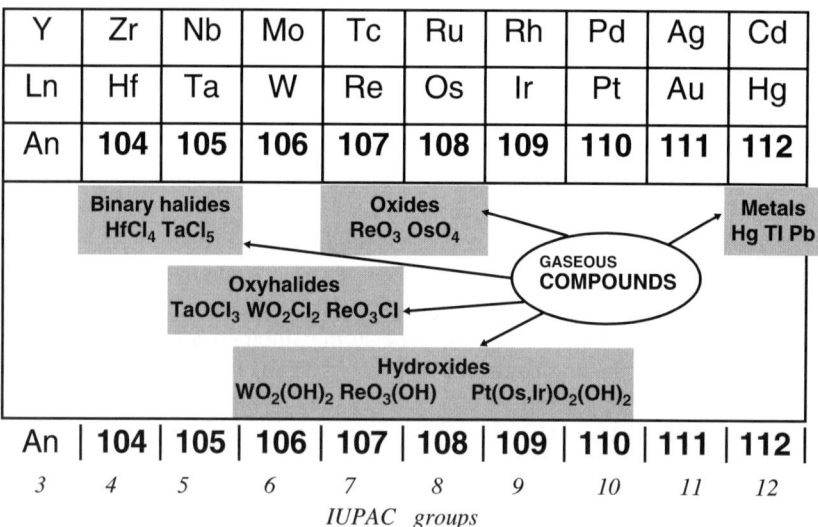

Fig. I.1 Volatile compounds and elemental forms characteristic of the lighter homologs of transactinoid elements.

as to the energies of the excited atomic levels. This is not very important for judging the differences in chemical properties between TAEs and their lighter congeners because the energies of molecular orbitals weakly depend on the detailed structure of the outer shells of the heavy element. The problem of finding the "relativistic effects in chemical properties" of a TAE is the next experimental goal after its chemical identification.

Figure I.1 shows some volatile compounds formed by the common elements of groups 4 to 11, as well as the relatively volatile metals of groups 12 to 14. The compounds of transactinoid elements with supposedly similar stoichiometry have been studied in works reported by now. A few studies have also been done with bromides and oxybromides.

References

1. Marshall JL, Marshall VR (2003) Bull Hist Chem 28:76
2. M. Curie (1955) Pierre Curie. Paris, Denoel
3. Kusaka Yu, Meinke WW (1961) Rapid radiochemical separations. Report NAS-NS 3104. Nat Acad Sci, Washington
4. Beard HC (1960), rev. by Cuninghame JG (1965) Radiochemistry of arsenic. Report NAS-NS 3002 (Rev). Nat Acad Sci, Washington
5. Herrmann G, Denschlag HO (1969) Rapid Chemical Separations. In: Ann Rev Nucl Sci 19: 1
6. Merinis J, Boussieres G (1961) Anal Chem Acta 25:498
7. Schädel M (2006) Angew Chem Internat Edit 45:368
8. Kratz JV (2003) Pure Appl Chem 75:103
9. Schädel M (2003) (ed) The chemistry of superheavy elements. Kluwer, Dordrecht
10. Kaldor U, Wilson S (2004) (eds) Theoretical chemistry and physics of heavy and superheavy elements. Kluwer, Dordrecht
11. Milligan WO (1970) (ed) Proceedings of Welch foundation conferences on chemical research. XIII. The transuranium elements – The Mendeleev centennial. Welch Foundation, Houston
12. (1990) Proceedings of Welch foundation conferences on chemical research. XXXXI. The transactinide elements. Welch Foundation, Houston
13. (1998) Proceedings of Welch foundation conferences on chemical research. XXXIV. Fifty years with transuranium elements. Welch Foundation, Houston
14. Eichler B, Eichler R (2003) Gas-phase adsorption chromatographic determination of thermochemical data and empirical methods for their estimation. In: Schädel M (ed) The chemistry of superheavy elements. Kluwer, Dordrecht, p 205
15. Pershina V (2003) Theoretical chemistry of the heaviest elements. In:Schädel M (ed) The chemistry of superheavy elements. Kluwer, Dordrecht, p 31
16. (2005) IUPAC periodic table of the elements. IUPAC, Research Triangle Park, NC. http://www.iupac.org/reports/periodic_table/IUPAC_Periodic_Table-3Oct05.pdf Cited 15 May 2007
17. Fluck E (1988) Pure Appl Chem 60:431
18. Koppenol WH (2002) Pure Appl Chem 74:787
19. Sargeson AM (for IUPAC Commission) (1997) Pure Appl Chem 79:2471
20. Corish J, Rosenblatt GM (for IUPAC Inorganic Chemistry Division) (2003) Pure Appl Chem 75:1613
21. Corish J, Rosenblatt GM (for IUPAC Inorganic Chemistry Division) (2004) Pure Appl Chem 76:2101

Chapter 1
Experimental Developments in Gas-Phase Radiochemistry

Abstract In the 1960s radiochemical separations of metallic elements were first accomplished by gas-solid chromatography of halides. Chlorides and oxychlorides in which both the oxidation state and coordination number of the metal is above three have mostly symmetric, three-dimensional molecular structures, which make the compounds relatively volatile molecular liquids. This fundamental property was soon successfully exploited for chemical identification of the very short-lived first transactinoids, Rf and Db. The continuing studies of new elements motivated further development of experimental techniques, which should ensure fast synthesis of the proper compounds, their transportation by gas flow or aerosol stream, chromatographic separation and efficient radioactivity measurements. Two somewhat different approaches have been developed. The first synthesizes necessary compounds in the target chamber with their subsequent separation by thermochromatography in columns made of radiation detectors. Otherwise, the thermalized recoils are transported by an aerosol flow to a distant setup to synthesize the compounds, perform isothermal chromatography and, using another aerosol, deposit the compounds for radioactivity measurements; it enables experiments with α-emitters when the working gas is corrosive. The concrete works described below illustrate the way to the present state of art in the field. Similarly reviewed are the techniques and prospective classes of compounds (other halides, sulfides, gaseous complexes) which, to date, have been employed only for common elements. Finally, there are some examples of separating mixtures of radioelements related by their origin, like fission or spallation products.

1.1 Early Gas-Solid Chromatography Studies

The 1960s were marked by increasing interest of nuclear industry in volatile compounds of metals. Not to mention UF_6, which had been exploited from the mid-1940s. Much effort was devoted to developments in the fluoride reprocessing of spent nuclear fuel. At that time, transition metals like Zr, Nb and Ta found many applications in nuclear industry. Some technologies for the extraction of these elements from ores and for the production of pure metals were based on the use of anhydrous halides, especially chlorides. It was hoped that such an approach might become universal.

The highest halides of the elements of groups four to eight are rather volatile molecular liquids or solids. It is not due to covalent bonding — the metal–halogen, M–X, bonds in the molecules of alkaline, alkaline earths, and rare earth halides are very polar, and this bonding character cannot greatly change in the halides of higher groups. The primary sources of the enhanced volatility are the 3D-structures and high symmetry of the molecules which contain four to six bonded halogen atoms. It holds even if the halogens are not all the same and if two of them are replaced by oxygen. Now the bond dipoles are effectively shielded by a tight shell of the halogens, and strong electrostatic interaction between touching molecules is prevented. The boiling points of these inorganic molecular liquids (and so their vaporization energies – cf. Trouton's rule) follow a remarkably tight correlation with certain universal solely geometric parameters, which takes into account only the coordination number and sizes of M and X [1].

Needs of the industrial technologies called for extensive studies and measurement of the physicochemical properties of halides and oxohalides: their thermodynamic characteristics, phase diagrams, reactions, complexing in gases and so forth. After two decades of growth, the intensity of these works waned; today, such studies are scarce, though the properties of a number of compounds are still known with low accuracy.

During the same period the fundamental studies in nuclear chemistry with increasingly powerful accelerators created a demand for radiochemical analyses of mixtures of nuclides, which were more and more complex as well as unusual in chemical composition. These tasks were tackled using solution chemistry techniques. These were time-consuming and resulted in loss of information about the relatively short-lived nuclides, which were originally present in the targets. Radiochemists and radioanalysts then, generally, did not keep pace with the progress in the nuclear chemical and refractory metal technologies. For a long time, adequate attention was not paid to the halides and other volatile compounds, which could help solve the new problems. Nevertheless, this attitude changed and the potentialities of the gas-solid chromatography techniques were recognized. They are both in separation of radioactive elements and in non-analytical studies, like measurement of the characteristics of adsorption of molecules or atoms. Developments of the gas-phase radiochemical techniques for heavy metals were mostly and primarily motivated by the quest for new transactinoid elements (TAEs) and superheavy nuclides. The studies focused on the chemical identification of TAEs, and on revealing the anticipated relativistic effects in their chemical properties. Below, we outline the major experimental steps which led to these goals. The works required sophisticated equipment to perform the necessarily long-lasting on-line experiments on intense beams of heavy ions. Another important requirement was high sensitivity of the measurements of particle radioactivity – high efficiency and extremely low background. These aspects are documented to help visualize the experiments. Presentation of these most difficult steps is followed by a more systematic outline of the gas phase studies made with relatively long-lived nuclides. They are considered

1.1 Early Gas-Solid Chromatography Studies

according to the classes and groups of elements, sorts of compounds and typical problems met in practice.

Understanding of some of the chemical processes, which take place in the experiments, is not yet satisfactory. It also concerns the rationales behind the observed regularities. Occasionally, empirical observations have served to hypothesize certain functional dependence, but the latter has not been substantiated by proposing a reasonable microscopic mechanism. This chapter emphasizes a few fundamental physicochemical problems, which were posed by the experimental works. Tentative solutions to some of them will be attempted in the later chapters.

Merinis and Boussières [2, 3] pioneered the method of thermochromatography (TC) in application to radiochemistry. With the equipment schematically pictured in Fig. 1.1, they investigated thermochromatographic behavior of some 20 elements, mostly in the form of chlorides. The elements were some alkaline, alkaline earth, rare earth, transition, noble, and actinoid (Th and Pa) metals. The authors' experimental technique was based on slow batch chemical volatilization. They obtained data on the shift in position of TC peaks as a function of the experiment duration.

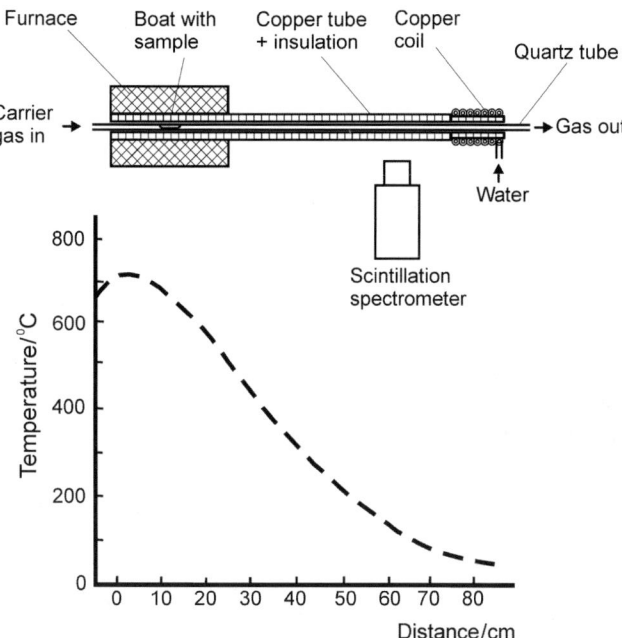

Fig. 1.1 Schematic of an experimental setup for thermochromatographic separations. The carrier gases Cl_2 or $Cl_2 + CCl_4$ also served as chlorination agents; the 90 cm long quartz column was of 0.5 cm i.d. Processing lasted several hours. Except for the initial and ending segments of the column, over some 60 cm, the temperature profile was nearly exponential.

Reproduced (adapted) from Radiochimica Acta, 12(2), Merinis J, Boussieres G, Etude de la migration des radioelements, 140–152, © 1969, with permission from Oldenbourg Wissenschaftsverlag.

1.2 Techniques for Isolation of Short-lived Accelerator Produced Nuclides

1.2.1 Off-line Simulation with Recoiling Fission Products

The basic idea in developing quick and efficient methods and techniques for TAEs was immediate, and continuous chemical processing of the nuclear reaction products in gaseous phase to obtain volatile compound(s) of the element under study. These had to be isolated from nonvolatile compounds of other elements and transported to detectors of radioactivity.

The very first simulation of the proposed gas-phase processing of the recoiling nuclear reaction products was done [4, 5] using the setup schematically depicted in Fig. 1.2. A supported layer of highly enriched $^{235}U_3O_8$, covered with a thin PTFE foil, was mounted on the inner wall of a heated cylindrical flow-through PTFE ampoule with 20 mm i.d. The target was bombarded with thermal neutrons from two closely attached 700-GBq Po-Be neutron sources, which were placed in a surrounding oil bath. The thicknesses of the target and its cover were such that all fission fragments penetrating into the ampoule volume thermalized in it; the activity did not get implanted into the opposite wall [6]. This made it possible to measure accurately enough (in separate experiments) the effective production rate of relatively long-lived nuclides by covering the target with a catcher — an Al-foil, thicker than the residual range of the fragments.

In one version of the experiments the target was bombarded with neutrons while the ampoule was continuously flushed with a gas. These were N_2, CO_2, Ar or Cl_2, containing about 1 mmHg partial vapor pressure of a carrier, such as $MoCl_5$, $ZrCl_4$, $SeCl_4$, $SnCl_4$, $TiCl_4$, $NbCl_5$ or $TaCl_5$. If solid at ambient temperature, the carrier was evaporated in the first furnace. Otherwise, the gas was bubbled through the liquid carrier and then fed into the device (not shown in Fig. 1.2). The target chamber was kept at different temperatures, not higher than 180 °C. The gas passed the distance from the target to a trap in about 15 seconds, and absorbed the

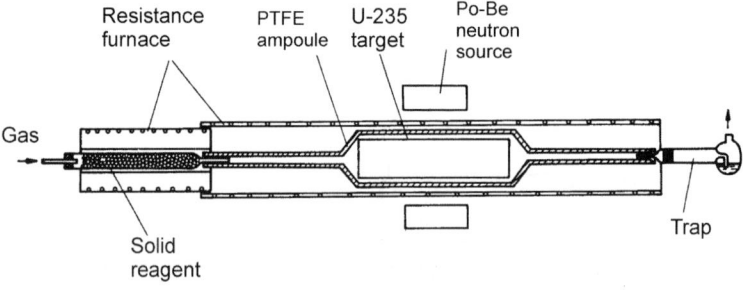

Fig. 1.2 First simulation of the fast on-line radiochemical method for heavy metals [5].

Adapted from Radiokhimiya, 8(1), Zvara I, Zvarova TS, Krivanek M, Chuburkov YuT Regularities in formation of volatile chlorides of ^{97}Zr and ^{101,102}Mo, 77–84, © 1966 with permission from Academizdat "Nauka" Publishers.

carrier and the transported activities. An experiment with the 17-h ^{97}Zr lasted about five hours. The content of the trap was analyzed with the standard radiochemical methods for Zr.

In other experiments the ampoule was filled with a pure inert gas at ambient temperature and was bombarded with neutrons for two half-lives of the radionuclide under study. Then, upon heating, the ampoule was flushed for a short time with the carrier containing gas, like in the experiments of the first kind.

The tests under certain conditions confirmed that ^{97}Zr, evidently in the state of ^{97}ZrCl$_4$, can be obtained and transported to the trap with a high chemical yield in less than seven seconds. The rare earths elements, which form nonvolatile chlorides, were not found in the trap. The results will be discussed in more detail in Sect. 3.2.1.

1.2.2 On-line Experiments with Spontaneously Fissioning Nuclides

The first on-line studies were performed by the present author and co-workers [7,8] on the intense circulating beam of the heavy ion cyclotron U-300 at JINR, Dubna. Extracted beams were then not available. To realize experiments inside the cyclotron chamber was not easy because of controversial technical requirements. The schematics of the setup [9,10] shown in Fig. 1.3 resulted from several trial-and-error attempts.

Before starting the tests aimed at element 104 with an expected lifetime of a second or less, two most important points had to be checked by producing Hf and Ln activities [7]. First, to verify that the recoil atoms can be chlorinated rapidly enough with the particular chemical agents, and at the working temperatures which do not cause severe technical problems. The second task was to check whether the behavior of tracer molecules in this processing, which was essentially gas-solid-chromatography, correlates with the characteristics of bulk volatility of the compounds involved – their energies of sublimation or vaporization. Specifically, it was verified that the retention time of Ln and An chlorides is much longer than that of hafnium compound, so that the IC column could serve as a sort of filter. The retention time of HfCl$_4$ was accurately measured with a special device attached to the exit of the tube-column. It proved to be some 0.5 seconds; details will be given in Sect. 3.2.1.

The construction of the target chamber assembly was described in Ref. [11]. It proved impossible to introduce the chlorinating reagents directly into the chamber because they corroded the target. An attempt failed to protect the target by a layer of the inert gas, which was supplied through a hollow frame with numerous directed pinholes. The layer was thinner than the range of recoils so that they were still thermalized in the presence of the reagents. However, the protection was not perfect enough to allow experiments lasting days with plutonium targets. It became necessary to flush the target chamber with pure nitrogen and to introduce the reagents only at its outlet. Along this short distance, about a third of the requested atoms

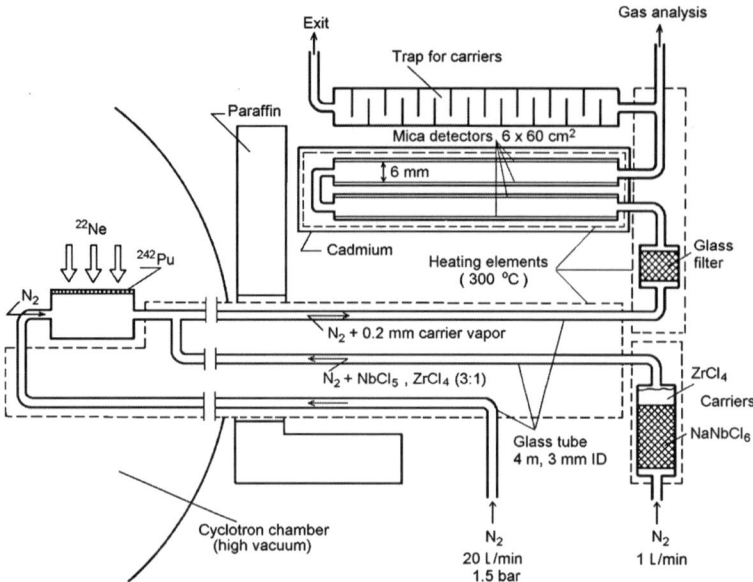

Fig. 1.3 Setup for first chemical experiments with element 104 – now Rf; Dubna, the mid-1960s [10]. The broken frames outline the placement of resistive heaters, paraffin and cadmium shielded the detectors from neutrons to prevent induced fission of uranium impurities in mica. Thermal decomposition of $NaNbCl_6$ was the source of $NbCl_5$ vapor. A Faraday cup was placed inside the target chamber (not shown).

Adapted from Radiokhimiya, 11(2), Zvara I, Chuburkov YuT, Caletka R, Shalaevskii MR, Experiments on chemistry of element 104 II. Chemical study of the spontaneous fission isotope, 163–174, © 1969, with permission from Academizdat "Nauka" Publishers.

were lost — by molecular and convective diffusion they reached the metallic walls of the target chamber and irreversibly adsorbed onto it. The majority of the atoms still stayed in the gas and got a chance to react. To exit the vacuum chamber of the accelerator, the volatile molecules passed through a 4-meter long tube; it functioned as an isothermal chromatography (IC) column. At its outlet, the gas flow was directed into a heated long, flat flow-through chamber, the walls of which were covered with mica sheets to detect fission fragments. The linear and time coordinates of the detected s.f. events of element 104 nuclei served to verify the half-life of the detected nuclide.

The fact that mica (and fused silica) can serve as "solid state track detector" (SSTD) of fission fragments was reported shortly before the final stage of development of the method for element 104 [12–14]. In the dielectric solids, fission fragments produce tiny tracks visible by electron microscopy. Mica and silica are very resistant to active chemical reagents and elevated temperatures. The tracks proved to stay in hostile conditions of real experiments for a reasonably long time. Thanks to this, after the end of bombardment (EOB), the mica sheets could be etched with hydrofluoric acid to enlarge the tracks to micrometer size; they were distinct in appearance and were searched out by scanning the surface of the detectors with an optical microscope. The target chamber accepting the heavy ion beam is the source of a considerably intense neutron flux. In view of this, each batch of mica was assayed for uranium impurities through exposing to much larger neutron fluences at a nuclear reactor and counting tracks from the induced fission events. The content of uranium in the batch selected for detectors was, by orders of magnitude, below the permissible level.

In these first experiments with Rf a few detectable atoms were isolated per day; now it is known that it was α-active 3.5-s ^{259}Rf, which has a few percent s.f. branching. A total of a 14 s.f. events [10] were observed with the column kept at 350 °C, while at 250 °C none survived because the chlorination or retention time of the nuclide was much longer than $t_{1/2}$. Independent of temperature, only about 1 percent of the produced atoms of actinoid elements passed to the detectors according to the measurements of α-active nuclides. Such a degree of separation proved good enough to exclude the possibility that some s.f. events on the detectors might originate from the actinoid nuclides — their total yield in the bombardment was known from independent physical experiments. Moreover, several identical chemical experiments were done with products of the target plus projectile combinations, which yielded only elements with atomic numbers less than 104. In these runs, the mica sheets did not detect any fission event. Thus, the careful considerations and experiments provided conclusive evidence for the observation of element 104 — in spite of the fact that spontaneous fission does not possess spectroscopic characteristics. The work was recognized as an independent discovery of element 104 [15, 16].

With a flow rate of the carrier gas of about 20 L min^{-1}, this technique of chemical identification is still a record as to its rapidness of a few tenths of a second, and to the almost quantitative chemical yield. Later on [17], a setup based on similar principals was mounted on an extracted beam of U-300 cyclotron, which greatly enabled the experiments: about 40 more atoms of Rf were isolated and detected.

The JINR, Dubna group also used thermochromatography in open columns for studies with s.f. nuclides. The then novel setup for identification of element 104 as ekahafnium, see Fig. 1.4, could also serve to compare the adsorbability of the tetrachlorides of these two congener elements. Especially for the latter purpose, thermochromatography has important advantages over isothermal chromatography. This time, the spontaneous fission of ^{259}Rf was detected by long mica sheets of a width equal to the column diameter. For the first time, the reagents were much more volatile than the compounds under study — their dew points were much lower than the temperature at the adsorption peak of HfCl$_4$ [18, 19].

The experimental data are shown in the bottom of Fig. 1.4. A filling in the initial section of the column served to enhance deposition of the nonvolatile chlorides of actinoid elements by disturbing the flow patterns. Success is demonstrated by the distribution of Sc activity as a marker for the nonvolatile species and by a few fission events (open circles) detected within the Sc zone. These are followed by one track over 100 cm of the isothermal section of the column. The thermochromatographic zone of Hf isotopes (measured by γ-activity) and that of Rf fission events, which were observed in the thermochromatographic section of the column, have

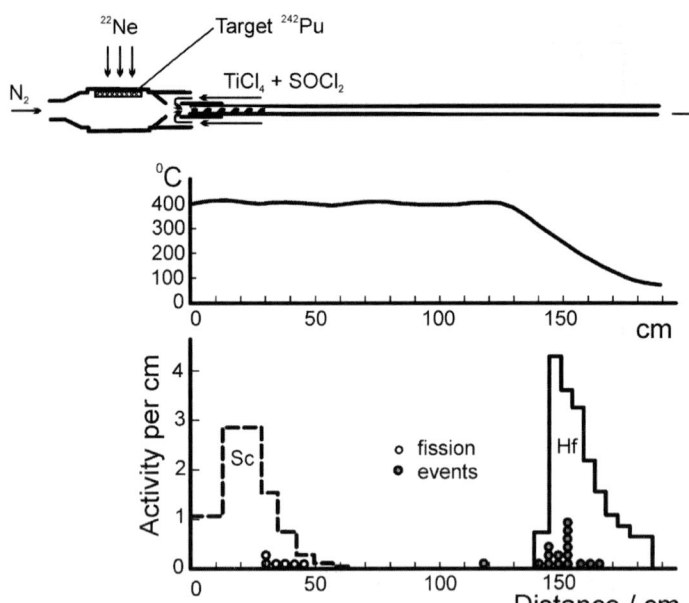

Fig. 1.4 Thermochromatographic identification of element 104 as ekaHf [18]. Extracted beam of the JINR cyclotron U-300.

Adapted from Radiokhimiya, 18(3), Zvara I, Belov VZ, Chelnokov LP, Domanov VP, Hussonois M, Buklanov GV, Korotkin YuS, Schegolev VA, and Shalayevsky MR, Chemical isolation of kurchatovium, 119–122, © 1972, with permission from Academizdatcenter "Nauka" Publishers.

characteristic TC profiles. Most probably, the solitary fission event at 120 cm distance was due to decay of the short-lived Rf in flight, rather than due to incomplete deposition of nonvolatile molecules.

An important achievement was chemical identification of element 106 as a homolog of tungsten. The aim was to produce an oxychloride of the element using gaseous $SOCl_2$, plus air as reagents. The experiment [20] had some distinct features. First, the 0.9-s ^{263}Sg, obtained via ^{249}Cf(^{18}O,4n), was to that date (and continues to be) the shortest chemically identified transactinoid nuclide. The thermochromatographic device had to be modified, compared with what was shown in Fig. 1.4. In particular, the temperature gradient took place along the entire column; cf. Fig. 1.5. Next, the column was made of fused silica and its walls served as SSTD. It required a special device, which allowed scanning the inner surface for the etched track through the column wall. The experiment remains unique in that the gaseous medium containing the molecules under study got in contact exclusively with the surface of silica. From the principal point of view it is very important for careful studies. A total of some 40 atoms of the element were detected; Figure 1.5 displays the experimental data. Interpretation of these and supplementary observations as to the chemical state of the homologous elements and to the mechanism of their thermochromatography is presented in Chapters 3 and 5.

1.3 Techniques for α-active Nuclides: Corrosive Reagents

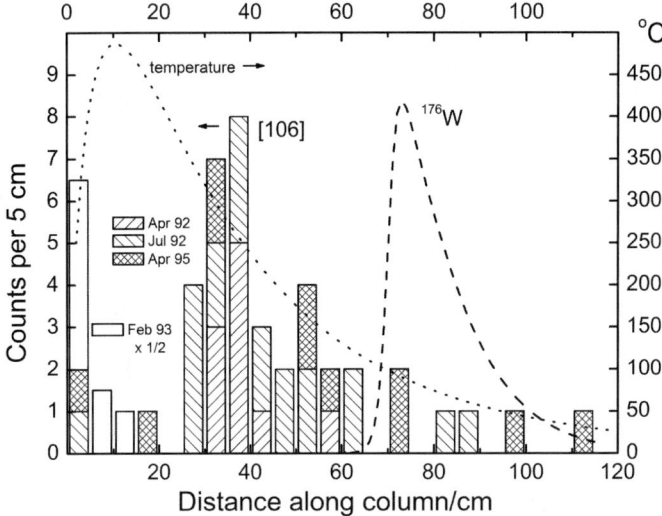

Fig. 1.5 Chemical identification of element 106, now seaborgium [20]; extracted beam of JINR U-300. The combined hatched histogram shows s.f. events from three separate runs. The white histogram shows fission events detected in an experiment with the inert carrier gas only. The thermochromatogram of simultaneously produced ^{176}W was traced through γ-radiation.

Reproduced from Radiochimica Acta, 81(4), Zvara I, Yakushev AB, Timokhin SN, et al., Chemical identification of element 106, 179–187, © 1998, with permission from Oldenbourg Wissenschaftsverlag.

1.3 Techniques for α-active Nuclides: Corrosive Reagents

The α-active transactinoid isotopes which possess half-lives suitable for chemical studies are quite numerous. However, the spectrometric detectors for measuring α particles or fission fragments are sensitive to the high neutron flux in the vicinity of the target. In addition, they do not withstand elevated temperatures and active chemical reagents. Hence, the bombardment products must be transported several meters away from the beam stop. This also enhances the technical problems of chemical experiments. Notice that in the first Dubna equipment depicted in Fig. 1.3, the bombardment products were transported to a distance of four meters from the target. However, the chemical processing was still done next to the accelerator and had to be remotely controlled from a shielded room, which was several more meters away. In the meantime, the radioactivity of the bombardment products is moderate, so that the necessary radiation protection measures are quite simple when the chemical equipment itself is placed behind a biological shield. In nuclear chemistry studies, researchers widely used a "helium jet" for long-distance transportation of short-lived nuclides from the targets. The techniques made use of the fact that vapors of various organic compounds in an inert carrier gas form molecular clusters when exposed to a particle beam. The recoils thermalized in such a flow are adsorbed to the clusters and can be transported over a long capillary through energetic pumping of its exit. In the late 1970s such transportation was realized also at ambient pressure at the capillary exit, with alkali halides as the cluster (aerosol) material; see more in Sect. 3.4. Using the target chamber like that in Fig. 1.3, Dubna researchers [21] were able to transport Pt+^{12}C bombardment products to about 12 meters with 70 percent efficiency. Hickmann, et al. [22] made an important step forward. They joined a gas-jet recoil-transport technique (with KCl aerosol) to a thermochromatographic system for continuous radiochemical separation of fission product; see also Sect. 1.5.3. Dincklage, et al. [23], still using mostly clusters of certain organics, were first to build a sophisticated equipment for continuous gas-solid isothermal chromatography with the possibility of measurement of α-active nuclides; see Fig. 1.6. The clusters were decomposed at the inlet of the chromatographic column in a quartz wool filter kept at 800 °C, while the transported bombardment products reacted with CCl$_4$ vapor, which was introduced into the flow shortly before. A nozzle-like end of the chromatographic column produced directed flow impinging onto a cooled copper wheel; the isotopes of Nb and Hf under study were deposited within a spot of about 1 cm in diameter. The wheel was rotated using a stepping motor to place the spot in front of the particle detectors. In the meantime, at the mouth of the chromatographic column, a portion of the gas was continuously taken aside and assayed for the activity in the same way as the gas exiting the column. Such measurements gave the survival yield of the particular nuclide after passing the column, and so the retention time of the appropriate analyte. The efficiency of cluster transportation was 30 percent; the overall efficiency of finding the long-lived products on the wheel was 10 percent.

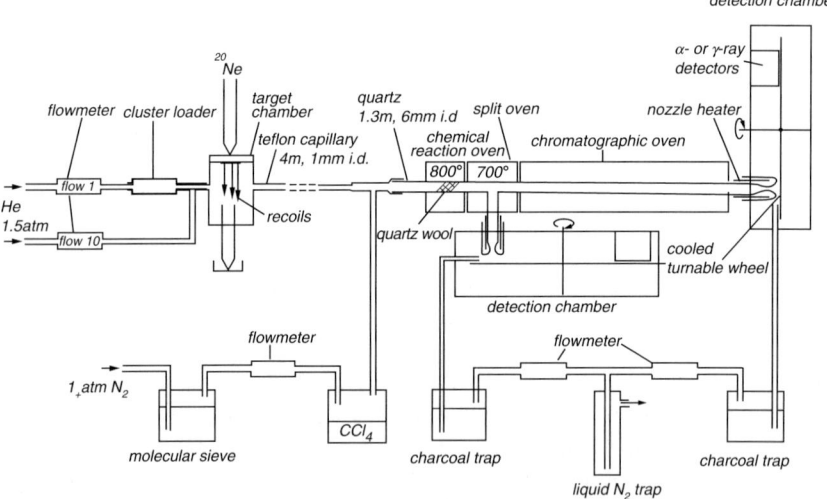

Fig. 1.6 Isothermal chromatography of short-lived cyclotron produced nuclides transported by helium (cluster) jet [23].

Reproduced from Nuclear Instruments and Methods, 176(3), von Dincklage RD, Schrewe UJ, Schmidt-Ott WD, Fehnse HF, Bachmann K, Coupling of a He-jet system to gas chromatographic columns for the measurement of thermodynamical properties of chemical compounds of ^{90m}Nb (18s), ^{160}Hf (12s) and ^{161}Hf (17s), 529–535, © 1980, with permission from Elsevier.

1.3 Techniques for α-active Nuclides: Corrosive Reagents

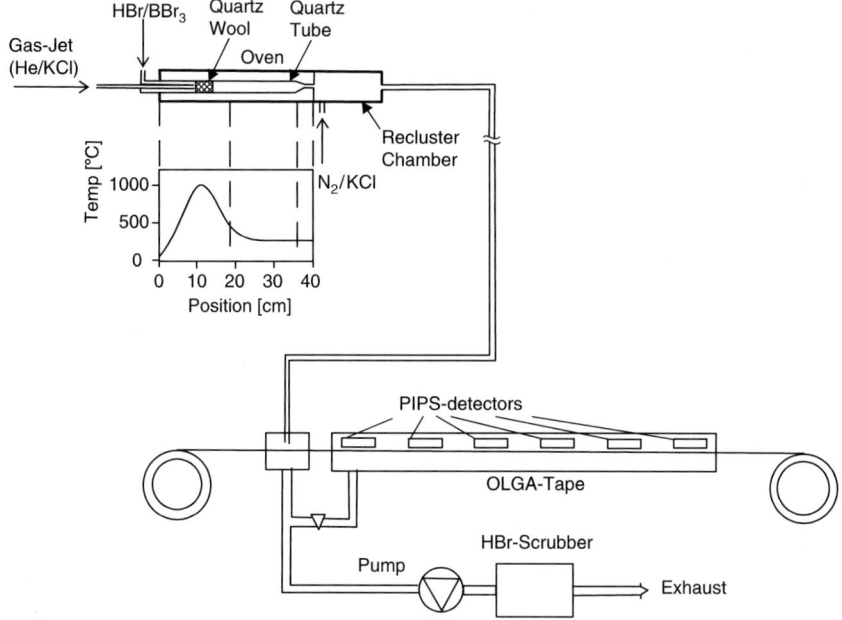

Fig. 1.7 Isothermal chromatography experiments (OLGA-type equipment) with volatile compounds of α-active nuclides [29].

Reprinted from Nuclear instruments and methods in physics research, A 309(1–2), Gäggeler HW, Jost DT, Baltensperger U, Weber A, Kovacs A, Vermeulen D, Türler A, Olga II, an on-line gas apparatus, 201–208, © 1991 with permission from Elsevier.

Later, using a similar approach, Gäggeler and co-workers [24–28] at PSI, Villigen built the first dedicated setup for isothermal chromatography experiments with short-lived, α- or s.f.-active isotopes of transactinoids. The essential parts of the equipment are shown in Fig. 1.7. A major concern was high efficiency at every step of processing. Several modifications of this OLGA setup have been built and employed. They differed in the parameters of the column and in the complexity, as well as quality, of the detection system. For example, the older detectors were replaced by the passivated implanted planar silicon (PIPS) detectors, which are much more resistant to elevated temperature and to chemicals.

The aerosol materials that have been exploited to date (not all of them in TAE studies) by the PSI and other groups are alkaline element halides, metal oxides (MoO_3), metals (Ag, Pb, C, and Pd) and carbon. The aerosols are introduced by mere evaporation of the material under experimentally found optimal conditions, or by spark discharge sputtering. After reaching the device for chemical experiments, the aerosol jet is mixed with gaseous reagents and passes through a hot filter (usually quartz wool plug). In this section the atoms of radionuclides are to react and yield the required compounds, while the particulates are destroyed and their matrix is either removed or (seldom) also chemically volatilized. The volatile tracer molecules are fed into an isothermal column; its temperature must be such as to make the retention time of the molecules comparable with the half-life of the particular nuclide. At the column outlet the gas is mixed with another (cold) aerosol stream to "recluster" the molecules containing the surviving atoms. Reclustering is a distinct feature of the novel equipment. It serves efficient transportation of the activity to the detection system, as well as complete deposition of the aerosol material as a thin layer suitable for the spectrometric measurements of particle radiation. To that end, the transport capillary is intensely pumped out from the exit to obtain a high velocity aerosol jet. The particulates are deposited by impact as a spot on the surface of a wheel [29] or tape [30]. These are stepped to make the sample successively face several detectors. If the foil supporting the deposit is very thin, the spectrometric measurements of the specimens can be done simultaneously from both sides. It allows achieving the ultimate detection efficiency and considerable probability of observing the "correlated" mother–daughter decay events. Correlation means consecutive detection from the same deposit of two α particles, which can be assigned to the expected pair by their energies, and by the time interval between them. It greatly enhances the problem of accounting for the possible background due to interfering α-activity of lighter elements.

A principal experimental value with OLGA is the column temperature at which 50 percent of the atoms decay during the retention time. The low production rate of TAEs does not allow simultaneous measurement of the incoming and exiting concentrations like in the above-mentioned studies [23] of common elements. Setups similar to that in Fig. 1.7 were constructed also in LBL, Berkeley [31] and in JAERI, Tokai [32]. That built in Radiochemistry Center, Dresden [33] allows column temperatures as high as 1,300 K; reclustering is omitted, and the exiting gas hits a cooled spot on the surface of a stepping wheel to deposit the tracer, like in Ref. [23].

Sample transactinoid data obtained by the Swiss group are presented in Fig. 1.8. The team pioneered the use of HCl, HBr, Cl_2 and Br_2 as the reagents. To date, with installations for isothermal chromatography, the above groups have reported adsorption studies of halides or oxohalides of rutherfordium [32, 34, 35], dubnium [36], seaborgium [37] and bohrium [38]. The data of Kadkhodayan, et al. [34] on $RfCl_4$ are distinguished for the best statistics in transactinoid research to date. Chemical identification of bohrium with HCl plus O_2 as the reagents producing BhO_3Cl is remarkable for the use of carbon aerosol, which could be removed on the quartz wool filter by oxidation to CO_2, rather than by absorption.

1.3.1 Relative Merit of Isothermal- and Thermochromatography

At this point, it seems appropriate to summarize the advantages and drawbacks of thermochromatography and isothermal chromatography as the tools for chemical identification and study of chemistry of the very heaviest elements:

1.3 Techniques for α-active Nuclides: Corrosive Reagents

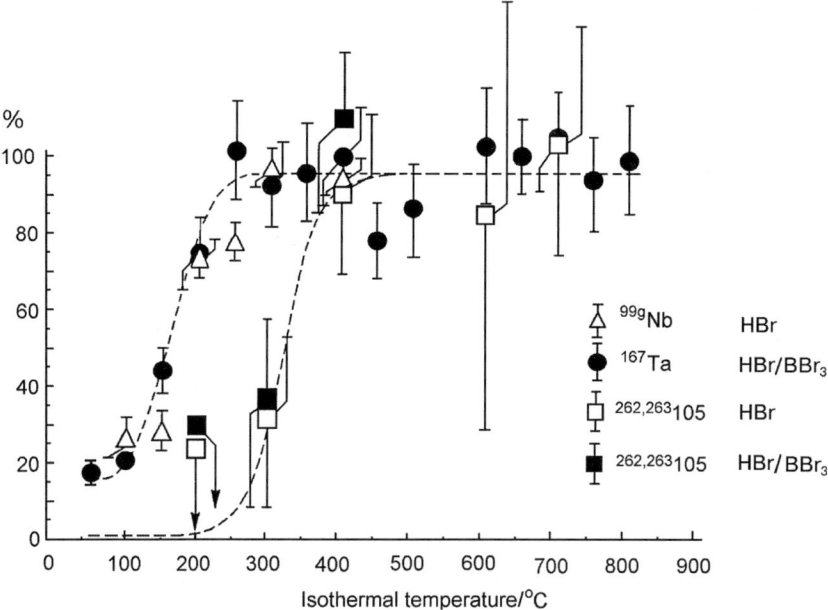

Fig. 1.8 Experiments with bromides of Db and its homologs. Survival yield of short-lived isotopes as a function of IC column temperature. The ≈0.5-min $^{262,263}105$ were produced through the reaction ^{249}Bk (^{18}O, 4, 5n) [26].

Reproduced from Radiochimica Acta, 57(2–3), Gäggeler HW, Jost DT, Kovacs J, Sherer UW, Weber A, Vermeelen D, Tuerler A, Gregorich KE, Henderson RA, Czerwinski KR, Gas phase chromatography experiments with bromides of tantalum and element 105, 93–100, © 1992, with permission from Oldenbourg Wissenschaftsverlag.

1.3.1.1 Thermochromatography

Disadvantages:

- When using solid state track detectors, results are not obtained in real time.
- If the reagents are corrosive and temperature elevated, α-active nuclides cannot be studied.
- Under the first two conditions, half-life of the measured nuclide cannot be measured or verified.

Advantages:

- All detectable nuclei reaching the column inlet contribute to the data.
- The new element and its known homolog, if produced simultaneously, are treated in identical conditions.
- The experiment, including chemical processing, is relatively simple and no mechanical devices are needed.
- In the case of noncorrosive reagents and very volatile analytes, the column can be made of spectrometric detectors, and the method also acquires the below-listed advantages of IC.

1.3.1.2 Isothermal Chromatography

Advantages:

- It is possible to study both s.f. and α-active nuclides.
- It is possible to identify nuclides by half-life and through spectrometry of its particle radiation.
- Data can be taken in real time.

Disadvantages:

- It is necessary to remove the aerosol matter by hot filters, which makes the chemical processes uncertain and difficult to control.
- The transportation and reclustering efficiencies strongly depend on the quality (concentration, size of particles) of the aerosol, which may change with time.
- Reclustering efficiency decreases with column temperature, which brings more uncertainty in the value of the 100 percent survival yield.
- It is necessary to perform several separate experiments at different column temperatures, while maintaining otherwise identical conditions may be difficult.
- Experiments with a lighter homolog of the TAE must often be performed in a different range of column temperatures.
- The temperature of the 50 percent survival yield can be found only at the expense of severe losses of otherwise observable nuclei.

The advantages of thermochromatography clearly prevail when the experiments involve only non-corrosive gaseous reagents, and when the compounds under study are so volatile that the "hot" end temperature of the TC column may be slightly above ambient temperature. This is evidenced by the works on chemistry of the transactinoids beyond seaborgium, which are cited in the following section.

1.4 Techniques for α-active Nuclides: Non-corrosive Reagents

1.4.1 Thermochromatography of Hassium Tetroxide

The Periodic Table suggests that hassium is a member of group 8, and thus akin to the lighter group members – Ru and Os. These are known to form unique highly volatile tetroxides. Chemical separation and characterization of hassium by thermochromatography of oxides was reported by Düllmann, et al. [39]; the design of the experiment is visualized in Fig. 1.9. The isotopes ^{269,270}Hs with the half-lives of several seconds were produced in bombardments of a ^{248}Cm target with ^{26}Mg. To obtain HsO_4, the recoiling nuclei were stopped in a mixture of He and O_2, and the gas exiting the target chamber was passed through a short quartz tube with a quartz wool plug heated to 600 °C. Quartz also absorbed many accompanying elements which form nonvolatile oxides.

1.4 Techniques for α-active Nuclides: Non-corrosive Reagents

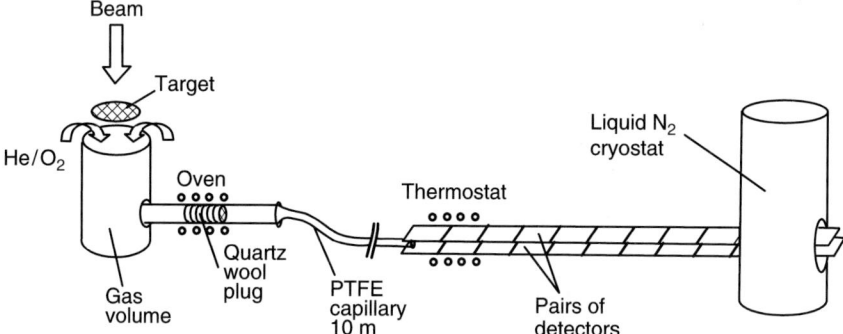

Fig. 1.9 Schematic drawing of the setup used to produce and isolate Hs isotopes in the form of volatile HsQ$_4$ [39]. Ref. [40] gives a detailed description of the target chamber and Ref. [41] — of the cryo-thermochromatographic separator.

Reprinted (adapted) by permission from Macmillan Publishers Ltd: Nature, 418(6900), Düllmann ChE, Brüchle W, Dressler R, Eberhardt K, Eichler B, Eichler R, Gäggeler HW, Ginter TN, Glaus F, Gregorich KE, Hoffman DC, Jäger E, Jost DT, Kirbach UW, Lee DM, Nitsche H, Patin JB, Pershina V, Piguet D, Qin Z, Schädel M, Schausten B, Schimpf E, Schott H-J, Soverna S, Südowe R, Thörle P, Timokhin SN, Trautmann N, Türler A, Vahle A, Wirth G, Yakushev AB, Zielinski PM, Chemical investigation of hassium (element 108), 862–859, © 2002.

The volatile oxides could be transported through a capillary by the gas flow at ambient temperature to an array of 24 PIPS detectors. These were arranged in 12 pairs to make a narrow rectangular channel, which also served as the thermochromatographic column. A downstream negative temperature gradient was established, the temperature changed from a slightly elevated to that of LN$_2$.

The seven Hs nuclei observed in detectors numbered 2 through 4 in Fig. 1.10 were identified owing to the detection of the correlated mother–daughter decay events. Simultaneously, α-active Os isotopes were produced and detected. The particular gaseous chemical system is very simple. It allows selective isolation of the group 8 elements Ru-Os-Hs from the great majority of the simultaneously produced α-active nuclides. Unfortunately, this is not true for group 18, namely for the short-lived isotopes of radon, which are carried by the gas all the way through the system and, when decaying in flight, they yield descendants with high energy α-radiation. This creates interfering background in the measurements of α–spectra of hassium.

Another fundamental problem is securing the necessary low residual water content in the gas. The supplied gas was reportedly dried with P$_2$O$_5$ to obtain the water content less than 10^{-3} mmHg. Reference sources give the residual water concentration in air dried over P$_2$O$_5$ as 10^{-5} mmHg; the corresponding dew point would be $-100\,°C$. In the conditions of the Hs experiments, it means some 10 mg of ice deposited per day on the surface of the detectors; fortunately, the region is beyond the peaks of Hs and Os; see Fig. 1.10. Such an amount would heavily deteriorate spectrometric measurements of α particles emerging from deposits of even more volatile compounds.

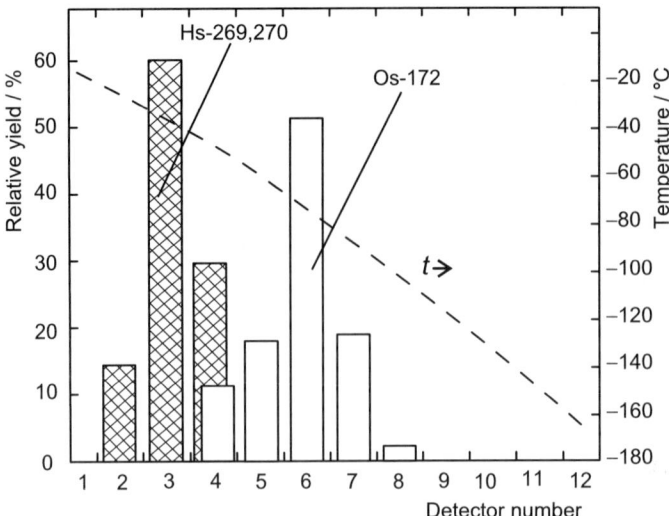

Fig. 1.10 Merged thermochromatograms of HsO$_4$ and OsO$_4$ [39]. Indicated are the relative yields of HsO$_4$ and OsO$_4$ for each of the 12 detector pairs. For Os, the distribution of some 10^5 decay events of ^{172}Os is given.

Reprinted by permission from Macmillan Publishers Ltd: Nature, 418(6900), Düllmann ChE, Brüchle W, Dressler R, Eberhardt K, Eichler B, Eichler R, Gäggeler HW, Ginter TN, Glaus F, Gregorich KE, Hoffman DC, Jager E, Jost DT, Kirbach UW, Lee DM, Nitsche H, Patin JB, Pershina V, Piguet D, Qin Z, Schädel M, Schausten B, Schimpf E, Schott H-J, Soverna S, Südowe R, Thörle P, Timokhin SN, Trautmann N, Türler A, Vahle A, Wirth G, Yakushev AB, Zielinski PM, Chemical investigation of hassium (element 108), 862–859, © 2002.

Generally, the work once more witnesses the advantages of in situ production of the desired compounds [40] and of thermochromatography as the most economic method of obtaining data when only a few atoms are produced.

1.4.2 Chemical Identification of Metallic Element 112

The conclusions presented above can be applied in designing experiments on chemical identification and studies of element 112. There is little doubt that the element is a congener of mercury, at least equally volatile and chemically inert. Then proper chemical environment can be even simpler than in the case of HsO$_4$. However, the experimental technique must allow for the possibility that element 112 is much *more* volatile and chemically inert than mercury. The problem is to guarantee the registration of the element (not to lose it) even if it resembles Rn, rather than Hg, in volatility and inertness. Atomic mercury in tracer quantities can be transported by inert gas at ambient temperature through tubes made of various materials. However, it adsorbs onto some metals, in particular, on gold.

1.4 Techniques for α-active Nuclides: Non-corrosive Reagents

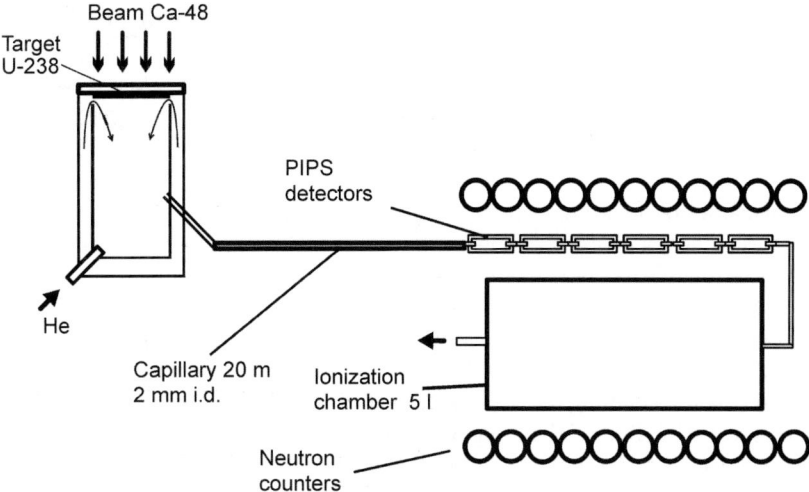

Fig. 1.11 Schematic of an experiment to distinguish between the mercury-like and radon-like behavior of element 112.

These expectations determined the principal schematics of the experiments in progress. The essential components of the experimental installation which were considered at Dubna are shown in Fig. 1.11. The recoiling α-active isotopes of element 112 and of simultaneously produced mercury are thermalized in streaming helium. The gas flows first through a flat chamber with several consecutive pairs of PIPS detectors coated with Au. Mercury is known to adsorb strongly on gold at ambient temperature. Hence, the chamber serves to deposit and register at least the simultaneously produced α-active isotopes of Hg. To detect spontaneous fission of element 112, if it *does not* get adsorbed on gold, the exiting gas is directed into a flow-through ionization chamber adjusted to register fission fragments. To assure reliable identification of the spontaneous fission events, both detecting devices are housed inside an assembly of neutron counters (^{3}H-filled) with polyethylene moderator. The device serves to detect the burst of prompt fission neutrons in coincidence with the triggering signal of fission fragment detection by the PIPSs, or by the chamber.

Experiments and attempts along the above lines, which have been reported up to now [42–44], are not completely consistent. More informative will be experiments using a TC column, like that utilized in the HsO_4 experiments. It offers a variety of ways to arrange the detectors, with or without the gold layer on the surface, to provide quantitative characteristics of the differences in volatility and reactivity within the range Hg–112–Rn.

1.5 Prospects for Future of Radiochemical Studies of Heavy Elements

1.5.1 Classes of Compounds

Several classes of binary and ternary compounds are potentially prospective for use in studies of heavy elements. Below are some remarks and observations about the merit of various classes of compounds — those already employed (like chlorides) as well as those investigated thus far only in off-line experiments with relatively long-lived nuclides. One can judge prospects for their use in radiochemistry of the very heaviest element from the tests with lighter homologs. The transactinoid elements reported to date, including those yet identified only by nuclear techniques, belong to groups 4 to 18; cf. Table I.1 and Fig. I.1. We also comment on the actinoids because they are always present in the experiments with TAEs.

1.5.1.1 Metals

Reports on isolation and interseparation of tracer metals appear in literature with increasing frequency. Long-standing interest comes from the ISOL installations. They need rapid extraction of (mostly) spallation products from bulk voluminous targets and supplying them to a mass separator. The research is largely empirical, and concrete tasks often bring novel problems with technology [45]. Meanwhile, more fundamental laboratory studies on separation of metals by gas-solid thermochromatography and by vacuum thermochromatography have been reported. The works also evaluated the adsorption energies of tracer metals on foreign metallic surfaces and as well as the most volatile metals. Some data and relevant references will be given in Sect. 4.4.

1.5.1.2 Oxides and Oxide Hydroxides

Gaseous oxygen usually reacts only at elevated temperatures. Moderately to considerably volatile oxides are characteristic of some heavy metals of groups 6 through 9. A thorough study of the oxides of group 7 to 10 elements by the programmed chromatography and thermochromatography was reported in Refs. [46,47]. Some of the data are presented in Figs. 1.12 and 1.13. More results can be found in Ref. [48].

As was illustrated in Sect. 1.4.1, the unique volatility of tetroxides makes their high selective isolation possible. Domanov, et al. [49, 50] reported observation of a volatile compound of plutonium, presumably the tetroxide. This unexpected result has not yet been independently confirmed and was questioned [51].

Of interest are also oxide hydroxides of some heavy elements of groups 6 to 10 [52–54]. They can seldom be obtained as bulk phases, but have been observed in tracer chemistry with oxygen, plus water vapor, as the reagents. Expecting that

1.5 Prospects for Future of Radiochemical Studies of Heavy Elements

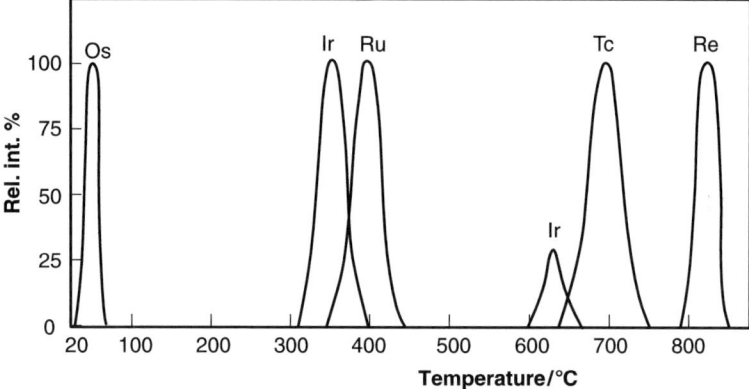

Fig. 1.12 Temperature-programmed chromatography of oxides [46]. Conditions: quartz column, length 100 cm, 8 mm i.d, filled with quartz granules 0.32–0.63 mm; heating rate 20 K min^{-1}; carrier gas and reagent oxygen, 3 cm^3 min^{-1}.

Reprinted from Talanta, 25 (10), Steffen A, Bachmann K, Gas chromatographic study of volatile oxides I, 551–556, © 1978, with permission from Elsevier.

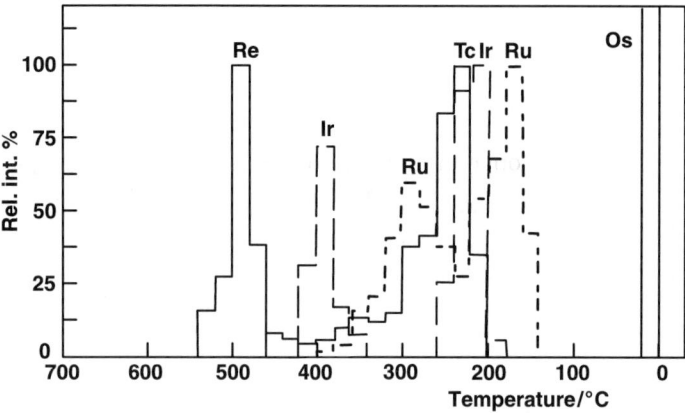

Fig. 1.13 Thermochromatogram of a mixture of oxides [47]. Conditions: quartz column, length 100 cm, 8 mm i.d, filled with quartz granules 0.32–0.63 mm; duration of run 30 min; carrier gas and reagent oxygen, 6 cm^3 min^{-1}.

Reprinted from Talanta, 25 (11–12), Steffen A, Bachmann K, Gas chromatographic study of volatile oxides II, 677–683, © 1978, with permission from Elsevier.

element 107 must form the compound similar to ReO$_3$(OH), the FLNR group once attempted the isolation of possible s.f. isotopes from the products of the bombardment of ^{249}Bk with ^{22}Ne [54]. The desired nuclides were not observed, either because of their small s.f. branching or because of an insufficient fluence. Later, Hübener, et al. [55, 56], making experiments with 21-s ^{266}Sg from the bombardment of ^{248}Cm with ^{22}Ne, succeeded in isolating two molecules of what was supposedly SgO$_2$(OH)$_2$. The temperature of the isothermal column had to be as high as 1,300 K.

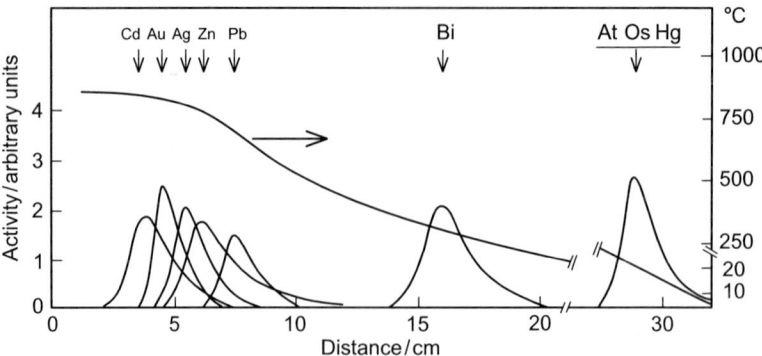

Fig. 1.14 Thermochromatography with sulfur vapor in the carrier gas [57].

Conditions: open silica column, length 100 cm, 3 mm i.d; duration of run 20 min; carrier gas argon, 10 cm^3 min^{-1}; reagent sulfur, 5 mmHg (effective concentration of monoatomic S).

Reprinted from JINR report P6-88-595, Korotkin YuS, Kim UJ, Timokhin SN, Orelowich OL, Altynov VA, Thermochromatography of sulfides of elements groups I–VIII, © 1988, with permission from Joint Institute for Nuclear Research.

1.5.1.3 Sulfides

Volatilized sulfides have not yet been utilized for "applied" separations. In the meantime, Korotkin, at al. [57] reported adsorption characteristics for a number of such compounds; a sample chromatogram is shown Fig. 1.14. The deposition temperatures of the investigated sulfides proved to correlate with the sublimation energies of the bulk compounds. Notice an enhanced volatility of the compounds with less than four sulfur atoms in the molecule; it seems to be attributable to less polar bonding than that in oxides. The paper offers an extended discussion of the potentialities of sulfides for isolation and radiochemical separations of elements.

1.5.1.4 Chlorides, Bromides and Corresponding Oxyhalides

Higher chlorides, higher bromides (to a less degree), oxychlorides and oxybromides have already been used in TAE studies, as was documented in previous sections of this chapter. In the pioneering works the chlorinating agents were $ZrCl_4$, $NbCl_5$, $TiCl_4$ and $SOCl_2$; later, in the tests with aerosol transportation, came HCl, Cl_2 and CCl_4. Carbon tetrachloride has some disadvantages: it only works on the filter at above 800 °C, when it decomposes to much less volatile C_2Cl_6 (sublimation point 184 °C) [23]. The isothermal chromatographic columns have usually moderate working temperatures, and C_2Cl_6 may condense and yield an aerosol. Then the particulates capture a fraction of the analyte, which is then retained in the column only for the gas hold-up time. This phenomenon may negatively affect the quality of experimental data.

1.5 Prospects for Future of Radiochemical Studies of Heavy Elements

Usually some mixtures of chlorinating reagents with oxygen are used to produce oxyhalides, though $SOCl_2$ alone can serve the purpose [58, 59]. The known oxyhalides in which the total number of the atoms attached to metal is four or more are volatile approximately like the binary halides with the same number of halogen atoms. It takes place because, in both cases, the positive charge of the metal atom is spatially shielded from electrostatic interaction with neighboring molecules [1]. The variety of oxyhalides of the metals of groups five to eight [60] is quite large. Occasionally, two such compounds of the same element (even in equal oxidation state) with different stoichiometry have similar volatility; it makes judging the formula difficult.

The brominating agents have been HBr, Br_2, as well as BBr_3 — a convenient and very powerful reagent. Bromides have been used less than chlorides, though they might have some advantages: deposition temperatures are similar, while the reagents are less corrosive than the chlorinating agents. Kim, et al. [61] compared thermochromatographic behavior of a number of elements as chlorides and bromides in laboratory batch experiments. The results are illustrated by Fig. 1.15. Differences are very small.

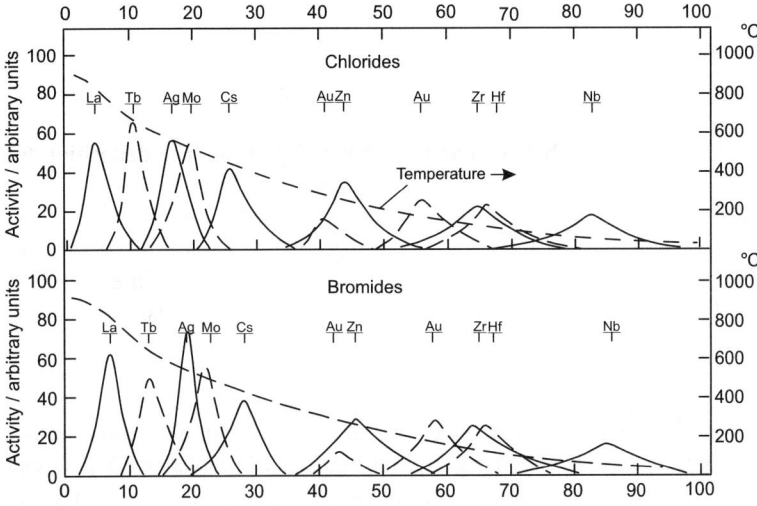

Fig. 1.15 Thermochromatograms of a mixture of radionuclides in the form of chlorides and bromides, obtained in otherwise similar conditions [61]. Conditions: open silica column, length 120 cm, 3 mm i.d; duration of run 20 min; carrier gas argon, $20\,cm^3\,min^{-1}$; reagents $SOCl_2$, 100 mmHg or BBr_3, 35 mmHg plus Br_2, 90 mmHg.

Reprinted from Environmental and Health Studies, 24(1), Kim UJ, Timokhin SN, Zvara I, The study of thermochromatography of bromides of various elements in relation to their chlorides (in Russian), 30–34, © 1988, with permission from Taylor and Francis Ltd., http://www.informaword.com.

It has been long known from macrochemical studies that $AlCl_3$ and $AlBr_3$, as well as the trihalides of other group 13 elements form gaseous complexes with the halides of the elements of groups 2 to 8 in various oxidation states [62–66]. Most of such complexes exist only in gas, in the presence of an excess of the complexing vapors of Al_2Cl_6 (dense vapors of all the group 13 trihalides are largely dimeric in gas). The number of attached ligands depends on the oxidation number of the metal and on the partial pressure of Al_2Cl_6. A simplest example is $La(AlCl_4)_3$ [67]. It proved that it is possible to chemically volatilize even the chlorides of Pd^{II} and Pt^{II}. Nevertheless, the complexes have not yet been exploited in radiochemistry, in spite of the successful chromatographic separation of lanthanoids and actinoids, which will be described shortly below.

1.5.1.5 Fluorides

Many heavy elements of group 5 and above form volatile pentafluorides and hexafluorides. In the same oxidation state, the fluorides are more volatile than the chlorides and bromides. What is more, in the fluorides or oxyfluorides, various elements (platinum metals, U–Cm and so forth) reach larger oxidation numbers than with other halogens. Then the highest fluorides are, of course, much more volatile. Widespread use of fluorides for radiochemical separations is prevented by the specific technical problems of handling the hazardous fluorination agents and by a limited choice of the materials which would withstand such reagents. Nevertheless, some French authors [68,69] (see also references in the papers) performed thermochromatographic studies of fluorides of some early transuranium elements. They observed a higher than expected microvolatility of these elements in TC experiments in nickel columns with HF, $HF + O_2$, F_2 or $BF_3 + F_2$ (the strongest oxidation reagent) as the carrier gas. Considering possible enhancement of the higher oxidation states, the authors of Ref. [69] conjectured that the previously unknown NpO_3F, NpF_7, PuF_5, PuO_3F, AmF_5, AmF_6, $CmOF_3$, CmF_6 and EsF_4 are partly responsible for the behavior of the elements. This assignment of the chemical states was based on a systematics of the deposition temperatures; the limitations of such an approach to identification of the chemical formulae will be considered in Chapter 5. The prospects for obtaining highest oxidation states, as well as compounds which are unstable as bulk phases, deserve more attention from radiochemists.

Hexafluorides of the elements of groups 6 to 8, not to mention similar compounds of U, Np and Pu, are very volatile. Nevertheless, their chromatographic separations are not easy because the volatilities (boiling points) of these compounds are quite close.

1.5.1.6 More Options

There are more exotic classes of volatile inorganic compounds like higher carbonyls, anhydrous nitrates or borohydrides. However, their use would probably bring severe experimental and technical problems: the appropriate chemical reagents and the compounds are not stable enough and are easily hydrolyzable or oxydizable (flammable).

1.5.1.7 Complexes with Heavy Organic Ligands

Many metallic elements form relatively volatile compounds with voluminous organic ligands. Well-known are hexafluoroacetylacenonates and heavier, highly fluorinated β-diketonates. Screening of the central atom by negatively charged ligands is, again, the source of considerable volatility of many chelates. Thermochromatography of hexafluoroacetylacetonates has been tested for use in radiochemistry of actinoids (from Th to Md) and for separation of some fission products by Fedoseev, Travnikov, et al. [70–72].

Gas chromatographic behavior of five dissimilar kinds of organic ligand complexes was studied in Ref. [73]. The complexing agents were various salicylaldimines, Schiff bases, a fluorinated β-diketone, a β-dithione and a hexadentate macrocycle. The metal ions included lanthanoids, transition metals, Pt, Pd and Zn. The monograph by Suglobov, et al. [74] gives more references related to organic ligands. Generally, chemical and thermal instability of the organic complexes constrain their applicability in gas-solid chromatography. In addition, kinetics of their formation has not yet been explored; it may be relatively slow. A careful study of the volatility and thermal stability of a number of β-diketonate chelates was reported by Berg and Acosta [75].

1.5.2 Groups of Related Elements

Refs. [45, 76] provide concise surveys of works on TC separations of common elements of all groups. They concern both compounds and elementary state. The present section quotes in more detail some works on separation of mixtures of the elements which have some common chemical properties, as well as of the elements which are related by their origin, like the fission products. The separations were performed by thermochromatography or by (eventually programmed) isothermal chromatography.

1.5.2.1 Lanthanoids and Actinoids

Clear interseparation of the lanthanoids or actinoids by gas chromatography of inorganic compounds is a challenge to experimental techniques. Thermochromatography does not have adequate resolving power for the separations, but the method proved very useful for adsorption studies of tracer metals. Hübener [77] was the first to realize the difficult experiments with some lanthanoids and the actinoids from Cf to Md; the tests included active metals of group 1 and 2 for comparison. The columns were made of Ti or Mo, the carrier gas was high purity He with some content of Ca vapor, which served to support the state of free atoms and to eliminate last traces of oxygen. The thermochromatograms are displayed in Fig. 1.16. Later work [78, 79] was done with extra pure He, in the presence of efficient solid getters

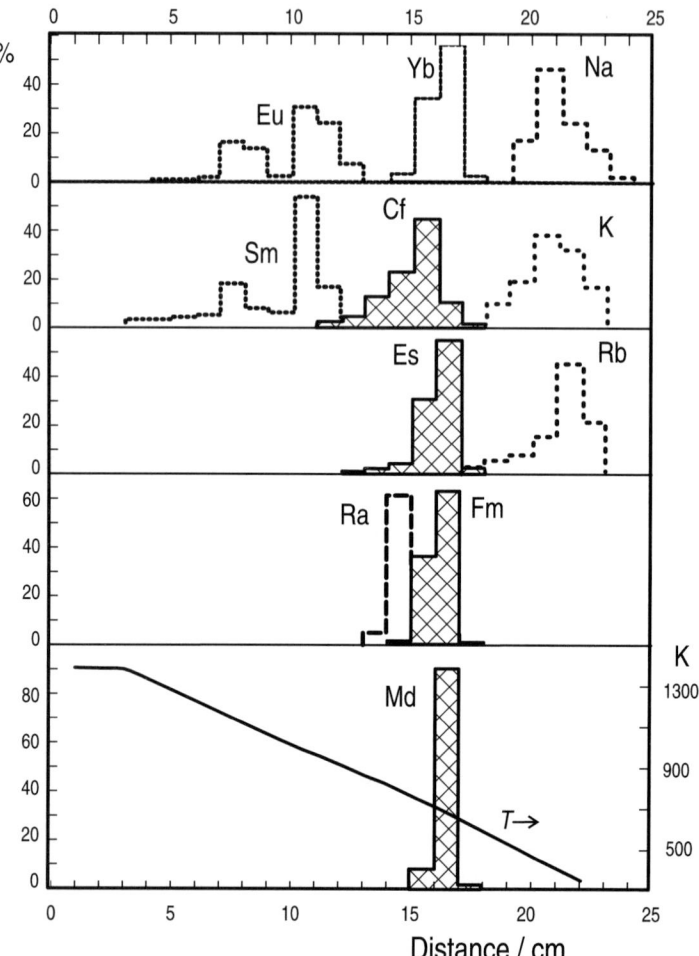

Fig. 1.16 Comparative thermochromatograms of some alkali, alkaline earth and rare earth metals in a titanium column [77]. Conditions: open titanium column, 4.5 mm i.d., average temperature gradient 57 K cm^{-1}, duration of run 15 min; carrier gas helium, 44 cm^3 min^{-1}; reagent 0.5 mg of Ca evaporated per run.

Reproduced from Radiochimica Acta, 31(1/2), Hübener S, Zvara I, Characterization of Some Metallic State Properties of Mendelevium and other Actinoids by Thermochromatography, 89–94, © 1982, with permission from Oldenbourg Wissenschaftsverlag.

1.5 Prospects for Future of Radiochemical Studies of Heavy Elements

for O_2; it provided adsorption data for the actinoids also on Fe, Nb and Ta surfaces, and achieved better resolution of the thermochromatograms.

The deposition temperature of known elements was found to correlate with their metallic valency; namely, the divalent metals were deposited at much lower temperatures than the trivalent ones. The data seen in Fig. 1.16 allowed deducing divalency of the metallic Fm and Md. This was the first experimental observation of this property; it enables some specific separations.

Successful clear separations of lanthanoids and actinoids by gas-solid chromatography were achieved owing to the volatile complexes of aluminium chloride with $LnCl_3$ [80] and $AnCl_3$ [81]. Unlike the trichlorides themselves, the complex molecules behave in the process of adsorption like the vapors of volatile molecular liquids. It is because they have composition and spatial structure of the kind shown in Fig. 1.17. Evidently, the outside of such a molecule looks very much like a cluster of the molecules of Al_2Cl_6. Because the complexes do not exist as condensed matter, an obligatory component of the carrier gas is a large concentration of Al_2Cl_6. Figures 1.18 and 1.19 present the results of elution chromatography on an isothermal or programmed temperature isothermal column. The separation of the halves of the lanthanoid series in one and one-half hours each is a very satisfactory achievement for a completely new method. Obviously, the method is not ready-to-use for any radiochemical problem. The separated elements exited the column mixed with a large amount of aluminium chloride and the chromatogram had to be detected by collecting fractions of the eluate, or by continuous deposition of the eluate on a moving belt.

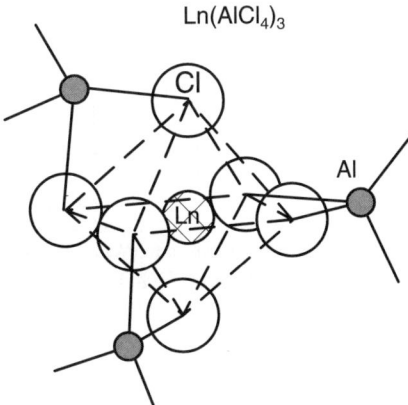

Fig. 1.17 Structure of $Ln(AlCl_4)_3$ molecules. The six "outer shell" chlorines attached to aluminium atoms are not explicitly shown.

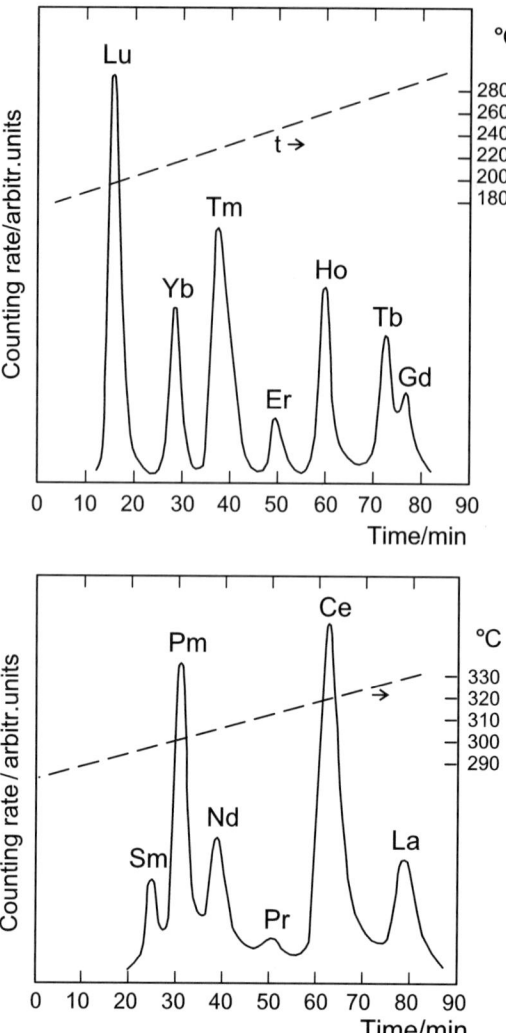

Fig. 1.18 Programmed temperature gas–solid chromatography of the complexes of lanthanoids with Al_2Cl_6 [82]. Conditions: Pyrex glass column, length 50 cm, 3 mm i.d., filled with 0.17 mm glass beads; carrier gas helium, 70 cm^3 min^{-1}; reagent 230 mmHg of Al_2Cl_6.

Adapted from Radiokhimiya, 16(5), Zvara I, Eichler B, Belov VZ, Zvarova TS, Korotkin YuS, Shalaevskii MR, Shchegolev VA, Hussonnois M, Gas chromatography and thermochromatography in the study of transuranium elements, 720–727, © 1974, with permission of Akademizdat "Nauka" Publishers.

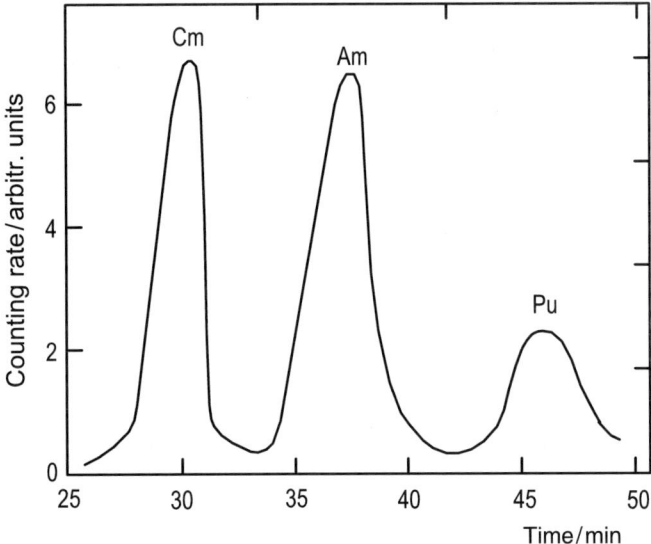

Fig. 1.19 Gas–solid IC chromatography of complexes of actinoids with Al_2Cl_6 [81]. Conditions: open glass column, length 250 cm, 1 mm i.d.; temperature 250 °C; carrier gas helium, 8 cm³ min⁻¹; reagent 100 mmHg of Al_2Cl_6.

Reprinted from Journal of Chromatography, 49(1), Zvarova TS, Zvara I, Separation of transuranium elements, 290–292, © 1970, with permission from Elsevier.

1.5.2.2 6p-elements of Groups 12 to 17

The problem of the superheavy nuclides with the atomic numbers around 114 stimulated TC separation studies of their expected chemical homologs — the elements of groups 13 to 18 in the sixth period of the Mendeleev system. With the use of H_2 as the reducing carrier gas, elements from Hg to Rn could be separated, mostly in their elemental state [83,84]. An illustration is given in Fig. 1.20. This approach was used in searching (yet unsuccessful) for SHEs in a uranium target after it had been bombarded with Xe ions [85].

1.5.2.3 Elements of Groups 7 to 10

Elements from Re to Pt are the expected homologs of elements 107 to 110. Chemistry of the halides in these groups is rather complicated; it is difficult to reveal any regularity which would allow reasonable extrapolations to the transactinoids. Meanwhile, the above closest congeners show an unexpected common behavior. In the TC experiments with either dry or moist air as a carrier gas, the tracer quantities of Re and platinum metals deposited at surprisingly low temperatures [53]. It indicated formation of gaseous oxides and hydroxides, which were mentioned above in this section. This group property is very specific and allows easy separation from the

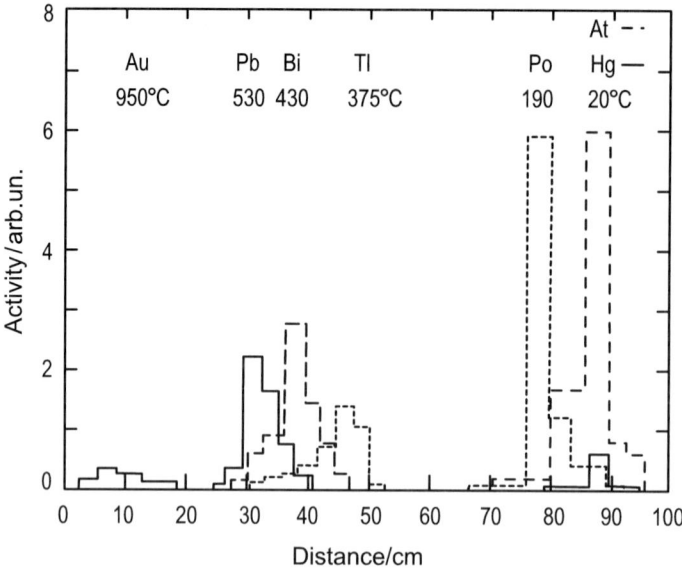

Fig. 1.20 Thermochromatographic separation of the expected homologues of SHE in metallic state [83]. The temperatures at the deposition zones maxima are indicated.

Reproduced (adapted) from JINR report, R12-6662, Eichler B. Thermochromatography of volatile metals, © 1974, with permission from Joint Institute of Nuclear Research.

elements of lower groups, which yield much less volatile hydroxides. The degree of separation from Hf, Ta and W was as high as 10^4. An attempt was made [85–87] to use this specific feature to search for s.f. isotopes in the products of an appropriate combination of the target and projectile; see Fig. 1.21. The range of detectable half-lives was from 0.5 s to 2×10^5 s. The obtained negative result (zero counts) indicated the effective cross-section less than 2×10^{-35} cm^2.

1.5.2.4 Fission Products

Rengan [88] reviewed the fastest gas phase isolations and separations of the fission products. Blachot, et al. [89] seem to be the first to separate the long-lived products of the neutron-induced fission of ^{235}U by thermochromatography of chlorides. Hickmann, et al. [22] made a careful study of fast continuous thermochromatographic separation of such fission products. They employed HCl, Cl_2, $SOCl_2$ and CCl_4 for chemical volatilization. Figure 1.22 shows their sample result.

Rudolph and Bächmann [90, 91] studied in detail isothermal chromatography of chlorides of several fission products. They used quartz columns filled with graphite, quartz or quartz coated with NaCl, KCl, $MgCl_2$ or CsCl. They made careful measurements of the retention time versus column temperature, and of the peak widths. The best separation was achieved by temperature-programmed chromatography. It is illustrated by Fig. 1.23.

1.5 Prospects for Future of Radiochemical Studies of Heavy Elements

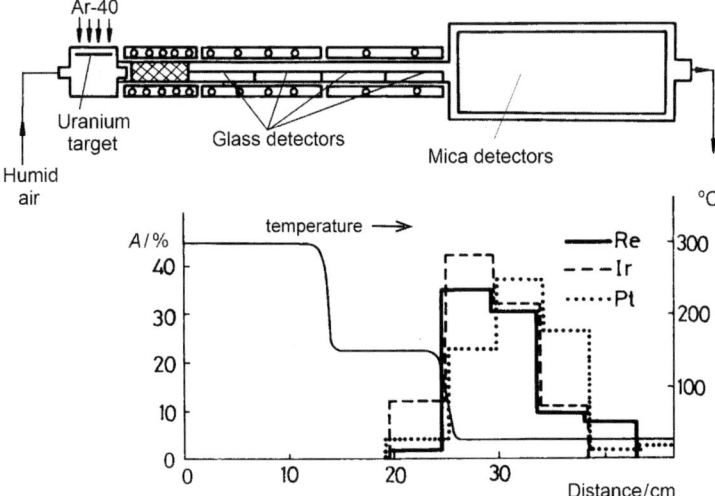

Fig. 1.21 Search for elements 107 to 110 as the products of the bombardment of ^{235}U with ^{40}Ar ions [86]. Quartz glass and mica served as SSTD of fission fragments.

Reproduced (adapted) from JINR Report R7-86-322, Domanov VP, Timokhin SN, Zhuikov BL, Chun KS, Eichler B, Chepigin VI, Zvara I, Search for spontaneously fissioning isotopes of elements 107–110 in the products of the interaction of ^{235}U $+^{40}$Ar, © 1986, with permission from Joint Institute for Nuclear Research.

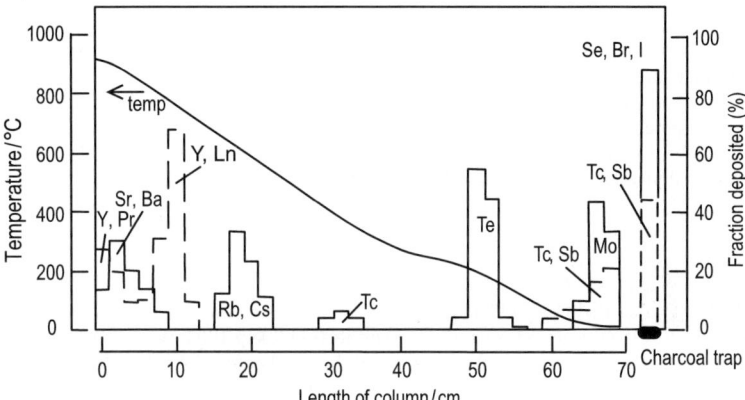

Fig. 1.22 Thermochromatographic separation of fission products in the form of chlorides [22]. Conditions: quartz column, 4 mm i.d., filled with quartz powder, 0.75 mm; carrier gas nitrogen, 1.2 L min^{-1}, reagent hydrogen chloride, 6 percent vol.

Reproduced (adjusted) from Nuclear Instruments and Methods, 174(3), Hickmann U, Greulich N, Trautmann N, Gäggeler H, Gäggeler-Koch H, Eichler B, Herrmann G, Rapid continuous radiochemical separations by thermochromatography in connection with a gas-jet recoil system, 507–513, © 1980, with permission from Elsevier.

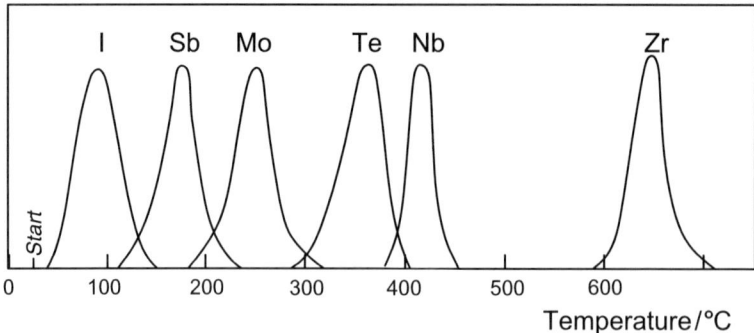

Fig. 1.23 Temperature-programmed separation of chlorides of fission products [90]. Conditions: quartz column, length 75 cm, 8 mm i.d., packed with 0.25 mm quartz particles coated with NaCl; heating rate 24.5 K min^{-1}; carrier gas nitrogen, 1 cm^3 min^{-1}; reagent CCl$_4$, 100 mmHg.

Reproduced from Chromatographia, 10(12), Bächmann K, Rudolph J, Inorganic gas chromatography Part I, 731–743, © 1977, with permission from author.

References

1. Zvara I (1996) J Radioanal Nucl Chem Articles 204:123
2. Merinis J, Boussieres G (1961) Analyt Chim Acta 25:498
3. Merinis J, Boussieres G (1969) Radiochim Acta 12:140
4. Zvara I, Zvarova TS, Krivanek M, Xu HG, Tarasov (1963) Dokl Akad Nauk SSSR 148:555
5. Zvara I, Zvarova TS, Krivanek M, Chuburkov YuT (1966) Radiokhimiya 8:77; (in French) Radiochimie 8:79
6. Zvara I, Zvarova TS, Krivanek M, Xu HG, Tarasov LK (1962) (in Russian) Report 1006. JINR, Dubna
7. Zvara I, Zvarova TS, Caletka R, Chuburkov YuT, Shalaevskii MR (1967) Radiokhimiya 9:231; Soviet Radiochem 9:226
8. Zvara I, Zvarova TS, Chuburkov YuT, Caletka R (1969) Radiokhimiya 11:154
9. Zvara I, Chuburkov YuT, Caletka R, Zvarova TS, Shalayevskii MR, Shilov BV (1966) At Energ 21:83; Soviet At Energ 21:709
10. Zvara I, Chuburkov YuT, Caletka R, Shalaevskii MR (1969) Radiokhimiya 11:163
11. Chuburkov YuT, Zvara I, Shilov BV (1969) Radiokhimiya 11:174
12. Price PB, Walker RM (1962) Phys Rev Lett 8:217
13. Price PB, Walker RM (1962) Phys Lett 3:113
14. Perelygin VP, Tretyakova SP, Zvara I (1964) Prib Tekh Eksp 1964:78
15. Greenwood NN, Hrynkiewicz A, Jeannin YP, Lefort M, Sakai M, Ulehla I, Wapstra AH, Wilkinson DH (1992) Progr Particle Nucl Phys 29:453
16. Wilkinson DH, Wapstra AH, Ulehla I, Barber RC, Greenwood NN, Hrynkiewicz A, Jeannin YP, Lefort M, Sakai M (1993) Pure Appl Chem 65:1757:1814
17. Zvara I, Chuburkov YuT, Belov VZ, Buklanov GV, Zakhvataev BB, Zvarova TS, Maslov OD, Caletka R, Shalayevsky MR (1970) Radiokhimiya 12:565; Soviet Radiochemistry 12: 530; J Inorg Nucl Chem 32:1885
18. Zvara I, Belov VZ, Chelnokov LP, Domanov VP, Hussonois M, Buklanov GV, Korotkin YuS, Schegolev VA, Shalayevsky MR (1972) Radiokhimiya 14:119
19. Zvara I, Belov VZ, Chelnokov LP, Domanov VP, Hussonois M, Buklanov GV, Korotkin YuS, Schegolev VA, Shalayevsky MR (1971) Inorg Nucl Chem Lett 7:1116
20. Zvara I, Yakushev AB, Timokhin SN, Xu HG, Perelygin VP, Chuburkov YuT (1998) Radiochim Acta 81:179

21. Zvara I, Domanov VP, Shalayevskii MR, Petrov DV (1980) (in Russian) Report 12-80-48. JINR, Dubna
22. Hickmann U, Trautmann N, Gäggeler H, Gäggeler-Koch H, Eichler B, Herrmann G (1980) Nucl Instrum Methods 174:507
23. von Dincklage RD, Schrewe UJ, Schmidt-Ott WD, Fehnse HF, Bächmann K (1980) Nucl Instrum Metods 176:529
24. Gäggeler H, Dornhofer H, Schmidt-Ott WD, Greulich N, Eichler B (1985) Radiochim Acta 38:103
25. Gäggeler HW, Jost DT, Baltensperger U, Weber A, Kovacs A, Vermeulen D, Türler A (1991) Nucl Instrum Methods Phys Res, Sect A 309:201
26. Gäggeler HW, Jost DT, Kovacs J, Sherer UW, Weber A, Vermeelen D, Türler A, Gregorich KE, Henderson RA, Czerwinski KR (1992) Radiochim Acta 57:93
27. Gäggeler HW (1994) J Radioanal Nucl Chem Articles 183:261
28. Türler A (1996) Radiochim Acta 72:7
29. Gäggeler HW, Jost DT, Baltensperger U, Weber A, Kovacs A, Vermeulen D, Türler A (1991) Nucl Instrum Methods Phys Res, Sect A 309:201
30. Türler A, Buklanov GV, Eichler B, Gäggeler HW, Grantz M, Hübener S, Jost DT, Lebedev VYa, Piguet D, Timokhin SN, Yakushev AB, Zvara I (1998) J Alloys Compd 271–273: 287
31. Kadkhodayan B, Türler A, Gregorich KE, Nurmia MJ, Lee D, Hoffman DC (1992) Nucl Instrum Methods Phys Res, Sect A 317:254
32. Sato TK, Tsukada K, Asai M, Akiyama K, Haba H, Toyoshima A, Ono S, Hirai T, Goto S, Ichikawa S, Nagame Y, Kudo H (2005) J Nucl Radiochem Sci 6(2):N1
33. Vahle A, Hübener S, Dressler R, Grantz M (2002) Nucl Instrum Methods Phys Res, Sect A 481:637
34. Kadkhodayan B, Türler A, Gregorich KE, Baisden PA, Czerwinski KR, Eichler B, Gäggeler HW, Hamilton TM, Jost DT, Kacher CD, Kovacs A, Kreek SA, Lane MR, Mohar MF, Neu MP, Stoyer NJ, Sylwester ER, Lee DM, Nurmia MJ, Seaborg GT, Hoffman DC (1996) Radiochim Acta 72:169
35. Sylwester E, Gregorich KE, Lee DM, Kadkhodayan B, Türler A, Adams JL, Kacher CD, Lane MR, Laue CA, McGrath CA, Shaughnessy DA, Strellis DA, Wilk PA, Hoffman DC (2000) Radiochim Acta 88:837
36. Türler A, Eichler B, Jost DT, Piguet D, Gäggeler HW, Gregorich KE, Kadkhodayan B, Kreek SA, Lee DM, Mohar M,, Hoffman DC, Hübener S (1996) Radiochim Acta 73:55
37. Türler A, Brüchle W, Dressler R, Eichler B, Eichler R, Gäggeler HW, Gartner M, Gregorich KE, Hübener S, Jost DT, Lebedev VY, Pershina VG, Schädel M, Taut S, Timokhin SN, Trautmann N, Vahle A, Yakushev AB (1999) Angew Chem, Int Ed 38:2212
38. Eichler R, Brüchle W, Dressler R, Düllmann ChE, Eichler B, Gäggeler HW, Gregorich KE, Hoffman DC, Hübener S, Jost DT, Kirbach UW, Laue CA, Lavanchy VM, Nitsche H, Patin JB, Piguet D, Schädel M, Shaughnessy DA, Strellis DA, Taut S, Tobler L, Tsyganov YS, Türler A, Vahle A, Wilk PA, Yakushev AB (2000) Nature 407:63
39. Düllmann ChE, Brüchle W, Dressler R, Eberhardt K, Eichler B, Eichler R, Gäggeler HW, Ginter TN, Glaus F, Gregorich KE, Hoffman DC, Jager E, Jost DT, Kirbach UW, Lee DM, Nitsche H, Patin JB, Pershina V, Piguet D, Qin Z, Schädel M, Schausten B, Schimpf E, Schott H-J, Soverna S, Südowe R, Thorle P, Timokhin SN, Trautmann N, Türler A, Vahle A, Wirth G, Yakushev AB, Zielinski PM (2002) Nature 418:859
40. Düllmann ChE, Eichler B, Eichler R, Gäggeler HW, Jost DT, Piguet D, Türler A (2002) Nucl Instrum Methods Phys Res, Sect A 479:631
41. Kirbach UW, Folden IIICM, Ginter TN, Gregorich KE, Lee DM, Ninov V, Omtvedt JP, Patin JB, Seward NK, Strellis DA, Sudowe R, Türler A, Wilk PA, Zielinski PM, Hoffman DC, Nitsche H (2002) Nucl Instrum Methods Phys Res, Sect A 484:587
42. Yakushev AB, Buklanov GV, Chelnokov ML, Chepigin VI, Dmitriev SN, Gorshkov VA, Hübener S, Lebedev VYa, Malyshev ON, Oganessian YuTs, Popeko AG, Sokol EA, Timokhin SN, Türler A, Vasko VM, Yeremin AV, Zvara I (2001) Radiochim Acta 89:743

43. Yakushev AB, Zvara I, Oganessian YuTs, Belozerov VA, Dmitriev SN, Eichler B, Hübener S, Sokol EA, Türler A, Yeremin AV, Buklanov GV, Chelnokov ML, Chepigin VI, Gorshkov VA, Gulyaev AV, Lebedev VYa, Malyshev ON, Popeko AG, Soverna S, Szeglowski Z, Timokhin SN, Tretyakova SP, Vasko VM, Itkis MB (2003) Radiochim Acta 91:433
44. Eichler R, Brüchle W, Buda R, Burger S, Dressler R, Düllmann ChE, Dvorak J, Eberhardt K, Eichler B, Folden III CM, Gäggeler HW, Gregorich KE, Haenssler F, Hoffman DC, Hummrich H, Jäger E, Kratz JV, Kuczewski B, Liebe D, Nayak D, Nitsche H, Piguet D, Qin Z, Rieth U, Schädel M, Schausten B, Schimpf E, Semchenkov A, Soverna S, Südowe R, Trautmann N, Thörle P, Türler A, Wierczinski B, Wiehl N, Wilk PA, Wirth G, Yakushev AB, von Zweidorf A (2006) Radiochim Acta 94:181
45. Novgorodov AF, Rösch F, Korolev NA (2003) Radiochemical separations by thermochromatography. In: Vértes A, Nagy S, Klencsár (eds) Handbook of nuclear chemistry Vol5. Kluwer, Netherlands, chap 7, p 227
46. Steffen A, Bächmann K (1978) Talanta 25:551
47. Steffen A Bächmann K (1978) Talanta 25:677
48. Zude F, Fan W, Trautmann N, Herrmann G, Eichler B (1993) Radiochim Acta 62:61
49. Domanov VP, Buklanov GV Lobanov Yu V (2002) J Nucl Sci Technol, Suppl 3, Nov:579
50. Domanov VP, Lobanov YuV (2004) In : Advances in nuclear and radiochemistry Qaim SM Coenen HH (eds) 6th InternatConf Nuclear and Radiochemistry NRC 6, Aachen 2004 Extended Abstracts, P 64
51. Hübener S, Bernhard G, Fanghanel Th, Advances in Nuclear and Radiochemistry Qaim SM Coenen HH (eds), 6th InternatConf Nuclear and Radiochemistry NRC 6, Aachen 2004 Extended Abstracts, P 66
52. Domanov VP, Eichler B, Zvara I (1984) (in Russian) Radiokhimiya 26:66
53. Domanov VP, Zvara I (1984) (in Russian) Radiokhimiya 26:770
54. Zvara I Domanov VP, Hübener S, Shalaevskii MR, Timokhin SN, Zhuikov BL, Eichler B, Buklanov GV (1984) Radiokhimiya 26:76; Soviet Radiochem 26:72
55. Vahle, A, Hübener, S, Funke, H, Eichler, B, Jost, DT, Türler, A, Brüchle, W, Jäger E (1999) Radiochim Acta 84:43
56. Hübener S, Taut S, Vahle A, Dressler R, Eichler B, Gäggeler HW, Jost DT, Piguet D, Türler A, Brüchle W, Jäger E, Schädel M, Schimpf E, Kirbach U, Trautmann N, Yakushev AB (2001) Radiochim Acta 89:737
57. Korotkin YuS, Jin KU, Timokhin SN, Orelovich OL, Altynov VA (1988) Report P6-88-595. JINR, Dubna
58. Hellas G, Hoffmann P, Bächmannl K (1977) Radiochem Radioanal Lett 30:371
59. Eichler R, Eichler B, Gäggeler HW, Jost, DT, Piguet D, Türler A (2000) RadiochimActa 88:87
60. Gartner, M, Boettger, M, Eichler, B, Gäggeler, HW, Grantz, M, Hübener, S, Jost, DT, Piguet, D, Dressler, R, Türler, A, Yakushev, AB: Radiochim Acta 78, (1997) 59
61. Kim UJ, Timokhin SN, Zvara I (1988) (in Russian) Isotopenpraxis 24:30
62. Schäfer H (1976) Z Angew Chem 88:775
63. Schäfer H (1981) Z Anorg Allg Chem 479:89,99
64. Schäfer H, Florke U (1981) Z Anorg Allg Chem 478:57
65. Papatheodorou GN (1982) Spectroscopy, structure and bonding of metal halide vapor complexes. In: E Kaldis (ed) Current topics in material sciences, vol 10. North Holland Amsterdam, chap 4, p 250
66. Boghosian S, Papatheodorou GN (1996) Rare earth halide vapors and vapor complexes. In: Gschneidner KA, Eyring LR (eds) Handbook on the physics and chemistry of rare earths. North Holland; Elsevier, Amsterdam
67. Oye HA, Gruen DM (1969) J Am Chem Soc 91:2229
68. Fargeas M, Fremont-Lamouranne R, Legoux Y, Merinis J (1986) J Less Common Metals 121:439
69. Boussieres G, Jouniaux B, Legoux Y, Merinis J, David F, Samhoun K (1980) Radiochim Radioanal Lett 45:121
70. Travnikov SS, Fedoseev EV, Davydov AV, Myasoedov BF (1985) J Radioanal Nucl Chem Lett 93:227

References

71. Fedoseev EV, Aizenberg MI, Timokhin SN, Travnikov SS, Zvara I, Davydov AV, Myasoedov BF (1987) J Radioanal Nucl Chem Articles 142:459
72. Fedoseev EV, Aizenberg MI, Travnikov SS, Davydov AV, Myasoedov BF (1989) J Radioanal Nucl Chem Lett 136:395
73. Robards K, Patsalides E (1999) J Chromatogr, A 844:181
74. Suglobov DN, Sidorenfo GV, Legin EK Letuchie organicheskie i kompleksnye soyedineniya f-elementov (Volatile organic and complex compounds of f-elements) (1987) Enorgoatomizdat, Moskva
75. Berg EW, Acosta JJC (1968) Analyt Chim Acta 40:101
76. Zvara I (1990) Isotopenpraxis 26:251
77. Hübener S, Zvara I (1982) Radiochim Acta 31:89
78. Hübener S, Eichler B, Schädel M, Brüchle W, Gregorich KE, Hoffman DC (1994) J Alloys Compd 213/214:429
79. Taut S, Hübener S, Eichler B, Gäggeler HW, Schädel M, Zvara I (1997) Radiochim Acta 78:33
80. Zvarova TS, Zvara I (1969) J Chromatogr 44:604
81. Zvarova TS, Zvara I (1970) J Chromatogr 49:290
82. Zvara I, Eichler B, Belov VZ, Zvarova TS, Korotkin YuS, Shalaevskii MR, Shchegolev VA, Hussonnois M (1974) Radiokhimiya 16:720; Soviet Radiochem 16:709
83. Eichler B (1972) (in Russian). Report R12-6662. JINR, Dubna
84. Eichler B (1973) J Inorg Nucl Chem 35:4001
85. Reetz T, Eichler B, Bruchertseifer H, Zhuikov BL, Belov VZ, Zvara I (1979) Radiokhimiya 21:877
86. Domanov VP, Timokhin SN, Zhuikov BL, Chun KS, Eichler B, Chepigin VI, Zvara I (1986) Search for spontaneously fissioning isotopes of elements 107–110 in the products of the interaction of U-235 + Ar-40 (in Russian). In: Heavy ion physics – 85. Report R7-86-322. JINR, Dubna, p. 16
87. Zvara I (1987) Radiochemical studies of new elements (in Russian). In: Proceedings of international school – seminar on heavy ion physics. Report D7-87-68. JINR, Dubna, p. 145
88. Rengan K (1990) J Radioanal Nucl Chem 142:173
89. Blachot LC, Carraz P, Cavallini A, Gadelle A (1969) In: 2eme Colloque sur la physique at la chimie de la fission. Report SM-122/88, IAEA, Vienna
90. Rudolph J, Bächmann K (1977) Chromatographia 10, 731
91. Rudolph J, Bächmann K (1979) Microchim Acta 71:477

Chapter 2
Physicochemical Fundamentals

Abstract Molecular kinetics provides initial interpretation of gas–solid chromatographic separations, which are based on the gas adsorption equilibrium. This chapter presents selected formulae and equations adapted for use later in the book. Their set contains the speed of gaseous molecules (also for two-dimensional gas), their rates of mutual collisions and chemical interaction, and the frequency of collisions with walls. The rate of desorption from surfaces is the Boltzmann factor multiplied by a frequency, which solely characterizes the crystal lattice of the adsorbent and can be calculated from semiempirical relationships. Some integrals of the Boltzmann factor times a power of temperature are presented in terms of the error or exponential integrals, as well as the appropriate asymptotic series. Because the data on molecular collision diameters and diffusion coefficients for many uncommon compounds are lacking, some semiempirical methods for their estimation are given. These diffusion coefficients, as well as those of aerosols, are needed to understand the rate of irreversible diffusional deposition of molecules and aerosols in channels, and to calculate the density of deposits and the penetration through ducts. The available formulae for the latter quantities were obtained by mathematical or engineering approaches. They deal with circular and rectangular channels, with the laminar and turbulent regimes, and with diffusionally and hydrodynamically developed flows; they are scarce for developing flows.

The specific problems discussed in this book require the use of fundamental concepts and equations from various fields like kinetic theory of gases, kinetics of chemical reactions, thermodynamics and mass transfer. This chapter presents some basic relationships relevant to these problems. From the very beginning, the studies of gas-phase radiochemistry of heavy metallic elements have been largely motivated by the quest for new man-made chemical elements. It necessitated experimentation with very short-lived nuclides on one-atom-at-a-time basis. We will pay much attention to this direction of research. Accordingly, we will consider microscopic pictures (at the atomic and molecular level) of the processes underlying the experimental methods and concrete techniques, and follow individual histories of the molecules.

Here and in later chapters, the formulae for some quantities are presented in two ways. One is a concise form obtained through all possible cancelling, as well as by combining numerical coefficients; it explicitly shows how the quantity depends on temperature, pressure, molar mass, geometric parameters, and so forth. The aim is to offer "ready-for-use" formulae, including the units of measurement of the result. In addition, typical values or orders of magnitude of the involved variables in common experimental conditions are sometimes recalled. The alternative is not to unveil quantities like the mean speed of molecules, which are involved in many equations of our interest. Such formulae help in estimating the order of magnitude of the result at a glance. In later chapters, quite often, we will need to restrict ourselves to the orders of magnitude because of the complexity of the tasks.

References are given in the present chapter only when the approach or treatment may be less known.

2.1 Molecular Kinetics

Most of the formulae below give the average values of quantities. For integer quantities, like concentration or frequency of collisions, in the case of poor statistics, the possible uncertainties in data evaluation are discussed when dealing with specific "applied" problems in Chapter 6.

2.1.1 Concentration and Speed of Gaseous Molecules

The average number of molecules per volume unit, the concentration, is:

$$n = N_A \frac{p}{RT} = \frac{p}{k_B T} \text{ cm}^{-3} \tag{2.1}$$

The number of molecules per cubic centimeter at standard temperature and pressure (STP), the Loschmidt number, is

$$L_N \equiv N_A \frac{p_0}{RT_0} = \frac{N_A}{22414} = 2.687 \times 10^{19} \text{ cm}^{-3} \tag{2.2}$$

so that

$$n = 2.687 \times 10^{19} \frac{p}{p_0} \frac{273.15}{T} \text{ cm}^{-3} \tag{2.3}$$

For different pressure units the appropriate numerical values are:

$$n = \frac{7.24 \times 10^{15}}{T} \frac{p}{\text{dyn cm}^{-2}} \text{ cm}^{-3}$$

$$n = \frac{7.24 \times 10^{16}}{T} \frac{p}{\text{Pa}} \text{ cm}^{-3}$$

2.1 Molecular Kinetics

$$n = \frac{9.66 \times 10^{18}}{T} \frac{p}{\text{mmHg}} \text{ cm}^{-3}$$

$$n = \frac{7.24 \times 10^{21}}{T} \frac{p}{\text{bar}} \text{ cm}^{-3}$$

The mean speed u_m of molecules with the mass m or the molar mass M is:

$$u_m = \sqrt{\frac{8k_B T}{\pi m}} = \sqrt{\frac{8RT}{\pi M}} \qquad (2.4)$$

When the numerical values of the constants and quantities are substituted in CGS units then, accurately enough, $u_m = 14550\sqrt{T/M}$ cm s^{-1}.

It holds for three-dimensional gas. On some special occasions we shall need the characteristics of two-dimensional gas; its mean speed is by about 22 percent smaller:

$$u_m^{(2)} = \sqrt{\frac{\pi k_B T}{2m}} = \sqrt{\frac{\pi RT}{2M}} \qquad (2.5)$$

For the two-dimensional pressure $p^{(2)}$ also holds that

$$n^{(2)} = \frac{p^{(2)}}{k_B T} \qquad (2.6)$$

so that at $p^{(2)} = 1$ dyn cm^{-2} and $T = 273.15$ K the concentration $n^{(2)}$ is 2.65×10^{13} cm^{-2}.

2.1.2 Number of Collisions with Wall

The mean speed serves to evaluate the rate of adsorption and the rate of chemical reaction of the adsorbed molecules with gaseous reagents. The average number of molecules striking unit area per unit time n_a is proportional to the concentration of the molecular entities and their mean speed:

$$n_a = \frac{n u_m}{4} = \frac{p}{\sqrt{2\pi m k_B T}} \qquad (2.7)$$

In CGS units it is numerically

$$n_a = 2.63 \times 10^{19} \frac{p}{\sqrt{MT}} \text{ cm}^{-2} \text{ s}^{-1} \qquad (2.8)$$

This number is usually of the order of 10^{23} cm^{-2} s^{-1} at STP (notice that 1 bar $= 1 \times 10^6$ dyn cm^{-2}).

Let us now consider a concrete molecule of a gas, which passes with the flow rate Q through an isothermal chromatographic column with the length z, the area per unit length a_z, and the free volume per unit length v_z. We shall soon need to know how many times each molecule hits the surfaces when moving through a column. Let Z_z be this number. It is obviously proportional to n_a, to the total surface za_z, and to zv_z/Q – the gas hold-up time; meanwhile, it is inversely proportional to the total number of molecules in the volume nzv_z. Hence:

$$Z_z = \frac{n_a z a_z}{n z v_z} \frac{z v_z}{Q} = \frac{n_a z a_z}{nQ} = \frac{u_m}{4} \frac{z a_z}{Q} \tag{2.9}$$

Thus, for a given sort of molecules and temperature, Z_z explicitly depends only on the total surface area of the column and on the true flow-rate. It is evidently a result of general validity — the size and internal geometry of the volume (including its eventual packing) are irrelevant. Notice that a_z must be the true surface area, which would take into account roughness and open porosity of real surfaces.

If Q_0 is the flow rate at an arbitrarily chosen standard temperature T_0 and pressure p_0, then

$$Z_z = 3640 \frac{z a_z T_0 p}{Q_0 T p_0} \sqrt{\frac{T}{M}} \tag{2.10}$$

and if Q_0 is measured at STP then:

$$Z_z = \frac{0.945 z a_z p}{Q_0 \sqrt{MT}} \tag{2.11}$$

2.1.3 Collisions in Gas and Rate of Chemical Interactions

In the first approximation the collision phenomena are described in terms of hard sphere molecular diameters, which are independent of temperature. Actually, the diameters decrease with higher temperature, approaching individual limits [1]. Let us consider a single molecular entity with the mass m_1 and the diameter $d_{m,1}$, which diffuses through a gas consisting mainly of more abundant dissimilar molecules of the mass m_2, the diameter $d_{m,2}$ and the concentration n_2. If the collision diameter is $\omega_{1,2} = (d_{m,1} + d_{m,2})/2$, the tracer molecule must collide each second with the host molecules contained in a volume of about $\pi \omega_{1,2}^2 u_m$. Because the host molecules also move, the mean relative speed $u_{1,2}$ is

$$u_{1,2} = \sqrt{\frac{8 k_B T}{\pi m_{1,2}}} \tag{2.12}$$

where $m_{1,2}$ is the reduced (harmonic mean) mass of the colliding particles:

$$m_{1,2} = \frac{m_1 m_2}{m_1 + m_2} \tag{2.13}$$

It is always smaller than the mass of the lighter partner; for identical molecules $m_{2,2} = m_2/2$.

2.1 Molecular Kinetics

The average number of collisions experienced by the tracer single per second is:

$$Z_{1,2} = \pi(n_1 + n_2)\omega_{1,2}^2 u_{1,2} \approx \pi n_2 \omega_{1,2}^2 u_{1,2} \tag{2.14}$$

Hence, in a common host gas at STP, the value of $Z_{1,2}$ is of the order of 10^9 s^{-1}; it is inversely proportional to the total bulk gas concentration. The mean free path between two collisions $\lambda_{m,1}$ is the mean speed divided by the above number of collisions:

$$\lambda_{m,1} = \frac{u_{1,2}}{n_{1,2}} \approx \frac{1}{\pi n_2 \omega_{1,2}^2} \tag{2.15}$$

It gives an order of 0.1 μm for the STP conditions. Evidently, $\lambda_{m,1}$ is proportional to temperature and inversely proportional to the total pressure.

In the experiments of our concern, the otherwise inert carrier gas usually contains a minor chemical active component. To estimate the rate of chemical interaction, we should take the concentration of the reagent for n_2, which is still by many orders of magnitude larger than n_1. Due to this, the reaction is formally of the first kinetic order, and a useful quantity is the mean time of interaction τ_r. It is obviously the reciprocal of the rate of "effective" collisions, the above n_{cz}, multiplied by the Boltzmann factor due to ε_{ea} – the energy of activation. It yields:

$$\tau_r = \frac{e^{\varepsilon_{ea}/k_B T}}{n_{1,2}} = \frac{e^{\varepsilon_{ea}/k_B T}}{\pi n_2 \omega_{1,2}^2 u_{1,2}} \tag{2.16}$$

The pre-exponential coefficient is taken here in its simplest form. A rigorous way to evaluate the collision diameters is offered by the fundamental, generally valid relationship between the dynamic viscosity of the gas μ_2 and its collision diameter:

$$\mu_2 = \frac{1}{\pi d_{m,2}} \sqrt{\frac{k_B T m_2}{\pi}} \tag{2.17}$$

The particular μ_2 and m_2 are taken here only to emphasize that the viscosity data are easily available only for major components of the carrier gas in gas–solid chromatography. Minor components of the gas and tracer species are less common compounds, and their gaseous viscosities are mostly unknown. The usual way to estimate the diameters is to assume that the molecules are hard spheres, and that they are closely packed in the condensed phase. Then the volume per molecule is $d_{m,1}^3/\sqrt{2}$ and the formula for the diameter is [2]

$$d_{m,1} = \left[\frac{\sqrt{2}M}{N_A \rho}\right]^{1/3} = 1.33 \times 10^{-8} \left(\frac{M}{\rho}\right)^{1/3} \text{cm} \tag{2.18}$$

where ρ is the density of the liquid or solid compound. The data in the upper lines of Table 2.1 are accurate values for 273.15 K obtained from the measurements of viscosity [1]. The lower part of the table contains the estimates based on Eq. 2.18, which are inherently of limited accuracy; they cannot be related to a particular

Table 2.1 Some Molecular (Atomic) Collision Diameters in Å

He	Ar	H_2	N_2	O_2	Cl_2	Br_2	HCl	H_2O	$SOCl_2$
2.17	3.66	2.73	3.76	3.62	5.55	6.16	4.50	4.32	5.6
$LaCl_3$	CCl_4	$TiCl_4$	$ZrCl_4$	$NbCl_5$	WCl_6	WO_2Cl_2	$ZrBr_4$	$TaBr_5$	OsO_4
5.3	6.0	6.3	5.7	6.0	6.3	5.2	6.0	6.4	4.9

temperature. We see that such data for the molecular compounds of the heavy elements fall in a rather narrow range. The differences are insignificant if we take into account other uncertainties involved in the microscopic picture of the physicochemical processes, which will be described in later chapters.

2.1.4 Diffusion in Gases

This process governs the rate of deposition of the molecules of nonvolatile compounds on the surface of gas ducts, and contributes to broadening of the chromatographic zones. Being of the order of 0.1 μm at STP, the mean free path of molecules, which is inversely proportional to pressure, reaches 1 cm only at about 0.01 mmHg. In dense enough gas, in the absence of convective flow, the macroscopic picture of migration of molecules (as well as of aerosol particulates) is described by the equations of diffusion. The mean squared diffusional displacement z_D^2 of molecules, the time of diffusion t and the mutual diffusion coefficient $D_{1,2}$ are related by:

$$z_D^2 = 2 D_{1,2} t \tag{2.19}$$

Molecular kinetic considerations of mutual diffusion [1] lead to the equation

$$D_{1,2} = \frac{u_{1,2}}{2\pi \omega_{1,2}^2 (n_1 + n_2)} \approx \frac{u_{1,2}}{2\pi \omega_{1,2}^2 n_2} \tag{2.20}$$

from which it follows that $D \sim T^{3/2}/p$.

The microscopic picture of diffusion in gases consists in that the molecule changes its velocity (vector) after each collision. The free paths are essentially random flights (displacements), which can be treated by the appropriate statistical theories. The latter conclude that $D_{1,2}$ is proportional to the frequency of random flights multiplied by their average squared length. Because the free molecular path lengths have exponential distribution [3], their average squared length equals the squared λ_m. Thus we come to

$$D_{1,2} = n_{1,2} \lambda_{m,1}^2 / 2 = u_{1,2} \lambda_{m,1} / 2 \tag{2.21}$$

which completely agrees with Eq. 2.20.

A well-known semiempirical equation for diffusion coefficient proposed by Gilliland [4] is

$$D_{1,2} = \frac{0.0044 \cdot T^{3/2}}{p \left[(M_1/\rho_1)^{1/3} + (M_2/\rho_2)^{1/3} \right]^2} \left(\frac{1}{M_1} + \frac{1}{M_2} \right)^{1/2} \text{ cm}^2 \text{s}^{-1} \tag{2.22}$$

2.1 Molecular Kinetics

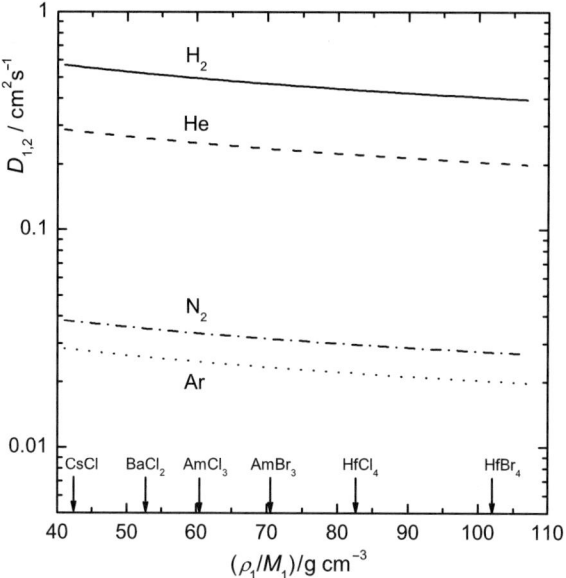

Fig. 2.1 Coefficients of mutual diffusion of tracers in various carrier gases at 1 bar and 500 K as a function of the molar volume of the condensed tracer; see text for explanations. Arrows indicate molar volumes of the particular compounds.

where p is in bars. The equation explicitly includes the reduced mass, cf. Eq. 2.13, and accounts for the theoretical dependence of $D_{1,2}$ on T and p. Evidently, it includes evaluation of the collision diameter in a manner consistent with Eq. 2.18.

We used Eq. 2.22 to calculate some $D_{1,2}$ values, which are related to the conditions and results of the experiments with the halides of heavy elements. The results are displayed in Fig. 2.1. In such studies the molar mass of the tracer is usually much larger than that of the carrier gas. Then the term $1/M_1$ in the parentheses in Eq. 2.22 becomes negligible, and the diffusion coefficients could be plotted in Fig. 2.1 as a function of only the molar volume. The $D_{1,2}$ values are given for 500 K; the figures for 273 K are less by a factor of 0.77, those for 700 K are obtained by multiplying with 1.18. We see that the combinations of common carrier gases with heavy voluminous molecules are characterized by $D_{1,2}$ of the order of 0.01–0.1 cm^2 s^{-1} at atmospheric pressure.

Some other proposed equations for $D_{1,2}$ better fit extended experimental data and point to a stronger temperature dependence of the coefficient, like $D_{1,2} \sim T^{1.75}$. A concise informative survey and critical discussion of them was presented by Giddings [5]. Unfortunately, these more sophisticated equations usually require knowledge of some molecular parameters, which are known only for a limited number of gaseous species. No values are available for the molecular compounds of metals. Thus, we are forced to adhere to the Gilliland's prescription and, if necessary, to estimate the unknown molar volumes based on the available data for the compounds of similar stoichiometry.

2.1.5 Elementary Adsorption–Desorption Event

Gas–solid chromatography is based on the fact that molecules migrating down the column experience numerous adsorption–desorption events on the walls of the tube and on the surface of an eventual packing. Let us assume that the surfaces are smooth and homogeneous. The simplest theory [6] suggests that at each collision with surface the molecule is adsorbed for a mean time interval τ_a obeying

$$\tau_a = \tau_0 e^{\varepsilon_d/k_B T} \tag{2.23}$$

where τ_0 is the period of vibration (perpendicular to the surface) in the adsorbed state, and ε_d is the energy of desorption per molecule. Due to the Boltzmann factor the random values of the adsorption sojourn time τ_a^* have an exponential probability density distribution:

$$\rho(\tau_a^*) = \frac{e^{-\tau_a^*/\tau_a}}{\tau_a} \tag{2.24}$$

The vibrations of adsorbed entities are induced by the vibrations of the adsorbent crystalline lattice. Hence, τ_0 is assumed to be a property of the adsorbent only and is supposed to be independent of temperature. Meanwhile ε_d characterizes the interaction between adsorbate and adsorbent. In the case of tracers, adsorbate – adsorbate interactions can be neglected.

The lattice vibrations have a complex spectrum; in various applications it is often replaced by an effective frequency ν_0, and τ_0 is taken as its reciprocal. Several approaches and formulae have been proposed for calculation of its value. In particular, Lindemann [7] proposed the relationship

$$\tau_0 = 4.75 \times 10^{-13} \sqrt{\frac{(M/\kappa_M)(M/\kappa_M \rho)^{2/3}}{T_{mp}}} \; s \tag{2.25}$$

where κ_M is the number of atoms in the molecular formula and T_{mp} is the melting point temperature. For example, in the case of SiO_2 ($M = 60$, $\kappa_M = 3$, $\rho = 2.21$ g cm^{-1}, $T_{mp} = 2001$ K) we obtain $\tau_0 = 1 \times 10^{-13}$ s.

The reciprocal of the Debye's [8] maximal allowed phonon frequency is

$$\tau_0 = \left(\frac{4\pi}{3}\frac{M}{\rho}\frac{1}{\kappa_M N_A}\right)^{1/3} \frac{1}{w_{sd}} \; s \tag{2.26}$$

where w_{sd} is the effective speed of sound — 5.9×10^5 cm s^{-1} for silica. This formula yields $\tau_0 = 0.7 \times 10^{-13}$ s. Notice that $\kappa_M N_A/M$ is the number density of atoms per unit volume of the adsorbent matrix.

2.1.6 Integrals Containing Boltzmann Factor

The mathematics of the problems in thermochromatography involves integration of the Boltzmann-type factor multiplied by a power of temperature. It is required to evaluate integrals of the general form

2.1 Molecular Kinetics

$$\int T^k e^{\frac{\varepsilon_d}{k_B T}} dT \tag{2.27}$$

with integer or half-integer k values in the range $[2, -2]$. By the substitutions $a_\varepsilon = \frac{\varepsilon_d}{k_B}$ and $u = \frac{1}{T}$, so that $dT = -\frac{du}{u^2}$, the integral is transformed into $\int \frac{e^{a_\varepsilon u}}{u^{k+2}} du$. Such integrals are expressed through the special functions erfi(x) and Ei(x). The functions can be expanded into asymptotic series [9]. Namely:

$$\text{erfi}(x) \approx \frac{e^{x^2}}{x\sqrt{\pi}} \left(1 + \frac{1}{2x^2} + \frac{1 \cdot 3}{4x^4} + \frac{1 \cdot 3 \cdot 5}{8x^6} + \cdots \right) \tag{2.28}$$

and

$$\text{Ei}(x) \approx \frac{e^x}{x} \left(1 + \frac{1!}{x} + \frac{2!}{x^2} + \frac{3!}{x^3} + \cdots \right) \tag{2.29}$$

The basic integrals are:

$k = -1$

$$\int \frac{e^{a_\varepsilon u}}{u} du = \text{Ei}(a_\varepsilon u) \approx \frac{e^{a_\varepsilon u}}{a_\varepsilon u} \left(1 + \frac{1!}{a_\varepsilon u} + \frac{2!}{(a_\varepsilon u)^2} + \frac{3!}{(a_\varepsilon u)^3} + \cdots \right) \tag{2.30}$$

and

$k = -3/2$

$$\int \frac{e^{a_\varepsilon u}}{\sqrt{u}} du = \sqrt{\frac{\pi}{a}} \text{erfi}(\sqrt{a_\varepsilon u}) \approx \frac{e^{a_\varepsilon u}}{a_\varepsilon \sqrt{u}} \left(1 + \frac{1}{2a_\varepsilon u} + \frac{1 \cdot 3}{4(a_\varepsilon u)^2} + \frac{1 \cdot 3 \cdot 5}{8(a_\varepsilon u)^3} + \cdots \right) \tag{2.31}$$

The rest of the integrals are obtained through integration by parts:

$k = -1/2 \quad \int \frac{e^{a_\varepsilon u}}{u^{3/2}} du = 2 \left[\sqrt{\pi a_\varepsilon} \, \text{erfi}(\sqrt{a_\varepsilon u}) - \frac{e^{a_\varepsilon u}}{\sqrt{u}} \right] \tag{2.32}$

$k = 0 \quad \int \frac{e^{a_\varepsilon u}}{u^2} du = -\frac{e^{a_\varepsilon u}}{u} + a_\varepsilon \, \text{Ei}(a_\varepsilon u) \tag{2.33}$

$k = 1 \quad \int \frac{e^{a_\varepsilon u}}{u^3} du = -\frac{e^{a_\varepsilon u}}{2u^2} - \frac{a_\varepsilon e^{a_\varepsilon u}}{2u} + \frac{a_\varepsilon^2}{2} \text{Ei}(a_\varepsilon u) \tag{2.34}$

$k = 2 \quad \int \frac{e^{a_\varepsilon u}}{u^4} du = -\frac{e^{a_\varepsilon u}}{3u^3} + \frac{a_\varepsilon}{3} \int \frac{e^{a_\varepsilon u}}{u^3} du \tag{2.35}$

In the experimental works discussed in this book, the real values of $a_\varepsilon u \equiv \varepsilon_d / k_B T$ have ranged from about 15 to 30. If so, just the first terms of the above power series 2.30 or 2.31 suffice to provide the precision that is required when treating experimental data. It can be verified that the series truncated this way yield a common approximate formula for the six integrals:

$$\int T^k e^{\frac{\varepsilon_d}{k_B T}} dT \approx \frac{k_B}{\varepsilon_d} T^{k+2} e^{\frac{\varepsilon_d}{k_B T}} \left[1 + (k+2) \frac{k_B T}{\varepsilon_d} \right] \tag{2.36}$$

2.2 Diffusional Deposition of Particles in Channels

In many experiments described and referenced in Chapter 1 an important role was played by diffusional deposition of nonvolatile molecular species (and of particulate matter) on the surface of gas ducts. First of all, in chromatographic columns, the nonvolatile molecules deposit within a short distance from the inlet and contribute to decontamination of the elements under study from interfering radioactive nuclides. Meanwhile, especially in chemical experiments with the aid of aggressive gases, unwanted aerosols occasionally form. The latter diffuse much more slowly, and when they catch and then carry a fraction of the nonvolatile molecular species, they make the depositional decontamination less efficient.

On the other hand, the aerosol "jets" have been widely utilized for primary transportation of the nuclear reaction products, from an accelerator to distant experimental equipment. Therefore, the major concern has been keeping the deposition rate of aerosols as low as possible.

All this substantiates our interest in understanding the regularities of deposition. In addition, the formulae for penetration of channels, and especially those for the deposition density profile, proved helpful when developing a Monte Carlo technique of simulation of chromatographic zones, which is described in Chapter 4.

2.2.1 Diffusion Coefficients of Aerosols

The values of D_{Bn} — the diffusivity for the Brownian motion of aerosol — are calculated from the Stokes–Einstein equation. For spherical particulates with the effective radius r_p, in a gas with the dynamic viscosity μ_2 (nearly constant for pressures about and less than one bar), the formula is:

$$D_{Bn} = \frac{k_B T c_C}{6\pi \mu_2 r_p} \quad (2.37)$$

Here c_C is the Cunningham slip correction factor given in Table 2.2. It should be included when r_p is comparable with the mean free path of the gas molecules.

Notice that the equation was evaluated by considering a particulate which moves through a fluid being pushed by the force resulting from impacts of many molecules of the (viscous) medium. In the same time, it experiences hydrodynamic resistance (friction). The dynamical viscosity of gases rather weakly depends on p and

Table 2.2 Cunningham Factor c_C [10] and the Dependence of Aerosol Diffusivity on Temperature and Pressure

λ/r_p	$r_p/\mu m$ in air at STP	Cunningham factor	$D_{Bn} = D_{Bn}(T, p)$
>2.6	<0.025	$1.66 \cdot \lambda/r_p$	$\sim T^{3/2} p^{-1}$
0.15–2.6	0.025–0.5	$2.61 \cdot (\lambda/r_p)^{1/2}$	$\sim T p^{-1/2}$
<0.15	>0.5	1	$\sim T^{1/2} p^0$

2.2 Diffusional Deposition of Particles in Channels

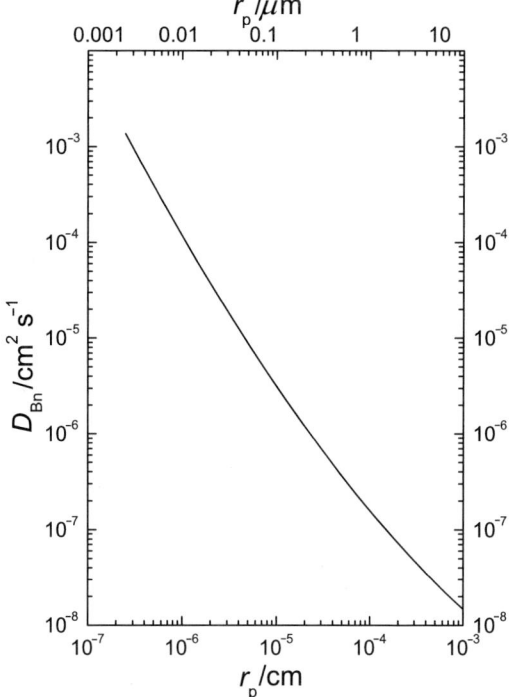

Fig. 2.2 Diffusion coefficients of the aerosol particulates as a function of their size at 300 K according to Eq. 2.37.

increases like $T^{1/2}$. Meanwhile, $\lambda_m \sim T/p$. The dependence of D_{Bn} on T and p observed for aerosol particulates greatly differs from $D \sim T^{1.5-1.75}/p$ for molecular species. D_{Bn} does not explicitly depend on the mass of the particulate; however, above certain mass, the velocity of gravitational fall out-performs the displacement due to diffusion (see Sect. 3.1.1).

The calculated functional dependence presented in Fig. 2.2 is roughly the same for helium, nitrogen and argon carrier gases, which at 300 K have similar viscosity — about 2×10^{-4} g cm^{-1} s^{-1} (= 20×10^{-6} Pa s). The viscosity of hydrogen is only one-half of the value so that diffusion in hydrogen is markedly faster.

2.2.2 Deposition from Laminar Flow

Several published works have been devoted to the mathematical solution of the problems of penetration of aerosols through a "diffusion battery," which is essentially a bundle of identical circular or rectangular channels in parallel [11]. The deposited matter is characterized only by its diffusion coefficient; it means that the derived formulae hold for aerosol particulates, as well as for molecular entities.

Neglecting the longitudinal diffusion, for the case of fully developed laminar flow in isothermal channels of the length z, the fractional penetration was found to be

$$F_p(z) = \sum_{i=0}^{\infty} a_i e^{-\beta_i \psi z} \qquad F_p(z_\psi) = \sum_{i=0}^{\infty} a_i e^{-\beta_i z_\psi} \qquad (2.38)$$

where $\psi \sim D/Q$; while the proportionality coefficient depends on the shape of the channel cross section (circular, rectangular, etc.), and $z_\psi \equiv \psi z$ is the "reduced" distance. The set of numerical coefficients a_i depends on the channel shape and on the distribution of the adsorbate concentration over the column section at $z = 0$. However, the sum of the infinite series obeys $\Sigma a_i = 1$. The eigenvalues β_i increase with i, their set depends only on the channel shape.

The longitudinal profile of the deposit density $\varphi_p(z_\psi)$ is obtained by differentiation:

$$\varphi_p(z_\psi) = -\frac{d}{dz_\psi} F_p(z_\psi) = \sum_{i=0}^{\infty} a_i \beta_i \, e^{-\beta_i z_\psi} \qquad (2.39)$$

In the true length units, it is convenient to present the distribution in the form

$$\varphi_p(z) = \sum_{i=0}^{\infty} \frac{a_i}{\eta_i} e^{-z/\eta_i} \qquad \eta_i = \frac{1}{\beta_i \psi} \sim \frac{Q}{D} \qquad (2.40)$$

to show that the density profile is the sum of weighted exponential terms characterized by steadily decreasing mean deposition distances η_i. Each such term is responsible for the fraction a_i of the total area under the profile.

Below, we will use appropriate mnemonic superscripts for the quantities related to circular or rectangular channels.

2.2.2.1 Circular Channels

If at $x = 0$ the concentration of the tracer over the column section is uniform, the numerical coefficients are now

$$a_0 = 0.819 \quad a_1 = 0.0976 \quad a_2 = 0.0325 \quad a_3 = 0.0154 \quad a_4 = 0.0088 \quad a_5 = 0.00573\ldots$$
$$\beta_0 = 3.65 \quad \beta_1 = 22.3 \quad \beta_2 = 56.9 \quad \beta_3 = 107.6 \quad \beta_4 = 174.3 \quad \beta_5 = 257\ldots$$

$$(2.41)$$

and

$$\psi^\circ = \frac{\pi D}{Q} \text{ cm}^{-1} \qquad \eta_i^\circ = \frac{1}{\beta_i} \frac{Q}{\pi D} \text{ cm} \qquad (2.42)$$

Hence

$$F_p^\circ(z_\psi) = 0.819 \, e^{-3.65 z_\psi} + 0.0976 \, e^{-22.3 z_\psi} + 0.0325 \, e^{-56.9 z_\psi} + \ldots \qquad (2.43)$$

and

$$\varphi_p^\circ(z_\psi) = 2.989 e^{-3.65 z_\psi} + 2.176 e^{-22.3 z_\psi} + 1.849 e^{-56.9 z_\psi} + \ldots \qquad (2.44)$$

2.2 Diffusional Deposition of Particles in Channels

As the values β_i rapidly increase in the series and so η_i values decrease, the shape of $\varphi_d^\circ(z)$ at sufficiently large z is determined mostly by the first term with

$$\eta_0^\circ = Q/3.65\pi D \tag{2.45}$$

On the other hand, near the column inlet, the deposition profile has a larger (negative) slope, which is described by the latter terms of the series. For small values of z_ψ, the infinite series by Eq. 2.44 converges very slowly, which makes the calculations cumbersome. There were simpler working formulae proposed for the penetration at small z_ψ's to avoid the evaluation of many members of the series. According to Ref. [12], near the inlet of a circular channel, at $z_\psi < 0.031$,

$$F_p^\circ(z_\psi) = 1 - 2.566 \cdot z_\psi^{2/3} + 1.20 z_\psi + 0.177 z_\psi^{4/3} \tag{2.46}$$

with the derivative

$$\varphi_p^\circ(z_\psi) = 1.71 z_\psi^{-1/3} - 1.20 - 0.24 z_\psi^{1/3} \tag{2.47}$$

At $z_\psi > 0.031$, the sums of just three first terms in Eqs. 2.43 and 2.44 closely approximate $F_d^\circ(z_\psi)$ and $\varphi_d^\circ(z_\psi)$, respectively.

2.2.2.2 Rectangular Channels

For the rectangular channels the diffusional deposition problem was solved only for an infinitely wide duct. Real channels have a finite ratio of the height to the width, b_h/b_w. If it is $\ll 1$, the formula allows one to use the flow rate rather than the flow velocity. The general form of the solution for such channels with

$$\psi^\square = \frac{8 b_w D}{3 b_h Q} \tag{2.48}$$

is in every respect similar to Eqs. 2.38 and 2.40. As many as 20 pairs of the values of numerical coefficients were reported [13]. The very first of them are:

$a_0 = 0.910 \quad a_1 = 0.0531 \quad a_2 = 0.0153 \quad a_3 = 0.0068 \quad a_4 = 0.0037, \quad a_5 = 0.0023\ldots$
$\beta_0 = 2.83 \quad \beta_1 = 32.2 \quad \beta_2 = 93.5 \quad \beta_3 = 186.8 \quad \beta_4 = 312.1 \quad \beta_5 = 469.5\ldots$

$$\tag{2.49}$$

Now we obtain:

$$\eta_0^\square = 0.375 \frac{b_h}{b_w} \frac{Q}{2.828 D} \tag{2.50}$$

A formula derived by Gormlay and Kennedy [12] for the initial part of the deposition profile, at $z_\psi < 0.013$ is

$$F_p^\square(z_\psi) = 1 - 1.54 z_\psi^{2/3} + 1.50 z_\psi \tag{2.51}$$

with the derivative:
$$\varphi_p^{\square'}(z_\psi) = 1.03\, z_\psi^{-1/3} - 1.50 \qquad (2.52)$$
Again, three first terms of the appropriate "rigorous" infinite series for $F_p^{\square}(z_\psi)$ and $\varphi_p^{\square}(z_\psi)$ suffice if $z_\psi > 0.013$.

The proportionality of η to Q/D in both cases and the absence of any dependence on the cross section area of the channel are easy to understand. First, the molecules which are initially near the tube walls reach the surface by diffusion in a short time; they are also carried by the laminar flow only a small distance. The layers near the surface are soon depleted, and the resulting steady *shape* of the concentration profile of the tracer over the radius no longer depends on that profile at the inlet of the channel. Then the probability of deposition per unit length gets constant and only one exponential term must characterize the density profile of the deposit. Second, the respective mean deposition length must be proportional to the flow velocity multiplied by the time of diffusion across the tube section: For circular channels we obtain $\eta_0^\circ \sim \frac{Q}{d_c^2}\frac{d_c^2}{D} \sim \frac{Q}{D}$ and for rectangular — $\eta_0^\circ \sim \frac{Q}{b_h b_w}\frac{b_h^2}{D} \sim \frac{b_h}{b_w}\frac{Q}{D}$. In the latter case the dimensions do appear in Eq. 2.50, but only as they ration, not the absolute values.

2.2.3 Diffusional Deposition — Engineering Approach

The engineering approach to the mass-transfer suggests full analogy between heat and mass-transfer when making use of dimensionless hydrodynamic and transport criteria. For our problems, and with the numerical subscripts to indicate the appropriate standard components of our chemical systems, we have:

$$Re = w d_c \rho_2 / \mu_2 \quad \text{(flow is laminar at Re} < 2300); \qquad (2.53)$$
$$Sc = \mu_2 / \rho_2 D \quad \text{is the analog of Pr};$$
$$Sh = k_m d_c / D \quad \text{is the analog of Nu};$$

Here k_m is the mass transfer coefficient and D stays either for $D_{1,2}$ or D_{Bn}. Notice that the engineering approach does not propose individual formulae for different geometric cross sections of the channels. Instead, the non-cylindrical channels are characterized by an effective diameter. For rectangular channels it is $2b_h b_w/(b_h + b_w)$, which approaches $2b_h$ for wide channels of small height.

Irreversible deposition of a tracer from the flow onto a channel wall is analogous to heat transfer at constant wall temperature. The formulae can be applied to our problems straightforwardly, because there are no gradients of macroscopic concentrations and no temperature fields to change the properties of the bulk fluid.

Below, we shall be interested in the formulae for penetration of channels and for deposition profiles in the case of laminar and turbulent steady flow regimes, and at various initial situations at the inlet of the channel — whether the flow is diffusionally and/or hydrodynamically developed. The engineering approach is not rigorous; various authors proposed different formulae which, nevertheless, yield reasonably consistent results. The author made a deliberate choice from the alternate formulae.

2.2 Diffusional Deposition of Particles in Channels

2.2.3.1 Diffusionally Developing Hydrodynamically Developed Laminar Flow

The title case for circular channels was characterized above by Eqs. 2.43 and 2.44. In the engineering approach [14, 15]:

$$\text{Sh} = 3.657 + \frac{0.19\,[\text{ReSc}\,d_c/z]^{0.8}}{1 + 0.117\,[\text{ReSc}\,d_c/z]^{0.467}} \qquad (2.54)$$

So, at large distances from the inlet, when $d_c/z \ll 1$, and in fully stabilized laminar flow, the Sh value is constant: $\text{Sh} = 3.657$. Then in agreement with Eq. 2.45 the deposition length is $1/\psi^\circ \text{Sh} = 1/3.657\psi^\circ$.

The formulae for penetration and for the deposit density can be obtained by the following way [16]. At a particular coordinate, the fraction of the tracer mass deposited per unit tube length in unit time is the mass flux to the surface, divided by the mass carried by the gas flow per unit time. In a cylindrical tube the deposition rate is $k_m c_g \times \pi d_c/\psi^\circ$ and the carried amount is Qc_g. The two quantities have the same dimension and their ratio is

$$\frac{\pi D}{Q} \frac{k_m d_c}{D\psi^\circ} = \text{Sh} \qquad (2.55)$$

In our notations

$$\text{ReSc}\frac{d_c}{z} = \frac{w d_c \rho}{\mu_f} \frac{\mu_f}{\rho D} \frac{d_c}{z} = \frac{4}{z_\psi} \qquad (2.56)$$

Then, at a given coordinate, the fraction of molecules or aerosol particulates staying in the gas, which is deposited per reduced length unit, is

$$\frac{dF_p^\circ(z_\psi)}{dz_\psi} \frac{1}{F_p^\circ(z_\psi)} = \text{Sh} = 3.657 + \frac{0.576}{z_\psi^{0.8} + 0.224\,z_\psi^{0.333}} \qquad (2.57)$$

However, it does not directly provide the formulae for penetration or deposition profile. The penetration obeys

$$\ln F_p^\circ(z_\psi) = -\int_0^{z_\psi} \text{Sh}\,dz_\psi \quad \text{and} \quad F_p^\circ(z_\psi) = \exp\left[-\int_0^{z_\psi} \text{Sh}\,dz_\psi\right] \qquad (2.58)$$

and so

$$\varphi_p^\circ(z_\psi) = -\frac{dF_p^\circ(z_\psi)}{dz_\psi} = -\text{Sh}\exp\left(-\int_0^{z_\psi} \text{Sh}\,dz_\psi\right) \qquad (2.59)$$

which is to be evaluated numerically.

2.2.3.2 Both Diffusionally and Hydrodynamically Developing Laminar Flow

This case seems much closer to most real situations near the channel inlet. Now [17, 18]

$$Sh = 3.657 + \frac{0.0677 \, [ReSc \, d_c/z]^{1.33}}{1 + 0.1 Sc \cdot [Re \, d_c/z]^{0.3}} \quad (2.60)$$

from which it follows that

$$Sh = 3.657 + \frac{0.422}{z_\psi^{1.33} + 0.152 \, Sc^{0.7} z_\psi^{1.03}} \quad (2.61)$$

and Eqs. 2.58 and 2.59 apply as well.

2.2.3.3 Developed Turbulent Flow

Under the conditions of a few transactinoid experiments, Re number was as high as 10^3–10^4; the value of Sc was about unity. A formula known and used for a long time is:

$$Sh = 0.023 Re^{0.83} Sc^{0.3} \quad (2.62)$$

It is reportedly valid over the range $0.6 < Sc < 100$.

More recent formulae for the heat transfer proposed by Gnielinski [19], when re-casted for the mass transfer, give

$$Sh = 0.12 \left(Re^{0.87} - 280\right) Sc^{0.4} \quad \text{if } 1.5 < Sc < 500 \text{ and } 300 < Re < 10^6 \quad (2.63)$$

as well as

$$Sh = 0.0214 \left(Re^{0.8} - 100\right) Sc^{0.4} \quad \text{if } 0.5 < Sc < 100 \text{ and } 10^4 < Re < 5 \times 10^6 \quad (2.64)$$

The mean deposition length derived from Eq. 2.62 is

$$\frac{1}{\psi \, Sh} = \frac{43.5}{\psi \, Re^{0.83} \, Sc^{0.3}} \quad (2.65)$$

It is an analogue of the values of η_0, obtained above from the analytical solutions to some problems with laminar flow.

Because Sc is inversely proportional to D, the values of Schmidt number for aerosols are much larger than unity. In practice one meets the problem of diffusional deposition of polydisperse aerosols. A solution to this problem for rectangular and circular tubes was proposed in Refs. [10, 20].

References

1. Moelwyn-Hughes EA (1961) Physical chemistry, 2nd revised edn. Pergamon, London
2. Emanuel NM, Knorre DG (1974) Kurs khimicheskoi kinetiki (A course in chemical kinetics), 3rd edn. Moskva, Vysshaya shkola, chap 3
3. Couture TL, Zitoun R (2000) Statistical thermodynamics and properties of matter. Taylor & Francis, London, chap 14
4. Gilliland ER (1934) Ind Eng Chem 26:681
5. Giddings JC (1965) Dynamics of chromatography, Part I, Principles and theory. Dekker, New York, chap 6
6. Frenkel, YaI (1948) Statisticheskaya fizika (Statistical physics). Izd Akad Nauk SSSR, Moskva
7. Lindemann FA (1910) Physik Z 11:609
8. Debye P (1912) Ann Physik 39:789
9. Wolfram Resources (2007). Wolfram Research, Champaign. http://integrals.wolfram.com/index.jsp/. Cited 15 May 2007
10. Lee KW, Kim SP (1999) Aerosol Sci Technol 31:56
11. Mayya YS, Kotrappa P (1982) J Colloid Interface Sci 90:509
12. Gormley PG, Kennedy M (1949 Proc R Irish Acad 52A:163
13. Tan CW, Thomas JW (1972) Aerosol Sci 3:39
14. Hausen H (1959) Allg Wärmetechnik 9:75
15. Frank-Kamenetskii DA (1967) Diffuziya i teploperedacha v khimicheskoi kinetike. Nauka, Moskva. (1969) Diffusion and heat transfer in chemical kinetics. Plenum, New York
16. Zvara I Unpublished results
17. Stephan K (1962) Chem Ing Tech 34:207
18. Stephan K, Preusser P (1979) Chem Ing Tech 51:37
19. Gnielinski W (1976) Int Chem Eng 16:359
20. Fuchs NA, Stechkina IB, Starosselskii VI (1962) Brit J Appl Phys 13:280

Chapter 3
Production of Transactinoid Elements, Synthesis and Transportation of Compounds

Abstract On-line chemical experiments with single short-lived atoms produced on accelerators involve a number of stages, besides the chemical partition processing itself. The newborn nuclei, ejected by recoil into a flowing gas, thermalize in heavily ionized environment; typical conditions are quantitatively characterized. A priori consideration shows that chlorinating (an example) must proceed rapidly, provided that it consists of several simple bimolecular steps — transfer of one or two chlorines — with a negative enthalpy change and so low activation energy in each of them. Fast synthesis of $HfCl_4$ was repeatedly confirmed in on-line elution-chromatography-like experiments. Analogous considerations helped to establish a safe upper limit for the concentration of interfering reactive impurities in the working medium. Briefly discussed are hot filters for retaining aerosols and for absorption of contaminants. When the gaseous molecules of the tracer should be transported a long distance, the concern is efficient deposition of unwanted aerosols in the channel. The presented universal graphs describe diffusional deposition as a function of certain reduced distance. Alternatively, molecular entities are intentionally attached to aerosol particulates for transportation. The size of particulates, which provides the highest possible penetration through a channel, corresponds to equal rates of gravitational settling and diffusional deposition; these two rates depend on the size in an opposite way. Concentration of the aerosol should be above a necessary minimum — diffusional deposition of the tracer on the particulates should exhaust the particular volume (like target chamber) during the gas hold-up time. Transportation of short-lived radionuclides by laminar aerosol flow has some prospective peculiarities.

The "on-line" (on an accelerator beam) experiments with gaseous compounds of short-lived heaviest elements, which were illustrated by the schematics in Figs. 1.3, 1.6 and 1.7, are most complicated. They involve numerous stages like:
- Production of the elements in heavy ion-induced reactions with radioactive targets
- Prompt isolation of the bombardment products from the target material using recoil energy
- Thermalization of the recoils in gas

- Fast synthesis of the desired compounds in situ
- Synthesis of the compounds after transportation of the thermalized recoils by aerosol flow to a distance (as alternative to the previous stage)
- Separation of compounds of different volatility (adsorbability)
- Detection of the characteristic radiation of the required, as well as interfering, radionuclides

In this Chapter we briefly consider the production and prompt isolation of the nuclear reaction products. The thermalization of recoils and the synthesis of compounds are discussed in more detail, as is the transportation of thermalized recoils by aerosol particulates to remote equipment for chemical experiments. The separations on chromatographic principles are discussed in Chapter 4. Some basic information about detection and measurement was given in Chapter 1 when describing concrete experiments.

3.1 Production of the Elements by Heavy Ion Accelerators

Today the only way to obtain TAEs, including their superheavy isotopes, is bombardment of high atomic number targets with proper heavy projectiles to achieve complete fusion of the two interacting nuclei. The compound nuclei are obtained with nearly geometric effective cross sections — around 10^{-24} cm^2 (one barn) for the lightest projectiles, but some orders of magnitude less with the heaviest ones. Because the newborn products of fusion are necessarily excited, a great majority of them undergo prompt fission. Still, some of them cool and reach ground state, thanks to evaporation of several neutrons; practically all these "evaporation residues" (EVRs) are radioactive. The general trend is that the effective production cross sections of transactinoids get smaller with higher Z. At present, discoveries and studies of the new elements are feasible down to picobarn level of cross sections. Because the numerous possible nuclear interaction channels take place in parallel, the bombardments also inevitably produce other kinds of radioactive recoils with total yield reaching some fractions of barns. In addition the beam hits various parts of the target chamber and nonselectively sputters their atoms. This way a small fraction of the target is also released into gas. It is necessary to account for the phenomenon if the target nuclide itself is radioactive.

To date, in the chemical experiments with TAEs, the targets have been certain isotopes of the actinoids Th to Cf, while stable isotopes of elements C to Ca have served as the projectiles. An optimal energy of the beam particles is slightly above the repulsive Coulomb barrier between the nuclei to be fused. Namely, it is about 5 to 6 MeV per nucleon of the projectile in the laboratory system; the corresponding velocity of the particles is about one-tenth of the speed of light.

Table 3.1 provides basic information on the production and radioactive properties of the transactinoid nuclides, including their superheavy isotopes. The data (as in the fall of 2006) are from various sources [1, 2]. Only nuclides with half-lives longer than 0.5 seconds are listed because, at present, chemical studies seem to be feasible

3.1 Production of the Elements by Heavy Ion Accelerators

Table 3.1 Reported Transactinoid Nuclides with Half-lives ≥ 0.5 s

Nuclide		Half-life, Decay mode(s)	Production	Nuclide		Half-life, Decay modes	Production
Z	A			Z	A		
114	289	2.6 s α	^{244}Pu(^{48}Ca, 3n) decay of ^{293}116	$_{106}$Sg	271	1.9 min α, SF	decay of ^{283}112, ^{287}114, ^{291}116
	288	0.8 s α	^{244}Pu(^{48}Ca, 4n) decay of ^{292}116		266	34 s	^{248}Cm(^{22}Ne, 4n)
	287	0.5 s α	^{242}Pu(^{48}Ca, 3n) decay of ^{291}116		265	7.1 s α	^{248}Cm(^{22}Ne, 5n)
113	284	0.5 s α	decay of ^{288}115		263	0.8 s SF, α	^{249}Cf(^{18}O, 4n)
112	285	29 s α	decay of ^{289}114, ^{293}116		259	0.5 s α, SF	^{207}Pb(^{54}Cr, 2n)
	283	3.8 s α	^{238}U (^{48}Ca, 3n) decay of ^{287}114, ^{291}116	$_{105}$Db[a]	268	1.2 d α, SF	decay of ^{284}113, ^{288}115
$_{111}$Rg	280	3.6 s α	decay of ^{288}115		267	1.2 h SF	decay of ^{283}113, ^{287}115
$_{110}$Ds	281	11 s SF	decay of ^{289}114, ^{293}116		266	22 min SF, EC	decay of ^{282}113
$_{109}$Mt	276	0.7 s α	decay of ^{288}115		263	27 s α, SF, EC	^{249}Bk(^{18}O, 4n)
	274	0.5 s α	decay of ^{282}113		262	34 s α, SF	^{249}Bk(^{18}O, 5n)
$_{108}$Hs	270	3.6 s α 22 s[b] α	^{26}Mg(^{248}Cm, 4n)	$_{104}$Rf[a]	267	1.3 h SF	decay of ^{283}112, ^{287}114, ^{291}116
	269	9.7 s α	^{26}Mg(^{248}Cm, 5 n)		263	15 min SF	EC decay of ^{263}Db
$_{107}$Bh	272	9.8 s α	decay of ^{288}115		262	2 s SF	^{244}Pu(^{22}Ne, 4n)
	270	1 min α	decay of ^{282}113		261	65 s α, SF, EC	^{248}Cm(^{18}O, 5n)
	267	17 s α	^{249}Bk(^{22}Ne, 4n)		259	3 s α, SF	^{242}Pu(^{22}Ne, 5n)
	266	1.7 s α	^{249}Bk(^{22}Ne, 5n)				

[a] dubnium 255, 256, 257, 258, 260, 261 and rutherfordium 253, 255, 257, 259 have half-lives in the range of 1.5–4.5 seconds
[b] calculated from the experimental decay energy [3]

only above this time limit. If the particular nuclide can be produced through a complete fusion reaction, the reaction is given in the Table 3.1. Some of the nuclides cannot be obtained this way. They were observed only as descendants in the chains of α-decays of the shown heavier mother activities. Of them, the nuclides $^{291-293}$116, ^{287,288}115, and ^{282,283}113 have half-lives shorter than 0.5 s. They were produced by the ^{48}Ca bombardments of ^{245,248}Cm, ^{243}Am, and ^{237}Np, respectively.

The effective production cross sections are not included in Table 3.1; their values are not known accurately enough. The cross sections for the superheavy isotopes in ^{48}Ca-induced reactions with heavy actinoid targets are not larger than a few picobarns. They reach 10s to hundreds picobarns for the reactions of ^{18}O and ^{22}Ne with the same targets, producing the first transactinoids. Generally, the larger the neutron number of the projectile at given Z, the larger the yield may be, and the longer the half-life of the complete fusion products. Hence, neutron rich radioactive nuclides would have great advantages as bombarding particles. However, intense enough beams of such projectiles will probably be available only in the distant future.

3.1.1 Recoil Separation from Targets

The recoil energy of an EVR is close to the energy of the compound nucleus. Let E_{HI} and A_{HI} be the energy (in MeV) and mass number of the heavy ion (projectile), while E_{CN} and A_{CN} are the same quantities for the compound nucleus. Then, from the law of momentum conservation, it follows that

$$E_{CN} = E_{HI} \frac{A_{HI}}{A_{CN}} = (5-6) \times \frac{A_{HI}^2}{A_{CN}} \text{ MeV} \qquad (3.1)$$

where the numerical coefficient comes from the optimal energy mentioned above. Thus, E_{CN} increases roughly as the squared projectile mass, and is of the order of 0.01 to 0.1 MeV per nucleon. The absorbed projectile momentum makes the angular distribution of the recoiling EVRs narrow. Hence, if the target is not thicker than the recoil range of EVRs in it, the residues are mostly ejected into the target chamber. Such "thin" target has a surface density of about 1 mg cm^{-2}. The targets for Ca projectiles can be thicker than in the bombardments with O ions. In the meantime, with the target thickness optimized for prompt isolation, one does not enjoy the maximum possible production yield of the required rare products. The reason is that the loss of the projectile energy in the optimal target is smaller than the width of the bell-shaped dependence of the cross section on E_{HI}. The production yield with the target thick enough to absorb the whole width of the excitation curve would be considerably larger.

In principle, the required fast (though not prompt) isolation from thick targets might be achieved by evaporation of the products from the target or a recoil catcher heated to high temperature. This has been realized with massive targets for low and medium Z elements in ISOL-type installations. The prospects of the ISOL-type approach for the elements 104 to 110, which are expected to be refractory metals, are dubious. On the other hand, SHEs number 112 to 118 must be relatively volatile; evaporating their atoms into vacuum would be a convenient way to supply them to a mass separator. Evaporation into gas and transportation by the flow also seems promising. Early attempts to apply such techniques in searching for SHEs in products of heavy ion-induced reactions were unsuccessful.

3.1.2 Thermalizing Recoils

The recoiling EVRs are let thermalize in a gas. When the target thickness is equal to (optimal) or larger than the nominal range of the recoils in the target material, the spectra of the residual energies, and so also ranges of the recoils ejected into the gas, span from zero to a maximum. In practice, the recoil separation efficiency under optimal conditions is high, but never 100 percent. It is so because of the fundamental phenomenon of "straggling" in the target, which results in wider (than might be expected) energy and angular distributions of the recoils, and shortens the projected ranges of the recoils. In addition, some local extra target thickness due to small variations can also cause some losses. If the target chamber volume is to be small, then the optimal chamber depth equals the nominal range of the recoils. For the TAE nuclei the range is some 5 cm in He or H_2, and not more than 1 cm in heavier gases.

In chemical experiments with the new elements, it is highly desirable to also produce, in parallel, their known and expected chemical (quasi)homologs. It requires simultaneous bombardment of an appropriate lighter target element; thanks to a larger effective cross section, the necessary mass of this additional target is relatively small. Such a two-component target can be realized in different ways. Generally, it is advantageous to use a homogeneous mixture of the two elements. It makes the ratio of the yields of the homologous nuclides practically independent of the uneven beam density profile over the target, which always changes with time. However, because the lighter recoils have longer range, it is not possible to guarantee that the conditions of thermalizing and extraction from the target chamber are completely identical for the recoiling TAE and its lighter homolog. Clearly, the larger range of the lighter homologous nuclides must be taken into account when optimizing the size of the chamber.

These minor disadvantages can be overcome by pasting a separate lighter target onto the heavy one down the beam. The supporting foil of this additional target can help to reduce the projectile energy to the individual optimum. From the chamber side, the target layer may be covered by another degrading foil to make the spectrum of the residual ranges in gas similar to that of the TAE. A serious drawback of this variant is that the ratio of the yields of the heavy and lighter activities varies with time because of the aforementioned unsteady beam density profile.

In the pioneering experiments on the circulating cyclotron beam (see Fig. 1.3), the depth of the target chamber was much smaller than the recoil ranges in gas discussed above. It was possible due to the strong magnetic field between the cyclotron poles, which made the charged and relatively low energy recoils move along a small-diameter spiral.

The ejected nuclei are thermalized under ionization of the gas by the beam penetrating the target chamber. It is still an open question whether this may affect the chemical fate of the recoils. The available beams of heavy ions (O, Ne, Ca etc.) have diameters in the millimeter to centimeter range, and intensities as high as a few "particle microamperes" — pμA. The latter is a frequently used unit which denotes a beam of 6×10^{12} particles per second; when singly ionized, they give a current of 1 μA. Table 3.2 presents the values of LET – the linear (per unit length) energy transfer – and the full residual ranges of various heavy ions in several gases at STP. When passing the target, the projectiles are stripped of almost all electrons, and they remain highly charged along a large part of their range. The LET of the projectiles is predominantly of "electronic" origin rather than due to atomic collisions. Its values increase with the atomic numbers of the projectile and gas. In a particular gas LET depends only on the atomic number and velocity of the particle. Hence, it is the same for isotopes of an element with the same ratio E_{HI}/A_{HI}.

It can be seen from Table 3.2 that in the target chamber, which is optimized for the few centimeter range of EVRs, a substantial portion of the beam energy is taken

Table 3.2 Radiation Effects and Absorbed Energy Per Cubic Centimeter of Stopping Gas at STP

Gas[a]	He		N$_2$		Ar	
C_p at STP (MeV K^{-1} cm^{-3})	5.77 × 10^9		7.86 × 10^9		5.77 × 10^9	
Heavy ion E_{HI} (MeV)	^{18}O 90	^{48}Ca 240	^{18}O 90	^{48}Ca 240	^{18}O 90	^{48}Ca 240
LET of HI in U (MeV mg^{-1} cm^{-2})	1.6	8	1.6	8	1.6	8
LET of HI in gas (MeV cm^{-1})	0.89	3.71	5.22	21.8	5.61	23.4
Range of HI in gas (cm)	73	52	12.8	9.6	7.3	8.9
Energy absorption rate in gas[b] (MeV cm^{-3} s^{-1})	0.534 × 10^{12}	2.23 × 10^{12}	3.13 × 10^{12}	13.1 × 10^{12}	3.38 × 10^{12}	14.0 × 10^{12}
Ratio of absorbed energy to C_p at STP	92	386	398	1670	586	2430
Energy deposited per electron - ion pair (eV)	31.7 [4]	–	36	39 [5]	26	–
Production rate of the pairs (cm^{-3} s^{-1})	0.17 × 10^{17}	–	0.87 × 10^{17}	3.4 × 10^{17}	1.3 × 10^{17}	–

The heavy ion projectiles are 5-MeV per nucleon ^{18}O and ^{48}Ca, beam density is 0.1 pμA per square centimeter ($\equiv 6 \times 10^{11}$ cm^{-2} s^{-1}). Beam power in watts is 0.1 × (E_{HI}/MeV).
[a] Values for H$_2$ are by 10 percent higher than for He, values for O$_2$ and air are close to those for N$_2$
[b] Equals 6 × 10^{11} s^{-1} × LET. The equivalents of 1 MeV are 1.602 × 10^{-6} erg and 1.602 × 10^{-13} J.

up by the gas. With a beam of 1 pμA per cm^2, the absorbed power is very large compared with the heat capacity of gases. Actually, there are several fundamental factors and experimental circumstances which prevent heating the gas to hundreds or thousands of degrees. First, not all the absorbed energy is directly converted into heat – it is also emitted as radiation of different wavelengths and so rapidly transferred to the (usually cooled) walls of the target chamber. Second, the gas expands upon heating which makes the projectile range longer. Last, the true volume penetrated by the beam within the chamber is often several times smaller than the total volume of the chamber. On the other hand, the target assembly also contributes to heating of the gas. Indeed, the target is typically a layer of a refractory oxide deposited onto a supporting metallic foil; it absorbs a significant portion of the beam energy and partly transfers it to the gas. Experience shows that the gas is considerably heated by beams of the order of 0.1 pμA, which seems to be close to the highest intensity in TAE gas phase chemical experiments performed to date. One more order of magnitude may bring novel technical problems.

The data of Table 3.2 show that, primarily, the beams of a "standard" intensity produce each second 10^{16} to 10^{17} electron–ion pairs per cm^3. It goes to about one percent of the 2.6 × 10^{19} cm^{-3} molecules present at STP. However, recombination of the electrons with the molecular ions is fast – typical values of rate constants for pairs produced in air are 10^{-7} to $10^{-6}\,\mathrm{cm^3\,s^{-1}}$ [6]. Evidently, the squared steady state concentration of the ions nearly equals the production rate divided by the recombination rate constant. In the present case, this concentration should be $\sqrt{10^{17}/10^{-7}}\,\mathrm{cm^{-3}} = 10^{12}\,\mathrm{cm^{-3}}$. However, the concentration of various chemically active atoms, radicals and molecules, including their excited and ionized states, may be higher [6].

To date, most of the discoveries of the heaviest elements or their particular nuclides have been done with sophisticated "on-line separators" performing on exclusively physical principles: charged particles move in magnetic and electrostatic fields, in vacuum or in rarefied gas. The task is efficient in-flight separation of the recoiling heavy evaporation residues from the beam of the bombarding particles, because they originally move along about the same lines. The separated rare products are directed to spectrometric detectors of α particles and fission fragments. The detectors are usually arranged in a high detection efficiency system, which is capable of tracing even long decay chains of mother nuclides. Incomplete isolation of the EVRs from the beam would deteriorate the detectors, and poor separation from the unwanted by-products would produce a hardly manageable background in counting. To achieve the necessary cleanness, the target should be thinner than the recoil range of the EVRs, because to be guided by the physical fields the recoils must possess energies above a certain minimum.

There are good prospects for using some simpler dedicated modifications or varieties of the above separators to help future studies of TAE chemistry. Such equipment will mainly serve to separate the recoils from the projectiles to avoid the technical and principal problems due to the heat and ionization produced in the target chamber by the heavy ion beam. A broader spectrum of recoil energies is allowed and so also thicker targets. An ultimately efficient separation from the

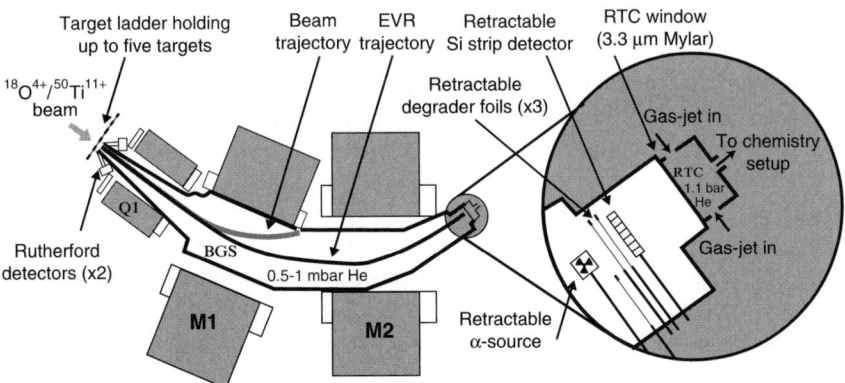

Fig. 3.1 Schematic of the Berkeley gas-filled separator (BGS) in combination with the Recoil transfer chamber (RTC) as used in a particular experiment [7].

Reproduced from Nuclear instruments and methods in physics research A, 551(2-3), Dullmann ChE, Folden CM, Gregory KE, et al., Heavy ion induced production and physical pre-separation of short lived isotopes for chemistry experiments, 528–539, © 2005 with permission from Elsevier.

radioactive by-products of the bombardment is not required because many eventual problems can be solved in chemical processing. Now, instead of hitting radiation detectors, the evaporation residues leave the low-pressure or vacuum chamber of the separator through a thin window. They enter a gas-filled chamber, being free from the ionizating beam and from a large fraction of the simultaneously produced interfering radionuclides. These important features open prospects for much broader diversity of chemical experiments. A schematic of such an experimental setup [7] is shown in Fig. 3.1.

Another project, TASCA – an experimental setup which includes a novel dedicated magnetic system — is being built at GSI Darmstadt [8].

3.2 Rapid Synthesis of Volatile Compounds

At the very beginning of the transactinoid radiochemistry, diverse functions of the gaseous reagents were of major concern. It was required that they:

- Rapidly react (chlorinate, oxidize etc.) with the nuclear reaction products to yield the desired volatile compounds of the new elements; it was not at all clear if it is possible only in the exceptional ionized medium, or also after the atoms become thermalized or even adsorbed on the walls
- Inactivate possible harmful impurities in the working gas, even those not detected by the available analytical techniques; the danger seemed to be in the formation of stable compounds which could interfere in obtaining the desired chemical form of the tracers
- Similarly to the previous point, block possible "active sites" with extra high energy of adsorption on the surfaces of the gas ducts and chromatographic columns

Problems raised by these requirements were like the following:

- It seemed that small bulk amounts of the "carrier," a compound homologous to that of the TAE, if introduced into the working medium, might take up at least inactivation of impurities and of the extra active adsorption sites. If so, can heavy losses of single atom tracers be avoided *only* by introducing almost "isotopic" carriers?
- If other, "nonisotopic" chlorination agents and carriers can also serve the purpose, then to what degree may their physicochemical properties differ from those of the compound of interest?
- Does the experimentally observed "effective volatility" – that is, possibility of transportation far from the target at certain temperature and gas composition – follow the bulk volatility of the compounds closely enough?

The anticipated short half-lives made the outlined functions and questions vital. Empirical studies of common binary compounds of the expected lighter homologs of TAEs usually did succeed in finding a suitable chemical reagent, its concentration, working temperature and other conditions under which the known volatile

3.2 Rapid Synthesis of Volatile Compounds

compounds were rapidly obtained and transported. However, mere homology – which, actually, is a rather vague term – cannot automatically guarantee that these conditions are also suitable for the new heavy elements. For example, the heavier homologs may not yield compounds of similar stoichiometry because of kinetic or thermochemical factors. Evidently, a deeper insight into the problem and the mechanisms involved is needed. This is of primary importance because the emphasis in chemical studies of TAEs is not just on chemical identification of still heavier elements. The interest naturally extends to the studies of both qualitative and quantitative differences in behavior of the particular TAE and of its nearest lighter homolog in possible detail. While chemical processes in gases are relatively simple, separation of tracers also involves interaction with surfaces, which is much more difficult to treat rigorously. Without understanding of the underlying mechanisms, the mere fact that the experiments gave results in the range consistent with reasonable expectations cannot bring any intellectual satisfaction.

The pioneering studies performed with chlorides at Dubna in the early 1960s had to answer at least some of the above-listed fundamental questions before proceeding to the studies on the inner circulating beam of the U-300 cyclotron. With a simple apparatus showed in Fig. 1.2, the experimenters [9] sought to achieve fast and efficient processing of single atoms in the "mildest" possible atmosphere (low concentration of weakly corrosive reagents), at the lowest possible working temperature. They investigated chlorination and transportation of the fission products – 17-h ^{97}Zr, ^{101,102}Mo and some Ln isotopes. The percentage of this activity transferred into the trap, the chemical yield, could be evaluated rather accurately. It was measured as a function of the gas phase composition and of the temperature regime. In addition, several time regimes of target bombardment and of switching of the carrier gas flow were tested. The recoiling fission products were thermalized in various carrier gases (air, N_2, CO_2, Ar or Cl_2) in the absence or presence of vapors of different molecular chlorides. These were, first, $ZrCl_4$ and $MoCl_5$, the truly "isotopic" carriers (in the traditional radiochemical terms) for ^{97}Zr and ^{101,102}Mo, respectively. Next were various "nonisotopic" carriers – $SeCl_4$, $SnCl_4$, $TiCl_4$, $NbCl_5$, $TaCl_5$ – as well as $MoCl_5$ for ^{97}Zr and $ZrCl_4$ for the isotopes of Mo. Thus, the physicochemical properties of the nonisotopic carriers differed from the properties of the isotopic carriers to various degrees.

The chemical yields were measured at $\leq 180\,°C$; higher working temperatures were not explored because of technical problems. The experimental regime mostly consisted in bombarding the target with neutrons, during several hours in the case of ^{97}Zr, while the ampoule was heated and continuously flushed with gas containing the reagents. The major results were the following: the employed gases, including Cl_2, did not transport Zr and Mo tracers. However, in the presence of the truly isotopic carriers (see above), the activities were efficiently transported. Moreover, the nonisotopic carriers $MoCl_5$, $NbCl_5$, and $TaCl_5$ were also effective in transporting ^{97}Zr, but $TiCl_4$, $SnCl_4$ and $SeCl_4$ were not. Later, with more sophisticated equipment, it was found that temperatures above $200\,°C$ allow using $TiCl_4$ and $SOCl_2$ for the Zr tracer. The efficiency of transportation of ^{97}Zr by $NbCl_5$ and other carriers did not depend on the nature of the supporting gas. Transportation of ^{97}Zr and ^{101,102}Mo depended on the temperature of the gas duct. Rare earth elements were not transported under any tested conditions.

Because the presence of free chlorine proved unnecessary, all the investigated carriers could, to some extent, play the role of chlorination agents. Even the thermalized atoms adsorbed on PTFE could be chlorinated. It was deduced that the carrier may be considerably more volatile than the transported compound, though the losses would increase, presumably due to adsorption. A strong temperature dependence of the processes was observed. Though these indications were necessarily rough, the behavior of various tracer elements seemed to correlate with the volatility of the expected chloride or oxychlorides.

3.2.1 Experimental Findings on Kinetics

One of the basic requirements is to synthesize the desirable compounds and to rapidly transport them to the equipment for chemical experiments. The time spent to accomplish such processes and its probability distribution can be properly determined only when a radioisotope of the element under study can be produced in a quantity that can be easily and accurately measured. It has never been the case for the transactinoid elements. Rough estimation of the chlorination time of Zr and Mo was done in the model experiments described in the above Section (see also Fig. 1.2). The ampoule was filled with pure inert gas, then closed and bombarded for some time with neutrons at ambient temperature. Thus, the thermalized fission products were accumulated on the walls. Then the ampoule was heated and flushed for a short time with the gas containing a reagent. Most of the activity got transferred into the trap in 30 seconds or so; it involved a mean gas hold-up time of 15 seconds, so that the actual upper limit of the chlorination time could be set as 15 seconds.

Advanced measurements of the sum of the mean hold-up, chlorination and net retention times – down to tenths of a second [10] – could be done by exploiting the basic equipment for the cyclotron experiments depicted in Sect. 1.2.2 and Fig. 1.3. The 4-m tube, which served for transportation of the volatile species from the target region to the detectors of radiation, also performed as an isothermal chromatographic column. At its exit there was mounted the device shown in the top of Fig. 3.2. Two heated valves served to commute the out-going gas between two traps for the de-sublimed reagent, which scavenged the transported radionuclide from the gas. To accumulate enough activity, an elementary elution experiment had to be repeated many times. To this end, the cyclotron beam was periodically switched on and off for equal time intervals, as shown by the first line in the bottom of Fig. 3.2. (This brought some technical problems in operating such a large machine). The valves were actuated in a similar regime – the trap A was opened and trap B kept closed when the beam was on, and *vice versa* when the beam was off, with just a short time shift to allow for the gas hold-up time in the tube. It is illustrated by the two bottom lines, which show the ideal and the expected actual profile of the activity concentration in the gas at the tube branching. Obviously, if the measured delay time

3.2 Rapid Synthesis of Volatile Compounds

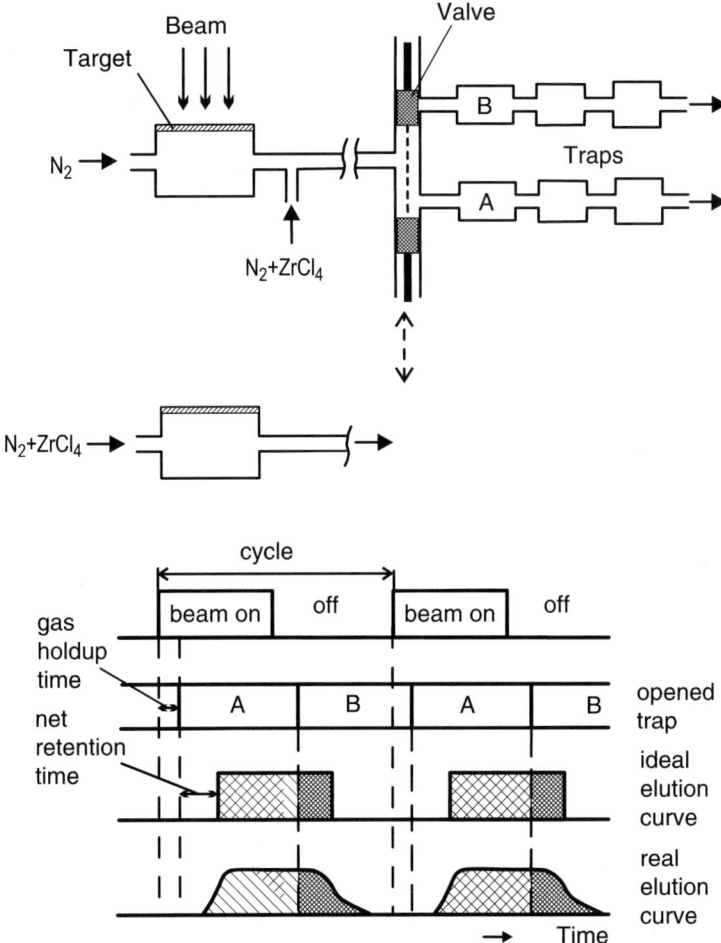

Fig. 3.2 Measurement of short elution time in an IC experiment on the accelerator beam by synchronous timing of the beam and two traps for exiting activity [10]; see text.

Reproduced (adapted) from Radiokhimiya, 9(2), Zvara I, Zvarova TS, Caletka R, Chuburkov YuT, Shalaevskii MR, Express continuous separation of transition elements of groups III and IV, 231–239, © 1967, with permission from Academizdatcenter "Nauka" Publishers.

is shorter than the beam on–off intervals, the activity must be found mostly in the trap A. The physicochemical conditions were identical to those in the later experiments on chemical identification of Rf. With 1.4 second on–off intervals and a gas hold-up time of 0.15 seconds in the 4-m tube, the ratio of activities of long-lived Hf isotopes found in the traps was $A/B = 3.5 \pm 0.5$. Assuming the ideal form of the elution curve, the sum of the mean chlorination and net retention times could be estimated as 0.3 seconds or less.

In the first experiments $ZrCl_4$ was one of the chlorinating agents. It was expected to be a nearly isotopic carrier, not only for $HfCl_4$, but also for $RfCl_4$. In the first thermochromatography of RfCl$_4$ [11], the reagents and carriers were SOCl$_2$ plus TiCl$_4$; they were obviously not that similar in properties to the main object of the study. Therefore, the elution time of Hf isotopes had to be checked again. To that end, the gas jet exiting the high temperature section of the column (cf. Fig. 1.4) hit the periphery of a cold rotated disk. HfCl$_4$ was deposited on the surface with high yield by adsorption, producing a ring. The much more volatile nonisotopic carriers were sucked off. In the particular tests the beam was on for 0.4 seconds and off for some 3 seconds. The cycle was again repeated many times; rotation of the disk and timing of the cyclotron beam were strongly synchronized again. This yielded a true elution curve shown in Fig. 3.3. It provides evidence for the high average rate of processing and indicates the extent of tailing in the total retention time.

The described techniques with blinking accelerator beam are quite cumbersome. Of course, mere observation that a short-lived radionuclide appears at the exit of the column indicates that the retention time is of similar order as the half-life. However, more precise measurements would require an independent determination of the initial activity released from the target. In the case of short-lived nuclides, the traditional simple catcher experiments can again be realized only by repeated bombardments and with the use of some technical tools.

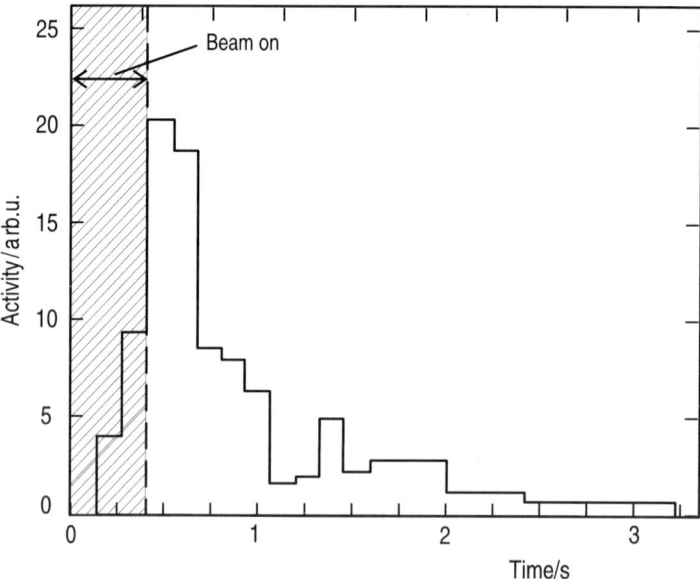

Fig. 3.3 Elution curve of Hf isotopes resulting from multiple cycles of a short bombardment followed by a longer beam-off interval [11]. The hatched area (the activity produced at beam-on) equals that under the histogram.

Reproduced (adapted) from Radiokhimiya, 14(1), Zvara I, Belov VZ, Chelnokov LP, Domanov VP, Hussonois M, Buklanov GV, Korotkin YuS, Schegolev VA, Shalayevsky MR, Chemical isolation of kurchatovium, 119–122, © 1972, with permission from Academizdatcenter "Nauka" Publishers.

3.2.2 Thermochemistry and Kinetics — Chlorination in Gas

The reactions of recoil atoms with chlorinating reagents (carriers) are considered in some detail to outline a general approach. It might be of use especially when it is desirable to avoid high working temperatures.

If the interactions proceed under the beam (in the target chamber), the atoms (ions) may still possess extra excitation energy and "hot" reactions at the initial stages cannot be excluded. However, the above-cited studies with the isotopes of Zr and Mo, which originated as fission fragments, proved the possibility to produce volatile compounds rapidly even after the recoil atoms have lost any extra energy. Similar observations were made in the experiments on the cyclotron beam; cf. the middle of Fig. 3.2. Here the recoil atoms were thermalized in pure nitrogen gas and in some hundredths to tenths of a second were transported to the point where the chlorinating agents were applied. On that way, typically, one-third of the atoms irreversibly adsorbed onto the walls; they were lost for experimentation. Beyond the point, all atoms were chlorinated. Strictly speaking, it is not known in what state they existed when interacting – whether mostly yet in gas or after getting adsorbed on the walls.

The synthesis of higher halides is easier to discuss if the interactions proceed in the gaseous phase [12]. From the point of view of thermodynamics, conversion of the atomic tracers through the reactions with halogen containing reagents must be practically complete, even at small equilibrium constants, because of the overwhelming concentration of the carrier. For example, almost any chlorine-containing gaseous compound is capable of chlorinating the recoil atoms of Zr, producing $ZrCl_4$. This is corroborated by a non-thermodynamic argument: in practice, the products of a reaction involving the tracer cannot meet again in a reasonable time to realize the reverse reaction. This argument will be quantitatively discussed in Chapter 6.

Generally, thermodynamics as such cannot say anything about the mechanisms of producing higher chlorides, and so about the duration of chlorination, which is of our major concern. Therefore, we need a plausible scheme of the interaction steps. Chlorination through a reaction like $^{97}Zr + TiCl_4 \rightarrow\ ^{97}ZrCl_4 + Ti$ seems unrealistic. It would require complex rearrangement of the participating atoms. Reactions proceeding through steps requiring triple and higher multiple collisions, for example $^{97}Zr + TiCl_4 \rightarrow\ ^{97}ZrCl_4 + 4TiCl_3$ as an extreme, are completely improbable. Our conclusion is that the mechanism yielding $^{97}ZrCl_4$ is a sequence of several bimolecular steps, each of which mostly results in transfer of only one chlorine:

$$^{97}Zr + TiCl_4 \rightarrow\ ^{97}ZrCl + TiCl_3$$
$$^{97}ZrCl + TiCl_4 \rightarrow\ ^{97}ZrCl_2 + TiCl_3$$
$$^{97}ZrCl_2 + TiCl_4 \rightarrow\ ^{97}ZrCl_3 + TiCl_3$$
$$^{97}ZrCl_3 + TiCl_4 \rightarrow\ ^{97}ZrCl_4 + TiCl_3$$

The transfer of two chlorines at a time like

$$^{97}Zr + TiCl_4 \rightarrow {}^{97}ZrCl_2 + TiCl_2.$$

might play some role, especially as the first step.

Reactions of the types listed above must be characterized by a relatively small entropy change. Hence, a thermodynamic equilibrium constant is small only if the enthalpy change is highly positive. On the other hand, the activation energy E_{act} of a reaction cannot be less than its enthalpy change: $E_{act} \geq \Delta_r H$. Then, given the half-life of the nuclide and the working temperature, any bimolecular step is inevitably slow for the purpose if $\Delta_r H$ is higher than certain value. When such a critical step cannot be bypassed, there is not enough time to obtain the volatile (weakly adsorbable) compound. The limit can be estimated, taking into account that in a bimolecular reaction, the concentration of non-reacted molecules of the tracer exponentially decreases with time, because the reaction has pseudomonomolecular kinetics. The mean duration of reaction is (cf. Eq. 2.14)

$$\tau_r = \frac{e^{E_{act}/RT}}{Z_{1,2}P} = \frac{e^{E_{act}/RT}}{\pi n_2 \omega_{1,2}^2 u_{1,2} P} \qquad (3.2)$$

where n_2 is now the concentration of the reagent (carrier) and $P \leq 1$ is the steric factor. In other terms, the rate constant of bimolecular reaction $k_{r1,2}$ is

$$k_{r1,2} = \pi \omega_{1,2}^2 u_\mu P e^{E_{act}/RT} \qquad (3.3)$$

and its maximum possible value, in the absence of any steric or energetic hindrance, is $\pi \omega_{1,2}^2 u_{1,2} \approx 10^{-10} n_2$. Because $n_{1,2} P \leq 10^{-10} n_2$, the shortest possible mean interaction time is:

$$\tau_r \geq \frac{10^{10}}{n_2} e^{E_{act}/RT} \qquad (3.4)$$

Hence the chlorination is fast enough, that is, its time does not exceed t_λ, provided that

$$\frac{E_{act}}{RT} < \ln(10^{-10} n_2 t_\lambda) \qquad (3.5)$$

For example, at $T = 300$ K, $t_{1/2} = 10$ s, and n_2 corresponding to 0.01 bar (see Eq. 2.3 and below it), only E_{act} values less than some 48 kJ mol^{-1} are "permissible;" the limit becomes 80 kJ mol^{-1} at 500 K.

Notice that the above limit holds only for the values of E_{act}: the inequality by Eq. 3.5, when rewritten for the value of the enthalpy change in reaction is only a necessary, but not at all sufficient, condition for ΔH_r.

Unfortunately, there is no broadly valid relation between the two quantities. The activation energies of the above chlorination reactions and analogous interactions of concern in this book have not yet been reported, and their calculation from the first principles seems difficult. Meanwhile, the above supposed chlorination steps involve atoms and molecules of lower halides – "the species with unpaired electrons on an otherwise open shell configuration" – which is exactly the IUPAC definition

3.2 Rapid Synthesis of Volatile Compounds

of free radicals. Numerous kinetic data are available in literature for the reactions of organic radicals abstracting atoms from molecules; the results of studying inorganic radicals are much fewer. The original findings about regularities in the E_{act} values were due to Evans and Polanyi [13]. Later, Semenov [14] proposed a general correlation between E_{act} and $\Delta_r H$:

$$E_{\text{act}} \approx 43 + 0.25 \Delta_r H \text{ kJ mol}^{-1} \qquad (3.6)$$

It is valid for both endothermic and exothermic bimolecular reactions between simple radicals and molecules (mostly organic) in gaseous phase. Semenov realized that the coefficients of the correlation formula may differ somewhat for individual classes of the reaction partners. Ref. [15] reviews the empirical models of radical abstraction. A rationale for the above correlation was provided by Morse or Lennard-Jones who considered the potentials of overlap in the reacting partners [16].

According to Eq. 3.6, all exothermic reactions have activation energies less than some 43 kJ mol^{-1}. From Eqs. 3.5 and 3.6 it follows that the reactions are fast, even at ambient temperature.

The experimental data obtained so far on reactions somewhat similar to ours are for the atoms of H, F, Cl, Br and B interacting with the molecules of fully or partially halogenated methanes. The rate constants of exothermic reactions of boron abstracting one halide atom from various (mixed) tetrahalides of carbon [17, 18], silicon and germanium [19] were found to be in the range 10^{-10}–10^{-14} cm^3 s^{-1} at ambient temperature (except for the case of CF$_4$ and SiF$_4$). It means that their activation energies are not higher than 22 kJ mol^{-1}. The transfer of two halide atoms was not expected to compete with the abstraction of just one. Below, we will assume that Eq. 3.6 approximately holds also for the chlorination reactions of our concern.

3.2.3 Synthesis of (Oxy)chlorides of Group 4 and 6 Elements

The two figures given below allow finding and analyzing conceivable paths of chlorination of group 4 and 6 elements by certain reagents. Included in the figures are the gaseous compounds for which the formation enthalpies are known or can be reasonably guessed. Figure 3.4 displays thermochemical data for the simple bimolecular reactions involving zirconium, which might take place when the reagents are SOCl$_2$ or TiCl$_4$, like in the thermochromatographic experiments with Rf chloride. In this case the two diagrams start with atoms and allow also for the reactions with oxygen. Notice from the data that many oxides of transition metals are rather reactive in gaseous state — the refractory properties and chemical inertness of the appropriate solids results from the extremely high energy of their ionic crystal lattices. Interactions with nitrogen were not considered despite that N$_2$ often serves as the carrier gas and is difficult to remove if present as an impurity. The reason is that the reaction

$$\text{Zr}(g) + \text{N}_2(g) = \text{ZrN}(g) + \text{N}(g)$$

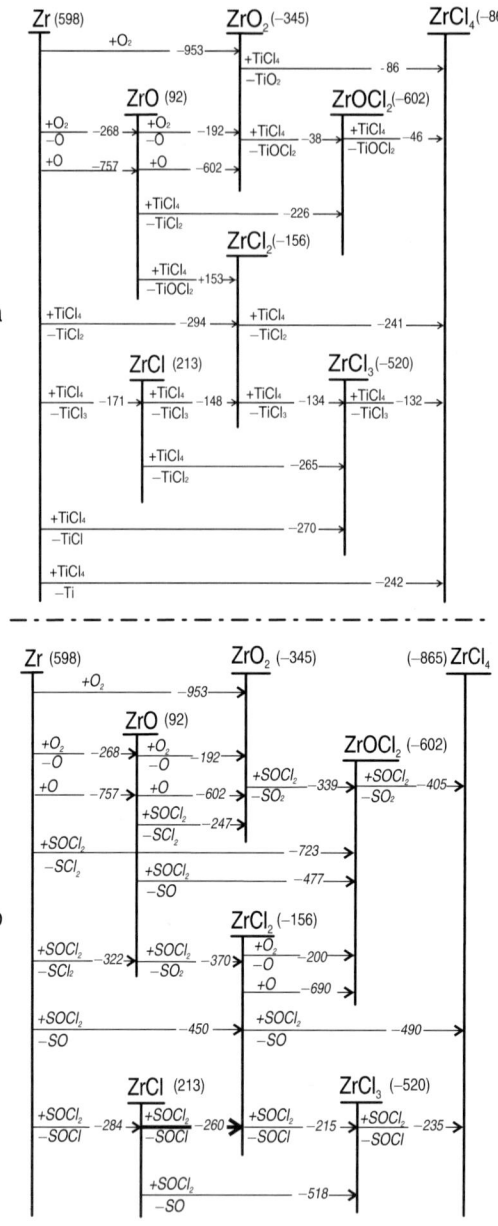

Fig. 3.4 Different imaginable ways of transformation of Zr atoms into ZrCl$_4$ molecules in the interactions with TiCl$_4$ (a) and SOCl$_2$ (b) in gaseous phase [12]. The reacting constituent of the gas and the resulting species are indicated above and below the arrow, respectively; the enthalpy of reaction in kJ mol^{-1} is given in the gap of the arrow. At the formulae of gaseous compounds of Zr are given their standard formation enthalpies (in the parentheses).

Reproduced from Pure and Applied Chemistry, 53, Zvara I, Problems in the identification of new elements, 979–995, © 1981, with permission from IUPAC.

has an extremely large enthalpy change of $+380$ kJ mol^{-1}. The figure for

$$Zr(g) + O_2(g) = ZrO(g) + O(g)$$

is considerably smaller, 267 kJ mol^{-1}.

The data for zirconium in Fig. 3.4 do reveal paths from an atom to the tetrachloride through bimolecular steps, all of which have negative enthalpies, involve radicals and do not require deep rearrangement of the atoms of reacting entities. In the first place, it is the sequence postulated in Sect. 3.2.2 for TiCl$_4$ as the reagent and an analogous one for SOCl$_2$. Some of the used thermochemical data may be of poor accuracy, which is evidenced by differences in the values recommended by renowned sources [20–25]. Nevertheless, the steps along these paths can by no means exceed the above limit $\Delta_r H > 80$ kJ mol^{-1} because the enthalpy change is mostly determined by the energies of the corresponding ruptured and newly formed bonds.

The interpretation might not be that straightforward with TiCl$_4$ as the only agent, taking into account the possible initial interaction of Zr atoms with oxygen. The problem might be clarified if the formation enthalpies of gaseous TiOCl or ZrOCl were known. The original diagram comes from Ref. [12] and contains thermochemical data taken from Refs. [20, 21]. Those for ZrCl$_2$ and ZrCl$_3$ today seem inconsistent – in Fig. 3.4 they are replaced by novel information from [24, 25].

Figure 3.5, using the data of the IvtanThermo base [22, 24], displays a part of $\Delta_r H$ data diagram for the reactions of gaseous compounds of Mo and W. The active components of the gas were nominally SOCl$_2$ + O$_2$ with a little excess of oxygen. According to the thermodynamic calculations using the same database, in the range of working temperatures 300 to 700 K, the two components must react to yield Cl$_2$, SO$_2$ and SO$_3$, leaving no untouched SOCl$_2$. However, a mass spectrometric analysis of the gas exiting the column revealed O$_2$, SOCl$_2$ and SOCl, plus

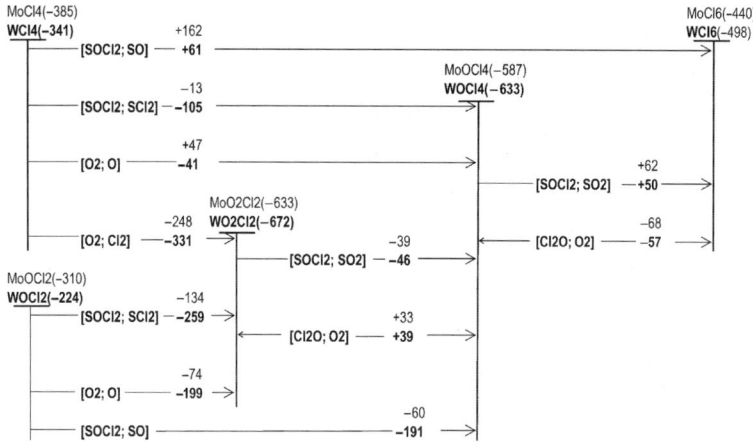

Fig. 3.5 Chlorination of group 6 elements from the oxidation state IV to state VI with SOCl$_2$ in the presence of O$_2$ [26]. See the legend of Fig. 3.4 except for the nuclear-reaction-like notation of the chemical interactions used in the present Figure.

a little of Cl_2, SO_2 and HCl. The hold-up time in the hottest part of the column (700 K) was probably too short to approach the equilibrium composition. All the actually present agents were included in the analysis.

In Fig. 3.5 the bimolecular reaction steps that yield compounds with the stoichiometry MCl_4, $MOCl_2$ and MO_2 are not shown; their thermochemical characteristics are similar to those for Zr compounds in Fig. 3.4. Notice that WCl_6, not to mention $MoCl_6$, could hardly come out, while there are many exothermic ways to convert W^{IV} and Mo^{IV} compounds into M^{VI} oxychlorides. Thus, from the point of view of the kinetics, both $W(Mo)O_2Cl_2$ and $W(Mo)OCl_4$ can be expected as the end products, because the reactions $W(Mo)O_2Cl_2 + SOCl_2 = W(Mo)OCl_4 + SO_2$ are exothermic with $\Delta_r H = -46(-39)$ kJ mol^{-1}. The M^V species are not discussed because $M^{IV, VI}$ compounds are preferable as to the thermochemical characteristics. These considerations will be taken into account in Sect. 5.7.4, when suggesting two possible chemical states and peculiar transport mechanisms to explain the results of comparative thermochromatography of W and Sg.

3.2.4 Chlorination in the Adsorbed State

As mentioned above, in real experiments the required volatile compounds could be obtained from the atoms, which became adsorbed on the walls. The number of the gaseous reagent molecules striking an adsorbed tracer entity per second differs from the collision rate of the gaseous tracer only slightly, by the factor $u_{1,2}/u_2$; cf. Eq. 3.3. However, the formation enthalpies of the tracer entities in adsorbed and gaseous state are very different because of the adsorption enthalpy. The latter must be highly negative, especially for the lower halides – the nonvolatile intermediates. This fact can strongly affect the enthalpy change in reactions.

Adsorption enthalpies of tracer entities on the surfaces of silica, metals, polymers and other materials contacted in experiments are not known. The values on truly bare surfaces may be very different for materials of a dissimilar nature. On the other hand, in the experimental studies with halides, at least the surface of columns seems to be deeply modified – densely covered by chemically bound fragments of the reagent molecules. This fact will be discussed in detail in Sections 5.3.4, 5.3.5 and 5.6. For example, most probable structures on the surface of silica modified by $TiCl_4$ seem to be $\equiv$Si–O–$TiCl_3$, $(\equiv$Si–O–$)_2TiCl_2$, as well as $\equiv$Si–Cl. If so, the adsorbed unsaturated tracer entity contacts mostly with these fragments rather than with the original surface of the adsorbent. Now the entity can realize its reactivity by also abstracting nearby chlorine atoms, which are directly or indirectly tied to the surface. On the surface densely covered by the grafted reagent, the tracer should "feel" nearly like in liquid $TiCl_4$. In particular, it may experience as many as 10^{11} to 10^{12} collisions with the neighboring chlorines per second — by some five orders of magnitude more frequent than in a carrier gas with 1 percent of $TiCl_4$ vapor.

The much higher collision rate considerably increases the tolerable activation energy of interaction compared with the reaction in gas. According to Eq. 3.5, the

3.2 Rapid Synthesis of Volatile Compounds

activation energy limit becomes higher by $RT \ln 10^5$. It is an important factor in favor of chlorination in the adsorbed state. However, the difference in the energetics of interactions because of the negative adsorption enthalpy is a strongly competing factor. In the specific case of the surfaces modified by halogenating reagents, there is a way to make some estimates of the adsorption enthalpy. They are based on an empirical correlation to be discussed in length in Sect. 5.6. Briefly, the adsorption enthalpy is close to the desublimation enthalpy of the solid adsorbate. We emphasize that such estimates are very rough, the more so as the available data on the sublimation energies of the lower halides are not always accurate.

As an example we will consider the reaction

$$\equiv Si-O-TiCl_3 + ZrCl_3(ads) = \equiv Si-O-TiCl_2 + ZrCl_4(ads)$$

Its enthalpy change must be more positive than in gas (cf. Fig. 3.4) by about $130\,\text{kJ mol}^{-1}$, which is the difference of the sublimation enthalpies of $ZrCl_3$ and $ZrCl_4$. With such a correction, the enthalpy of this step of chlorination of Zr atoms, which happens to be $-132\,\text{kJ mol}^{-1}$ in gas (see Fig. 3.4), may be about zero on the surface. Provided that Eq. 3.6 is still valid, chlorination on the surface is fast enough for the purpose. It is illustrated by the data of Table 3.3.

So far, we have left aside the question whether the change in the structure of the fragment modifying the surface might also bring a contribution. We would expect it to have an opposite sign than that due $ZrCl_3$–$ZrCl_4$. Let us consider an obviously extreme situation. It would be the reaction

$$TiCl_4(ads) + ZrCl_3(ads) = TiCl_3(ads) + ZrCl_4(ads).$$

The enthalpy of this reaction would evidently gain one more, this time negative contribution, approximately equal to the difference between the sublimation energies of the two Ti compounds. In the particular case, this would markedly compensate the contribution from the zirconium chlorides. Thus, when a priori neglecting the possible contribution from the surface, we obtain a safer margin to judge whether the reaction on surface is fast enough.

Let us go to the allowable reaction enthalpies. We have seen from the concrete example that the $\Delta_r H$ value for a reaction on surface can be much less negative than for the essentially similar reaction in gas. Very probably, with other combinations of the reaction partners, some of the halogenation steps may have positive $\Delta_r H$. Hence, it is not any more guaranteed that they are fast enough. Evidently, $\Delta_r H$

Table 3.3 Limits for Quantities which Govern the Kinetics of Halogenation of Thermalized Recoils

	Units	Reactions in gas			Reactions on surface		
Temperature	K	300	500	700	300	500	700
$RT \ln 10^5$	kJ mol^{-1}				34	57	80
Allowed [a]/ E_{act}	kJ mol^{-1}	≤ 48	≤ 79	≤ 108	≤ 82	≤ 136	≤ 188
Allowed $\Delta_r H$	kJ mol^{-1}	≤ 1	≤ 9	≤ 16	≤ 10	≤ 23	≤ 36

[a] In gas — from Eq. 3.5; on surface — that in gas plus $RT \ln 10^5$

must simultaneously obey the correlation 3.6 and the condition $E_{act} \geq \Delta_r H$. Then the allowed values are $\Delta_r H \leq (E_{act}^{max} - 43)/4$. Table 3.3 gives some figures related to typical temperatures of the experiments.

Needless to say, this section has presented a largely mental picture, not sufficiently substantiated by direct experimental data on the energetics of the processes involved. Posed questions call for carefully designed experiments with various reagents at different temperatures, concentrations and so forth. They could provide important evidence in favor or against the proposed picture. Studies along these lines also seem to be important in view of the prospective use of the compounds with the organic ligands, which may bring problems with kinetics.

3.2.5 Chemistry on Hot Aerosol Filters

The experiments with aerosol transportation of nuclear reaction products to the equipment for gas-phase chemistry include the use of filters, which serve to remove and absorb the aerosol matter from gas, as well as to enhance synthesis of the desired compounds of tracers. To date, the aerosols of various alkali metal halides, MoO_3 and elemental metals (Ag, Pb, C, Pd) have been utilized. The filters have been mostly varieties of silica (wool plug, powder), but also some crystalline salts. The synthesis is performed by introducing some gaseous reagents into the aerosol flow right ahead of the filter. Its temperature is chosen empirically; it is typically 1,300 K or somewhat higher. Ref. [27] reports 800 °C as the lower limit of satisfactory performance of a quartz wool filter for organic clusters carrying nuclear reaction products. Vahle, et al. [28] measured the efficiency of a quartz wool filter in absorbing barium activity carried by MoO_3 aerosol. Chemical reactions yielding the desired volatile species have not been discussed in any detail.

The high temperature of filters is an essential difference with the in situ synthesis of the required compounds at moderate heating, which was considered in the previous section. Occasionally, high temperature might enhance using the molecular halogens and hydrogen halides as halogenating agents. It can be discussed only for the reactions in gaseous phase; possible influence of the solid matter of the aerosol and filter is not clear. Here are some examples of the reactions for which the data are available (enthalpy is given for molar amounts):

- $ZrO(g) + Cl_2(g) = ZrCl_2(g) + O(g)$ $\Delta_r H = 49$ kJ
- $ZrO(g) + HCl(g) = ZrCl(g) + OH(g)$ $\Delta_r H = 315$ kJ
- $Zr(g) + HCl(g) = ZrCl(g) + H(g)$ $\Delta_r H = -20$ kJ
- $ZrCl_3(g) + HCl(g) = ZrCl_4(g) + H(g)$ $\Delta_r H = -45$ kJ

Hickmann, et al. [29] investigated how different chlorination agents (HCl, Cl_2, $SOCl_2$ and CCl_4) affect the appearance of the thermochromatogram of fission products of uranium; see Sect. 1.5.3 and Fig. 1.22 herein. The reagents, as well as Ln and Zr among the separated fission products, are of interest in the context of the present discussion. These important data deserve a detailed analysis of what concerns the chemical mechanisms involved.

3.3 Scavenging of Gaseous Chemically Active and Radioactive Impurities

A principal advantage of working with the short-lived nuclides in gaseous phase is a safe upper limit for the permissible concentration of any chemically active contamination. Indeed, the chemical fate of the recoil atom cannot be affected when its collisions with the contaminations are improbable because of the short lifetime. Equation 3.4 shows that the safe concentrations are smaller than about $10^9/t_{1/2}$ cm^{-3}.

3.3.1 Removing Water and Oxygen

In the first experiments with Rf and Db, which were characterized by huge flow rate of the carrier gas consisting mostly of nitrogen, the halogenating compounds fulfilled all three functions indicated in the introduction to Sect. 3.2. In particular, they scavenged the possible minor contaminations of nitrogen by water vapor and oxygen. These were supposed to be the most harmful trivial impurities, in view of the optimal path, and so the kinetics of chlorination. An example is the study of niobium bromide [30, 31] as a model of the behavior of dubnium. With helium and Br$_2$ vapors only, it proved possible to obtain volatile compounds of Ge, As, Se, Ga and Zn, but not such of Zr, Nb or Ta. As soon as BBr$_3$ was added to the gas, volatile compounds of Hf, Nb and Ta were produced with high yields. The two reagents were present in the carrier gas already at the inlet of the target chamber made of nickel. Thanks to their low corrosivity, it was possible to realize true in situ synthesis of the required tracer compounds.

BBr$_3$ is a well-known active brominating agent in macrochemistry, capable of converting solid oxides of transition and other metals into higher bromides. In principle, the gaseous molecules of oxides, or even those adsorbed on the surface, might be brominated. However, it is not evident that such reactions are fast enough. The authors of Refs. [30, 31] were inclined to believe that it was namely thorough purification of the gas from water and oxygen, which made fast synthesis of the higher bromides of Nb, Db possible when BBr$_3$ served as an active component of the gas. The reaction

$$2\text{BBr}_3(g) + 3\text{H}_2\text{O}(g) = \text{B}_2\text{O}_3(c) + 6\,\text{HBr}(g)$$

has a free enthalpy change of $\Delta_r G = -364$ kJ so that an equilibrium constant is as large as $\approx 10^{40}$. The reaction

$$\text{BBr}_3(g) + 3\text{H}_2\text{O}(g) = \text{H}_3\text{BO}_3(c) + 3\text{HBr}(g)$$

is characterized by $\Delta_r G = -211$ kJ.

At some typical concentrations of the reagents, straightforward equilibrium calculations yield a residual content of free water and oxygen molecules by several orders of magnitude smaller than the permissible one (vide ante). The above gross reactions very probably also proceed through some simpler steps like

$$\text{BBr}_3(g) + \text{H}_2\text{O}(g) = \text{HBr}(g) + \text{HOBBr}_2(g),$$

which must be exothermic by $\Delta_r H = -60\,\text{kJ mol}^{-1}$.

Similarly, in the experiments on chlorides, with $SOCl_2$ and $TiCl_4$ vapors in the carrier gas, water and oxygen impurities must be rather efficiently removed through the reactions

$$SOCl_2(g) + H_2O(g) = SO_2(g) + 2HCl(g);$$
$$SOCl_2(g) + O_2(g) = SO_3(g) + Cl_2(g); \text{ or}$$
$$TiCl_4(g) + 2H_2O(g) = TiO_2(c) + 4HCl(g).$$

They are characterized by $\Delta_r H$ equal to 67, -176, and $-1001\,\text{kJ mol}^{-1}$, respectively. The detailed kinetics of the reactions is undoubtedly fast.

Notice that, under intense beam, the steady concentration of the products of radiation-induced reactions like (excited) molecules, atoms and radicals can exceed the limit for the contaminations; see Sect. 3.1. Whether this might be a real danger has not yet been discussed in the literature on heavy element chemistry. In any case, one must keep in mind the above estimates when introducing reactive (in broad sense) gases into the target chamber. To date, the only reagents directly applied into the target chamber for real in situ chemical conversion or stabilization have been bromine and BBr_3 to produce bromides, hydrogen to support the metallic state, oxygen to produce higher oxides and water vapor to produce oxide hydroxides.

3.3.2 Chemical Filter After the Target Chamber

Experiments with the most volatile elemental tracers or their volatile oxides do not require aerosol transportation. Thanks to the non-corrosive environment, it is possible to place some chemical filters at the exit of the target chamber to remove the elements which might interfere in the measurements of radiation of the desired products of nuclear reactions. The problem is not only decontamination from the unwanted heavy nuclides which emit alpha particles and undergo spontaneous fission. The reactions of nucleon transfer between the projectiles and the target, or the carrier gas, yield many relatively light radionuclides "around" the projectile or the nuclei of the carrier gas. Some of these activities are produced with highly effective cross sections. They mostly emit β particles and may negatively affect spectrometric quality of the charged particle detectors, when deposited on their surface. This can be an acute problem when using argon as the carrier gas, as well as with the projectiles like the isotopes of calcium.

Zhuikov [32] was concerned with a broad task — the use of chemical volatilization, active filters in gases and thermochromatography for separation of elements in all regions of the Periodic Table. He investigated behavior of the elements at high temperature (1,100 °C) in hydrogen or oxygen flow, as well as their interaction with silica (tubes) and deposition on chemical filters. Great parts of his data, which are summarized in Ref. [33], are of interest for experimenters in heavy element studies. Some of his observations are outlined in Table 3.4.

3.3 Scavenging of Gaseous Chemically Active and Radioactive Impurities

Table 3.4 Chemical Volatilization of Elements and their Deposition on Silica Tube or Chemical Filters

Atmosphere	Behavior	Elements
Hydrogen	Volatilized	halogens, Zn, Se, Cd, In, Sb, Te, Hg, Tl, Pb, Bi, Po
	Slightly volatilized	Mn, Cu, Ga, Ge, As, Ag, Au, Sn
	Volatilized, react with silica	alkali metals, alkaline earth metals, "divalent" Ln and An: Sm, Eu, Tm, Yb, Cf, Es, Fm, Md
Oxygen	Volatilized, do not react with silica	halogens, Zn, Ag, Cd, Os, Ir, Au, Hg
	Slightly volatilized	(Pd), (Re), (Pt)
	Volatilized, react with CaO filter	Se, Mo, Tc, Ru, Rh Te, W, Re Pt, Bi, Po

3.3.3 Diffusional Deposition of Nonvolatile Species in Gas Ducts

The mathematics of the irreversible diffusional deposition in circular and rectangular channels was considered in Sect. 2.2. These regularities are related to the experimental problems of two kinds. First, when the radioelement wanted for study is converted into a volatile compound, either for the purpose of chemical studies or for easy transportation by the gas flow away from the target, some of the initially present interfering radionuclides yield compounds, which are strongly adsorbable on the walls of the channel; they get deposited in the duct, providing a profitable preliminary decontamination. Second, the carrier gas inevitably contains some hidden aerosol particulates (or molecular clusters) of various origin, concentration and size. These can nonselectively adsorb nonvolatile molecular entities from the gas and transport them a long distance. This way they may seriously hamper achieving the highest possible decontamination factor from truly nonvolatile species. The aerosols can also simulate a higher than actual volatility of compounds. For example, in the isothermal chromatography experiments on $RfCl_4$ [34] with CCl_4 as the reagent, a nonzero survival yield of ^{261}Rf persisted, even at very low temperatures of the column. It was obviously due to the aerosols from the thermal destruction of CCl_4.

The mathematics of diffusional deposition is the same for particles ranging from atoms to aerosol particulates. Because of smaller diffusion coefficients, tiny aerosols deposit more slowly than do the molecular entities.

Figure 3.6 displays graphs of the pertinent formulae for the laminar flow regime from Sect. 2.2. They describe the profile of the deposit density $\varphi_p(z_\psi)$ and that of penetration $F_p(z_\psi)$ – the fraction of the adsorbable particles still staying in gas at the exit of the channel. Notice that the reduced distance also equals the ratio of the doubled gas hold-up time t_g in a tube of the length z to the average time needed by the particles to diffuse across the channel diameter d_c:

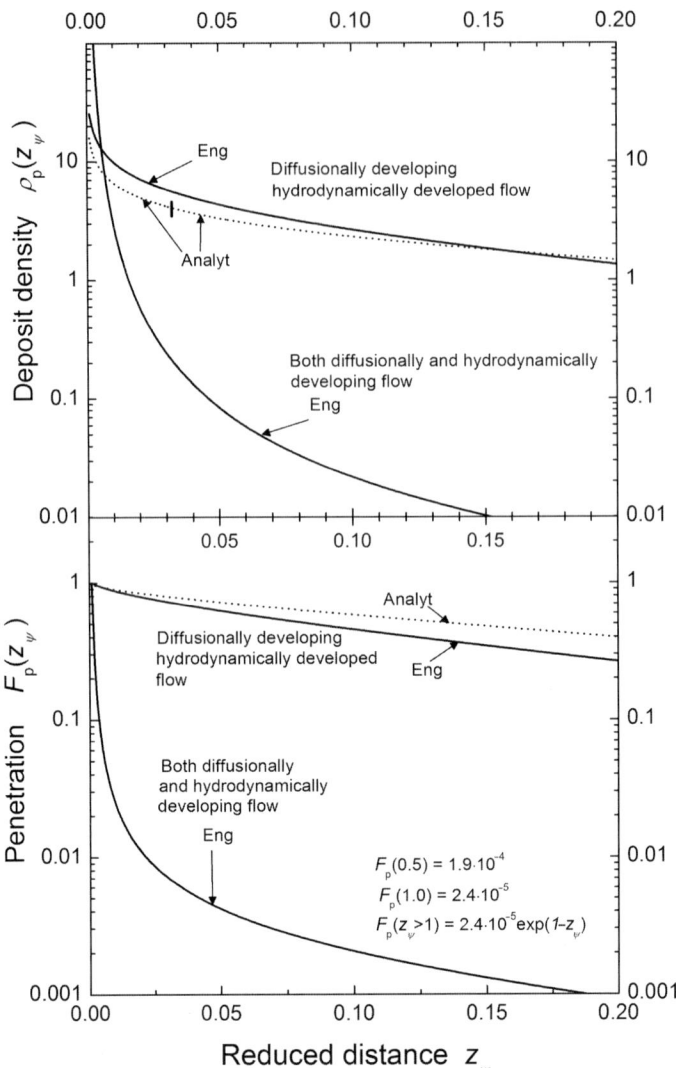

Fig. 3.6 Top: Density profile of an irreversibly adsorbed species deposited from laminar flow in circular tubes. The reduced distance is $z_\psi = \psi^0 z = \pi Dz/Q$. Analytical solution for hydrodynamically developed flow is given by Eqs. 2.47 (near inlet) and 2.44 – corresponding parts of the curve are divided by the thick mark; the engineering approach is Eq. 2.59 with Sh from Eq. 2.54. In the case of hydrodynamically developing flow, ψ taken from Eq. 2.61 was substituted into Eq. 2.59.

Bottom: Penetration of an irreversibly adsorbable species through circular tubes in laminar flow. Analytical solution for the developed flow is Eq. 2.43, that engineering is Eq. 2.58 with Sh from Eq. 2.57. Engineering formula for the developing flow is Eq. 2.57 with Sh substituted from Eq. 2.61; penetration values for some reduced distances larger than 0.5 are shown in the right bottom corner.

3.3 Scavenging of Gaseous Chemically Active and Radioactive Impurities

$$\psi^{\circ} z = \frac{\pi D z}{Q} = \frac{\pi d_c^2 z}{4Q} \frac{4D}{d_c^2} = 2 t_g \frac{2D}{d_c^2} \qquad (3.7)$$

In practice, it is difficult to introduce the radioactive nuclides into streaming carrier gas without disturbing the hydrodynamic patterns of the flow. Still, the analytical and engineering solutions to the problem of hydrodynamically developed flow proved useful in mathematical simulations of the gas-solid (thermo) chromatography (see Sect. 4.4). The two solutions are compared in the top of Fig. 3.6; their agreement is reasonable.

The case of both diffusionally and hydrodynamically developing flow is directly related to most experimental situations. For example, this regime takes place in the initial section of the tubing attached to a target chamber. The initial disturbance intensifies deposition of nonvolatiles and requires some distance and time to relax. Figure 3.6 shows that a very large fraction of adsorbable particles is deposited just behind the tube inlet: in particular, deposition of 99 percent of them within $z_\psi \leq 0.05$ is guaranteed. Sooner or later, the profiles of the penetration function and of the deposit density become almost strictly exponential; they are characterized by the mean reduced deposition length $1/3.657$, and the mean deposition length $1/3.657\psi^{\circ}$.

There are several engineering solutions available for both hydrodynamically and diffusionally developed *turbulent* flow. The most common is that by Eq. 2.53 in Sect. 2.2.3. This criterion relationship can be deciphered in terms of the involved variables to elucidate the corresponding dependencies [10]. We come to Eq. 2.56 from which it follows that

$$\eta_0 \sim \frac{w^{0.17} d_c^{1.17} v_2^{0.30}}{D^{0.56}} \sim \frac{Q^{0.17} d_c^{0.83} v_2^{0.30}}{D^{0.56}} \qquad (3.8)$$

where v_2 is the kinematic viscosity of the carrier gas, which is about $1 \text{ cm}^2\text{s}^{-1}$ for H_2 and He, being $0.15 \text{ cm}^2\text{s}^{-1}$ for N_2 and Ar. We can see an explicit strong dependence on d_c, but much weaker dependence on D and especially on Q compared with the laminar flow for which

$$\eta_0 \sim \frac{w d_c^2}{D} \sim \frac{Q}{D} \qquad (3.9)$$

Turbulent flow seldom occurs in radiochemical studies. A rule of thumb is that the constant mass transfer coefficient is established starting from a distance of about $50 d_c$ from the tube inlet. With radioactive tracers it is equally easy to measure either the penetration of the irreversibly adsorbed molecular entities or the density of their deposit. Zvara, et al. [10] compared the deposits of chlorides of three tracer elements from a turbulent flow; the results are seen in Fig. 3.7. The mean deposition lengths observed for NaCl, $ScCl_3$ and $LnCl_3$ were 15.4, 21.3 and 26.2 cm, respectively, with the error bars (1σ) ± 1.5 cm; hence, the differences are significant. When the diffusion coefficients were estimated from Eq. 2.19, the values calculated from Eq. 2.56 were 18, 24 and 23 cm, respectively. The agreement seems satisfactory. In any case it provides evidence for completely nonselective deposition mechanism of different nonvolatile chlorides. Initial parts of the curves evidence gradual stabilization of the flow patterns.

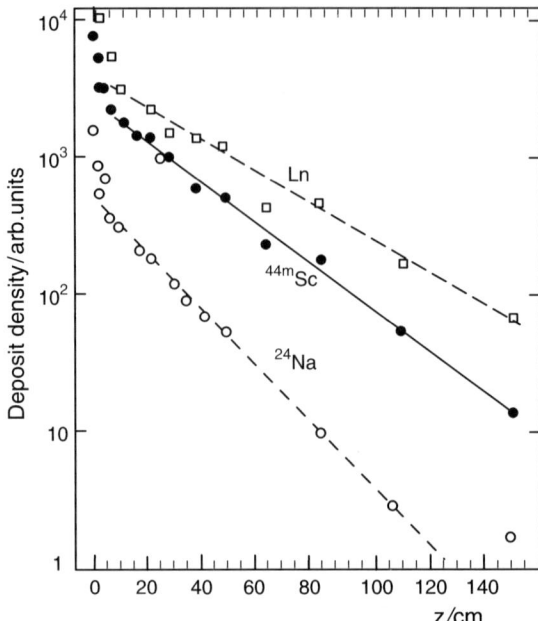

Fig. 3.7 Deposition profiles of nonvolatile chlorides [10]. Experimental conditions: Tube: stainless steel, i.d. 0.28 cm, length 400 cm. Gas: nitrogen + 0.2 mmHg of NbCl$_5$ (chlorinating agent), Q_0 335 cm^{-3} s^{-1}, $T_c = 530$ K, $p = 1.6$ bar at the inlet, 1 bar at the outlet, Re $\approx$ 8000, Sc $\approx$ 1.

Reproduced from Radiokhimiya, 9(2), Zvara I, Zvarova TS, Caletka R, Chuburkov YuT, Shalaevskii MR, Express continuous separation of transition elements of groups III and IV, 231–239, © 1967, with permission from Academizdatcenter "Nauka" Publishers.

3.3.4 Deposition of Heat

A problem somewhat analogous to those discussed in previous sections is deposition of heat (more illustratively, deposition of temperature of the gas) in thermochromatographic columns. Indeed, such columns have sometimes very steep temperature profile, or the flow is fast. Thus, we cannot be a priori sure that the heat exchange between the gas and the surface of the column is intense enough to ensure their equal temperatures at any given point. To our best knowledge, the problem has not yet been analyzed in detail. The fused silica columns are usually placed into thick outer metallic tubings to smooth and stabilize the temperature profile. It seems wise to measure the temperature profile inside a column by a possibly thin Pt-Pt(Rh) thermocouple, the wires of which come from opposite ends to compensate for the heat conductivity of the metal. The results of measurements in flowing gas (laminar regime) and with the flow stopped are usually identical. Still, this approach might not be perfect — the thermocouple slightly disturbs the patterns of the flow, enhancing heat exchange.

Simple estimates of the rate of heat transfer corroborate equal temperatures of the flow and of the column surface. Let us call the "thermal diffusivity," which is the heat conductivity divided by the heat capacity of the unit volume of the matter. The quantity plays the same role as the diffusion coefficient in mass transfer, and its dimension is also $cm^2\ s^{-1}$. The values for H_2, He, N_2 and Ar at 500 K and one bar are about 1.3, 3.8, 0.6, and $0.5\ cm^2\ s^{-1}$, respectively; they are roughly proportional to the absolute temperature. Comparison with the typical $0.1\ cm^2\ s^{-1}$ or smaller diffusion coefficients of the heavy voluminous molecules of our interest indicates that the thermal equilibrium is reached considerably faster. Hence, the heat transfer probably cannot negatively affect quantitative interpretation of the thermochromatographic data.

3.4 Transportation of Molecular Entities by Aerosol Stream

The previous section touched the problems which arise in the presence of unwanted aerosols in the gas-phase chemistry experiments. On the other hand, intentionally produced aerosols have found important application in nuclear chemical and radiochemical experiments as a tool for transportation of the accelerator produced tracers.

The 1970s saw intense development of the so-called He-jet technique of fast, nonspecific transportation of nuclear reaction products to counting area. The recoils were thermalized in helium containing an organic gas or vapor. The atoms stuck to the large molecular clusters of the organics emerging under the beam. Through intense pumping, the clusters could be sucked through a capillary as long as hundreds of meters away. At the exit, a special nozzle shaped supersonic jet, which helped to achieve high collection efficiency of the clusters impacting a catcher foil. The field was reviewed in Refs. [35, 36].

The above technique could not continuously supply radionuclides for chemical experiments. Catching the activity, for example by letting the gas bubble through a solution, requires ambient pressure at the capillary exit. Kosanke, et al. [37] seem to have pioneered such modification of the method. They still used organic (benzene) clusters and succeeded in efficient collection after transportation to 14 meters in a second. Other experimenters used ethane or ethylene gas. However, the high partial pressure of organic compounds required to produce clusters of necessary size and concentration, brought difficulties in the subsequent chemical experiments with gaseous halides.

The next logical step was to abandon organic clusters in favor of inorganic aerosols produced by controlled generators. In Ref. [38], a chamber with ^{235}U target was irradiated in a beam hole of the Mainz Triga reactor to produce fission products. Five different alkali halides were tried as aerosols. With 0.1-μm KCl particulates of the concentration $2 \times 10^5\ cm^{-3}$, the products could be transported eight meters away from the target in some tenths of a second with an efficiency of 70 percent. In the subsequent work [29], the aerosol particulates were removed by a quartz wool plug kept at 920 °C while some halogenating reagents were introduced at this point to accomplish an on-line thermochromatographic experiment; cf. Sect. 1.5.3 for the results.

At the JINR heavy ion cyclotron, an isotope of Po was produced in the target chamber bombarded by the internal circulating beam, and was transported by helium containing NaCl aerosol [39]. A sophisticated aerosol generator produced 0.2-μm NaCl particles with a concentration of 3×10^6 cm^{-3}. The transportation efficiency to 12 meters in some two seconds was as high as 70 percent; in the meantime, 20 percent of the activity was deposited in the target chamber and 10 percent – in the long capillary. Because the gas carrying aerosols is heavily ionized when passing through the target chamber, see Table 3.2, the electrical charging could affect deposition of the aerosols. The two latter experimental figures provided indirect evidence that the influence, if any, is not negative — similar losses are observed in the case of formation and transportation of volatile molecular compounds.

3.4.1 Optimal Parameters of Aerosol

When using aerosols for transportation of atoms or molecules thermalized in gas, the concern is achieving fast and highly efficient deposition of the entities on the surface of particulates, and low losses of the aerosol due to deposition on the walls of gas duct. This imposes requirements of certain concentration and size of the particulates. Electrically neutral particulates reach the duct wall by diffusion and gravitational settling. The rates of these processes depend in different ways on the effective radius r_p and the density of the particulate matter ρ, as well as on p and T. Diffusion is to be as slow as possible, which points to larger aerosol particles; however, these fall faster with larger size and density. For the laminar flow regime, there is a straightforward way to estimate the optimal size of the particles. The Brownian diffusion coefficients are given by Eq. 2.37, while the required Cunningham factor in the particle size range of present interest is $c_C = 2.61(\lambda_m/r_p)^{1/2}$; see Table 2.2. On the other hand, the terminal gravitational settling velocity is

$$w_{ps} = \frac{2g_g \rho r_p^2 c_C}{9\mu_2} \qquad (3.10)$$

where g_g is the gravitational constant.

Now, the optimal size is evidently such that the particles are displaced across the channel by the two independent deposition processes in the same time. Approximately (without integrating over the channel cross section), the condition can be written like

$$\frac{d_c}{w_{ps}} = \frac{d_c^2}{2D_{Bn}} \qquad (3.11)$$

so that the optimal radius is:

$$r_p^{opt} \approx 4 \times 10^{-7} \sqrt[3]{\frac{T}{d_c \rho}} \qquad (3.12)$$

Under common conditions, the calculated value of the particulate diameter is of the order 0.1 μm; see also Ref. [40]. It means that $D_{Bn} \approx 10^{-6}$ cm^2 s^{-1} and $w_{ps} \approx 10^{-4}$ cm s^{-1}; the latter velocity is established within a very short relaxation time of $w_{ps}/g_g = 10^{-7}$ s. According to Eq. 2.45, the steady deposition length $\eta_{0,p}$ is:

$$\eta_{0,p} = \frac{w d_c^2}{14.6 D_{Bn}} \qquad (3.13)$$

Hence, at the above D_{Bn} value, in ideal conditions, with a 0.3 cm i.d. capillary and $w = 10$ cm s^{-1}, the theoretical length would reach kilometers.

The next step is estimation of the number concentration of the particulates necessary to guarantee that all tracer molecules deposit on the aerosol within a certain time. Porstendorfer [41] found that the volume v_p "exhausted" by a spherical particulate per second is

$$v_p = \frac{4\pi r_p D_{1,2}}{\frac{4 D_{1,2}}{u_{m,1} r_p} + \frac{1}{1 + \lambda_{m,1}/r_p}} \text{ cm}^3 \text{ s}^{-1} \qquad (3.14)$$

(here again $D_{1,2}$ characterizes the molecular entities). The formula is an approximation – very appropriate for the purpose – of the coagulation frequency function [42]; the dependence on D_{Bn} and $d_{m,1}$, is neglected here.

Experimental measurements [41] of the uptake of ^{212}Po atoms by a latex-aerosol in the range of radii from 0.05 to 1 μm gave v_p values between 1×10^{-6} to 6×10^{-5} cm^3 s^{-1}, respectively, in agreement with the formula. Charged atoms were attached with a very similar rate. It means that a particulate concentration of the order of 10^6 cm^{-3} or higher would be needed to uptake the recoils in a typical target chamber in a second or so. It would prevent them from reaching the chamber wall by molecular diffusion, which is characterized by mean displacements of ≈ 0.3 cm during the first second. With such concentration and the 0.1 μm size of NaCl particulates, the content of the salt in gas is some 15 μg L^{-1}. It corresponds to the vapor saturated at 670 °C so that the necessary temperature of the aerosol generator based on evaporation cannot be lower.

At the above optimal concentration, the rate of coagulation of the aerosol happens to be rather small. According to Smoluchowski's equation, the particle number concentration decreases like

$$\frac{dc_p}{dt} = -k_{pr} c_p^2 \qquad (3.15)$$

where k_{pr}, the rate constant of the binary coagulation, equals:

$$k_{pr} = 8\pi r_p D_{Bn} = \frac{4 k_B T c_C}{3\mu_2} \qquad (3.16)$$

For 0.1-μm particles, the constant is of the order of $10^{-9} - 10^{-8}$ cm^3 s^{-1}; the concentration would then halve only in thousands of seconds [43,44].

The above-recommended parameters of the aerosol stream cannot be fully realized. The particles always have some size distribution, and it is difficult to maintain the optimal parameters of the aerosol during many hours or days. The isothermal chromatography experiments with TAEs, as a rule, require "reclustering" of the molecules at the column exit [45] to transfer and deposit the radionuclide onto a catcher for measurements. The recluster chamber was schematically shown in Fig. 1.7. There have been problems with the efficiency of reclustering — it decreases with higher column temperature and so brings additional uncertainty in measurement of the retention time as a function of column temperature. The origin of this phenomenon seems understandable. In a target chamber swept by the optimized aerosol stream, each thermalized recoil atom stops not further than ≈ 0.01 cm away from the nearest particulates. Meanwhile, the gas flow exiting the column contains free molecules and mixes with aerosol stream of about equal flow rate. Then the minimal required concentration of the particulates becomes higher. In addition, ideal mixing of the two streams takes some time. Forced mixing can hardly help because it enhances deposition of the particulates on walls.

3.4.2 Peculiarities in Aerosol Transportation of Short-lived Activities

Taking into account the flow rates, tube diameters and other parameters of real experiments, provided that the surface is smooth, the channel is straight and temperature gradients are absent, the diffusion of the aerosol particulates of optimal or larger size can be neglected. Then the stream can be seen as independent thin annular layers, which move with different velocities according to the parabolic law; so they pass the channel in different times. As a result, there can be envisaged some nontrivial peculiarities [46]. Some of them are discussed below.

3.4.2.1 Internal "Chromatogram" and "Elution Curve" of Aerosol Spike

When a nonadsorbable gaseous activity is introduced into laminar flow at $t = 0$ as a momentary spike of an infinitely thin disk, there develops a nearly Gaussian profile along the flow, with the center at wt and the dispersion $tw^2 d_c^2 / 94D$ (cf. Sect 4.2.3.1 and references herein). Its profile weakly depends on the initial radial density distribution of the tracer in the spike. Similar spike of an aerosol yields a prolonged longitudinal distribution of the particulates in the interval $0 \leq z \leq 2wt$; it is uniform in the case of the homogeneous disk spike. The top of Fig. 3.8 schematically displays the expected distributions for the reduced coordinate z/wt. The elution curves from a column of the length $l_c \equiv wt_g$ are shown in the bottom of Fig. 3.8 for the reduced time t/t_g.

The situation strongly changes when the particulates are injected right along the tube axis or, on the other hand, just released from the wall. In the first case they are

3.4 Transportation of Molecular Entities by Aerosol Stream

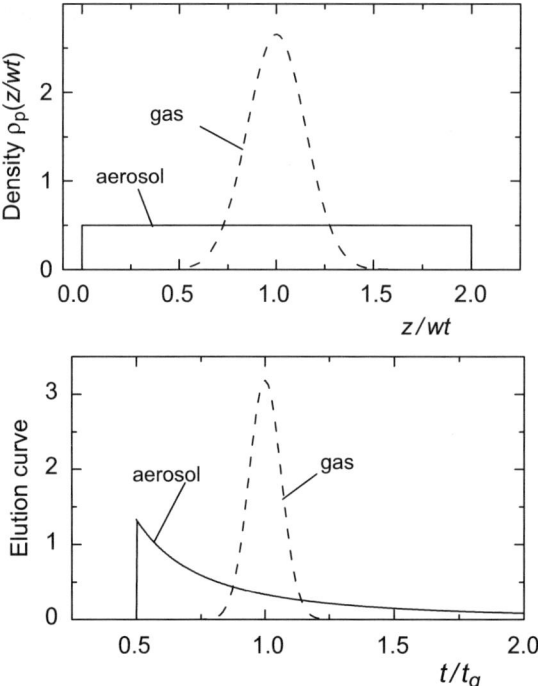

Fig. 3.8 Top: Distribution of a nonadsorbable gaseous tracer and that of the tracer carried by aerosol stream in a straight open cylindrical tube some time after injection of an infinitely thin plug. Hydrodynamically developed, diffusionally developing flow. Bottom: Character of the resulting elution curves.

carried twice as fast as the gaseous tracer molecules; in the second case they would not reach the exit in a reasonable time.

3.4.2.2 Transportation Efficiency of Rapidly Decaying Nuclides by Aerosol

Because of the dissimilar elution curves shown in the previous figure, the nuclei which have t_λ about equal to t_g or less also survive at the tube exit to a different degree. The transportation time of the nonadsorbable molecules is not much dispersed around t_g, and the surviving fraction is $\exp(-t_g/t_\lambda)$. It can be shown that the nuclei carried by aerosol survive to a greater extent, given by:

$$(1 - t_g/2t_\lambda)\exp(-t_g/2t_\lambda) - (t_g/2t_\lambda)^2 \mathrm{Ei}^*(-t_g/2t_\lambda) \qquad (3.17)$$

The graph of the two efficiencies and of their ratio (enhancement) is presented in Fig. 3.9. For nuclides with t_g/t_λ equal 4, 3, 2, 1 and 0, the ratio is 3.31, 2.28, 1.62, 1.20 and 1, respectively. The enhancement is larger with shorter t_λ, with longer l_c or with slower w. It does not compensate for decay losses, which increase the same way; still, generally; the aerosol jet transportation is preferable.

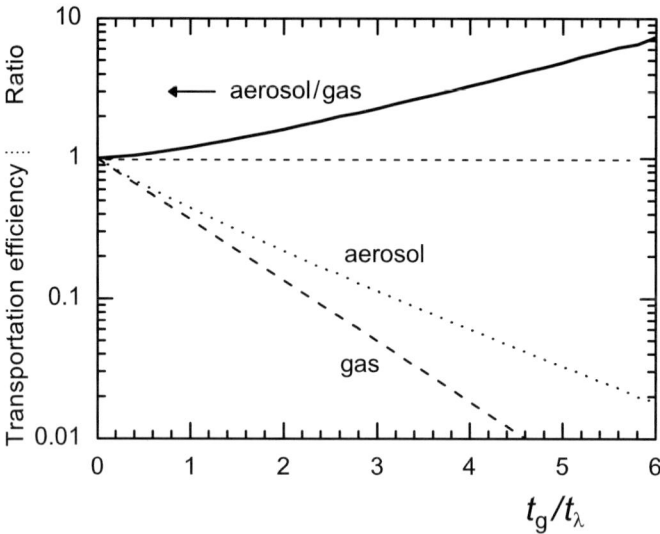

Fig. 3.9 Comparison of the transportation efficiencies (including the survival yields), measured at the exit of a straight open cylindrical tube, for relatively short-lived tracers, which are carried either as gaseous molecules or as aerosols. The values are plotted versus the hold-up time of the gas divided by the mean lifetime of the particular nuclide.

Strictly speaking, the optimal size from Eq. 3.12 holds only for straight horizontal tubes. Curvatures enhance deposition of the particulates owing to inertia, but the effect is small for r_p values less than some 0.5 μm. In the meantime, to the best of our knowledge, the peculiarities in behavior of the "horizontally" optimized aerosols in non-horizontal channels have not yet been discussed. In a straight, strictly vertical tube, even the particulates much larger than those earlier optimized will still be efficiently transported. It follows that the steady settling velocity is proportional to r^2 (see Eq. 3.10) and so even at $r_p \approx 30$ μm reaches only 3 cm s^{-1}, which is much less than common values of the velocity of the carrier gas. Other phenomena like the gradient of gas velocity across the capillary section can pose a limit on the size.

References

1. Gupta M, Burrows TW (2005) Nucl Data Sheets 106:251
2. Magill J, Pfennig G, Galy J (2006) Karlsruher Nuklidkarte (Chart of the nuclides) 7th edn. Rep EUR 22276 EN-DG, Joint Research Center ITU, Luxembourg
3. Dvorak J, Brüchle W, Chelnokov M, Dressler R, Düllmann ChE, Eberhardt K, Gorshkov V, Jäger E, Krucken R, Kuznetsov A, Nagame Y, Nebel F, Novackova Z, Qin Z, Schädel M, Schausten B, Schimpf E, Semchenkov A, Thörle P, Türler A, Wegrzecki M, Wierczinski B, Yakushev A, Yeremin A (2006) Phys Rev Lett 97: 242501
4. Knoll GF (1979) Radiation detection and measurement. Wiley, New York
5. Stuglik Z, Apel PYu (1998) An attempt of experimental determination of differential energy of the ion pair formation in air bombarded with 240 MeV Ca-40 ions, 300 MeV Co-59 ions, and

200 MeV Fe-56 ions. In: Oganessian YuTs, Kolpakchieva R (eds) Heavy ion physics. World Scientific, Singapore, p 796
6. Pikaev AK (1986) Sovremennaya radiacionnaya khimiya. Radiooliz gazov i zhidkostei. (Modern radiation chemistry. Radiolysis of gases and liquids), vol 1. Nauka, Moskva
7. Düllmanna ChE., Folden CM, Gregorich KE, Hoffman DC, Leitner, Panga GK, Sudowe R, Zielinski PM, Nitsche H(2005) Nucl Instrum Methods Phys Res, Sect A 551:528
8. Schädel M, et al. (2007) TASCA at GSI. GSI, Darmstadt. http://www-win.gsi.de/tasca/. Cited 15 May 2007
9. Zvara I, Zvarova TS, Krivanek M, Chuburkov YuT (1966) Radiokhimiya 8:77; (in French) Radiochimie 8:79
10. Zvara I, Zvarova TS, Caletka R, Chuburkov YuT, Shalaevskii MR (1967) Radiokhimiya 9:231; Soviet Radiochem 9:226
11. Zvara I, Belov VZ, Chelnokov LP, Domanov VP, Hussonois M, Buklanov GV, Korotkin YuS, Schegolev VA, Shalayevsky MR (1972) Radiokhimiya 14:119
12. Zvara I (1981) Pure Appl Chem 53:979
13. Evans MG, Polanyi M (1938) Trans Farad Soc 34:11
14. Semenov NN (1954) O nekotorykh problemakh khimicheskoi kinetiki i reaktsionnoi sposobnosti (On some problems of chemical kinetics and reactivity). AN SSSR, Moskva
15. Denisov ET (1997) Usp khim 66:953; Russ Chem Rev 66:859
16. Masel RI (2001) Chemical kinetics and catalysis. Willey, New York
17. Tabacco MB, Stanton CT, Sardella DJ, Davidovits P (1985) J Chem Phys 83:5595
18. McKenzie SM Stanton CT, Tabacco MB Sardella DJ, Davidovits P (1987) J Chem Phys 91:6563
19. Stanton CT, McKenzie SM, Sardella DJ, Levy RG, Davidovits P (1988) J Chem Phys 92:4658
20. Karapetiants MKh, Karapetiants ML (1968) Osnovnye khimicheskie konstanty neorganicheskikh soyedinenii (Basic thermodynamical constants of inorganic compounds). Khimiya, Moskva
21. Krasnov KS (1979) (ed) Molekulyarnye postoyannye neorganicheskikh soedinenii (Handbook of molecular constants of inorganic compounds). Khimiya, Leningrad.
22. Gurvich LV, Veitz IV, Alcock ChB (1989) (eds) Thermodynamic properties of individual substances, 4th ed. Hemisphere, New York. (1982) (in Russian) Nauka, Moskva
23. Gurvich LV (1989) Pure Appl Chem 61:1027
24. Gurvich LV, Iorish VS, Chekhovsioi DV, Yungman VS (1993) IVTAN-THERMO Users Manual. CRC Press, Boca Raton
25. NIST Chemistry WebBook, NIST standard reference database (2005) *http://webbook.nist.gov/chemistry/* Cited 15 May 2007
26. Zvára I, Yakushev AB, Lebedev VYa, Timokhin SN (1999) In: Extended abstracts of TAN-99, Seeheim 1999. Abstract O-15
27. von Dincklage RD, Schrewe UJ, Schmidt-Ott WD, Fehnse HF, Bächmann K (1980) Nucl Instrum Metods 176:529
28. Vahle A, Hübener S, Dressler R, Grantz M (2002) Nucl Instrum Methods Phys Res, Sect A 481:637
29. Hickmann U, Greulich N, Trautmann N, Gäggeler H, Gäggeler-Koch H, Eichler B, Herrmann G (1980) Nucl Instrum Methods 174:507
30. Zvara I, Keller OL, Silva RJ, Tarrant JR (1975) J Chromatogr 103:77
31. Zvara I (1976) In: Muller W, Lindner R (eds) Transplutonium Elements, Proceedings of 4th international transplutonium element symposium Baden Baden. North Holland, Amsterdam, p 12
32. Zhuikov BL (2005) J Radioanal Nucl Chem 263:65
33. Zhuikov BL (1982). Report P12-82-63. JINR, Dubna. Zhuikov BL (1984) Report ORNL-TR-5152. ORNL, Oak Ridge
34. Kadkhodayan B, Türler A, Gregorich KE, Baisden PA, Czerwinski KR, Eichler B, Gäggeler HW, Hamilton TM, Jost DT, Kacher CD, Kovacs A, Kreek SA, Lane MR, Mohar MF, Neu MP, Stoyer NJ, Sylwester ER, Lee DM, Nurmia MJ, Seaborg GT, Hoffman DC (1996) Radiochim Acta 72:169

35. Macfarlane RD, McHarris WC (1974). In: Cerny J (ed) Nuclear spectroscopy and reactions, vol A. Academic, New York, p 243
36. Wollnik H (1976) Nucl Instr Methods 139:311
37. Kosanke KL, McHarris WC, WarnerW RA, Kelly WH (1974) Nucl Instrum Methods 115:151
38. Stender E, Trautmann N, Herrmann G (1980) Radiochem Radioanal Lett 42:291
39. Zvara I, Domanov VP, Shalayevskii MR, Petrov DV (1980). Report 12-80-48, JINR, Dubna
40. Wu ChYu (2005) University of Florida, Gainesville. *http://www.ees.ufl.edu/homepp/cywu/ENV6130/Diffuse.pdf*. Cited 15 May 2007
41. Porstendorfer J (1968) Z Phys 217:136
42. Rogak SN, Baltensperger U, Flagan RC (1991) Aerosol Sci Technol 14:447
43. Hofmann W (2003) Aerosolfysik-I (Physics of aerosols). Queensland university of technology, Brisbane. *http://www.sbg.ac.at/ipk/avstudio/pierofun/transcript/aerosol1.pdf*. Cited 15 May 2007
44. Morawska L (2003) Environmental aerosol physics. Queensland University of Technology, Brisbane Queensland. *http://www.sbg.ac.at/ipk/avstudio/pierofun/transcript/aerosol2.pdf*. Cited 15 May 2007
45. Gäggeler HW, Jost DT, Baltensperger U, Weber A, Kovacs A, Vermeulen D, Türler A (1991) Nucl Instrum Methods Phys Res, Sect A 309:201
46. Zvara I (1995) In: Pustylnik BI (ed) Heavy Ion Physics JINR FLNR, Scientific Report 1993–1994. JINR, Dubna, p 168

Chapter 4
Gas–Solid Isothermal and Thermochromatography

Abstract A molecular-kinetic picture of the processes underlying adsorption chromatography in open columns allows deducing formulae for the characteristics of internal chromatograms; the emphasis is on thermochromatography. The retention time or the mean migrated distance is expressed as a function of the involved variables, like desorption energy, for the ideal case of homogeneous surface and fast adsorption equilibrium. The attempts to derive formulae for the dispersion or profile of zones have not yet been very successful, but the employed approaches are prospective. Meanwhile, mathematical simulations of the individual histories of migrating molecules have been quite successful. The corresponding Monte Carlo technique is versatile not only in accounting for different experimental conditions — it can also simulate special features of the experiments with short-lived nuclides, for example registration of the coordinates of the decay events which happen in the course of chromatographic processing. The simulated (as well as experimental) profiles can be well-fitted by the "exponentially modified Gaussian." It seems to offer rational parameterization of the shapes. Thermochromatography-like separations (but not isothermal chromatography) prove possible also in evacuated columns, in spite of the unique origin and essence of the Knudsen flow. This molecular flow drastically slows with decreasing temperature and the mathematical problem is the diffusion in the particular temperature field. It can be also treated as a random flight problem. Monte Carlo simulations of the diffusion are easy and can give zone profiles.

4.1 Characteristics of Methods

In the isothermal chromatography (IC) experiments, one usually measures elution curves – the time-dependent relative concentration of each of the analytes in carrier gas at the column exit. In principle, one could also fix the internal chromatogram (e.g., by rapid cooling of the column well below the working temperature); actually, it is impractical except for the compounds of similar volatility. In the thermochromatographic (TC) separations of mixtures of differently volatile tracers, given a proper stationary temperature profile, each of the components finally comes to practical rest somewhere within the column. The temperature gradient g may be constant or vary along the column; this possibility contributes to the versatility of TC technique. Even one or more isothermal sections (usually, near the column inlet) are eventually included. Zero gradient over the whole column brings one back to isothermal chromatography. Examples of the real temperature profiles of TC columns were presented in Sect. 1.2. A kind of hybrid between IC and TC is the temperature-programmed gas chromatography [1]. In this method, an isothermal column is operated in the elution regime, while its temperature is steadily raised, so that less and less volatile species become mobile. It facilitates obtaining a compact (in time) elution chromatogram of compounds with very different volatility. Obviously, the technique is inherently batchwise. Its gas–solid variety was tested for some compounds of present interest in Ref. [2].

The TC techniques are less widely known and used than the IC ones. They have some peculiar features, while sharing others with IC. In typical batch TC separations, the elements or compounds are volatilized at the inlet of the column by a thermal spike. It allows approaching δ-function-like input profile, longitudinal as well as temporal. Unlike this, in long lasting on-line experiments at accelerators, we meet a sort of frontal regime. In both cases, after the end of bombardment (EOB), the thermochromatogram is measured by detecting the distribution of radioactivity along the column.

We will discuss the quantitative characteristics of IC and TC, which are of importance when the techniques are used in studies of radiochemistry of the heaviest elements. Most interesting are the fast TC separations in open columns, which would yield reasonable resolution of mixtures and information on the adsorption–desorption energetics of the species involved. It is reasonable to start with what has been learned about isothermal chromatography and can be applied to thermochromatography. In the latter field, empirical works aimed at "useful" separations have prevailed. Here, we will pay interest primarily to the scarcer fundamental studies and theoretical developments which dealt with the characteristics of the resulting adsorption zones. Special conditions of the studies of TAE chemistry require attention to both the elution and frontal regimes of TC processing.

The simplest mechanism of migration of an unaltered molecule down the column is a repeated sequence of the adsorption on the surface for some sojourn time, desorption, and convective plus diffusional displacement in gaseous phase. In principle, surface diffusion can take place during the adsorption residence time, resulting in minor displacements. The major microscopic factor governing the effective migration rate is the positive desorption energy, which makes the elementary adsorption sojourn time grow strongly with inverse temperature.

When considering the "ideal" IC or TC, one implies that the adsorption equilibrium distribution of the tracer is immediately established at any point of the column, while longitudinal diffusion is absent. In the elution regime, with its input profile $\delta(z) \times \delta(t)$, both IC and TC would yield $\delta(z_A)$-like internal chromatograms. The coordinate of an ideal TC zone is shown in Fig. 4.1 by T_A^{id}, which corresponds to the ideal z_A. In the *frontal* regime of ideal IC, the elution curve has the form of continuous uniform distribution in time, and the internal chromatogram is an analogous longitudinal distribution. In the meantime, frontal TC results in a peak along z or

4.2 Theory

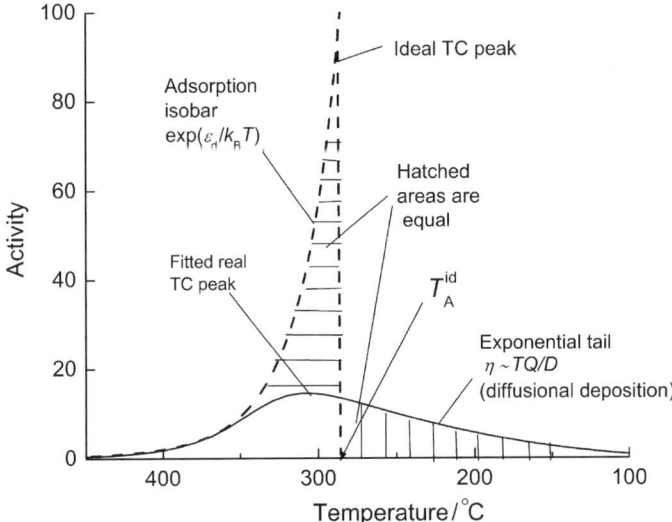

Fig. 4.1 Ideal (dashed curve) and real thermochromatograms in frontal regime [3]. The T_A^{id} marks the position of the ideal, $\delta(z)$-like elution thermochromatogram.

Reproduced from Physics Atomic Nuclei, 66(6), Zvara I, Accuracy of the chemical data evaluated from one-atom-at-a-time experiments, 1161–1166, © 2003, with permission from Pleiades Publishing.

the derived temperature coordinate, like that shown by the broken line in Fig. 4.1. Its left wing must be the adsorption isobar, while from the low temperature side it abruptly terminates at the temperature corresponding to T_A^{id} for the given processing time. Deviations from adsorption equilibrium due to the high velocity of the gas, exponentially distributed lifetimes of the nuclei and other factors broaden IC and TC peaks, compared with the ideal case. A specific characteristic of the real TC peak is the exponential (with distance) tail towards low temperatures; it stems from the diffusional, (practically irreversible) deposition of the non-equilibrated portion of molecules on the column walls, which takes place beyond T_A^{id}. The characteristic length of the exponential is proportional to Q/D (see Sect. 4.3.1).

4.2 Theory

4.2.1 Ideal Isothermal Chromatography

The first approach to understanding the regularities of gas–solid IC and TC was based on the laws of molecular kinetics. Within this framework, the time which a molecule spends in the adsorbed state when migrating down the column is seen as a sum of the random elementary adsorption sojourn times which accompany each adsorption event. The number of adsorption events is individual; of major interest is, of course, the mean value of the sum – the "net retention time" t_R. It is believed that the molecular entities get adsorbed upon every collision with the surface, though it is easy to account for a "sticking coefficient" less than unity. The desorption energy ε_d is assumed to be constant at any point on the surface; in other words, the wall is thought as being smooth in every respect. The next assumption is that the adsorbed entity experiences oscillations perpendicular to the surface. These are driven by the vibrations of the adsorbent lattice with a characteristic frequency ν_0, so that the period of the adsorbate oscillations τ_0 is the reciprocal of ν_0; see Sect. 2.1.5. It means that any structure of the surface is disregarded. From the point of view of thermodynamics, the above assumptions are equivalent to postulating two-dimensional gas as the state of the adsorbed molecules. We will consider the thermodynamic approach in Sect. 5.1.3. In this chapter, we shall adhere to molecular kinetics.

Let t_R^{IC} be the mean time necessary to travel the distance z_A down an IC column with the temperature T_c. It is the mean desorption time τ_d from Eq. 2.23 multiplied by the number of adsorption–desorption events on the way; the latter is Z_z calculated from Eq. 2.9. Then:

$$t_R^{IC} = \frac{z_A a_z}{Q} \frac{u_m}{4} \tau_0 e^{\frac{\varepsilon_d}{k_B T_c}} \tag{4.1}$$

If z_A is substituted by the IC column length l_c, the equation yields the net retention time in the column. We again stress that we discuss the internal chromatograms rather than the elution curves, unless stated otherwise. It means that now the total retention time equals the duration of the experiment. The latter is preset by the experimenter, thus characterizing experimental conditions rather than the results. Therefore, we need to know how z_A depends on t_R^{IC} rather than vice versa, and a more logical form of Eq. 4.1 would be:

$$z_A = \frac{t_R^{IC} Q}{a_z} \frac{4}{u_m} \frac{1}{\tau_0} e^{-\frac{\varepsilon_d}{k_B T_c}} \tag{4.2}$$

On the other hand, relationships of the form of Eq. 4.1 have been used for a long time in publications on gas-phase radiochemistry, while the corresponding formulae for z_A might look uncommon. It especially concerns the most important equations for TC, which are discussed in the present and next chapters; these, unlike Eq. 4.2, usually do not explicitly involve z_A. Therefore, it seems reasonable to adhere to the traditional presentation of the equations.

Here and below, we assume no pressure drop along the column and a total pressure of one bar. It means that we account only for the temperature dependence of the appropriate quantities. Expanding Eq. 4.1 we obtain

$$t_R^{IC} = z_A \frac{a_z T_0}{Q_0} \sqrt{\frac{k_B}{2\pi m T_c}} \tau_0 e^{\frac{\varepsilon_d}{k_B T_c}} \tag{4.3}$$

and for an open smooth circular column:

$$t_R^{IC,O} = z_A \frac{\pi d_c T_0}{Q_0} \sqrt{\frac{k_B}{2\pi m T_c}} \tau_0 e^{\frac{\varepsilon_d}{k_B T_c}} = z_A \frac{4 T_0}{d_c w_0} \sqrt{\frac{k_B}{2\pi m T_c}} \tau_0 e^{\frac{\varepsilon_d}{k_B T_c}} \tag{4.4}$$

4.2 Theory

where w_0 is the velocity of the carrier gas at an arbitrary chosen standard T_0. The net retention time t_R^{IC} is the duration of the experiment less t_g – the gas hold-up time; the latter equals the free volume of the column section of length z_A divided by the flow rate. For an open circular tube:

$$t_g^{IC,0} = \frac{z_A T_0}{w_0 T_c} = z_A \frac{\pi d_c^2 T_0}{4 Q_0 T_c} \tag{4.5}$$

4.2.2 Ideal Thermochromatography

The formulae of the above section give the retention time as a function of desorption energy or, vice versa, they allow evaluating the energy from the measured time, provided that τ_0 is known. We emphasize that it holds for an imaginary ideal situation, when the input spike and the resulting chromatographic zone are infinitely narrow.

In the TC processing the average molecular speed and the flow rate vary with z because of changing temperature. The retention time and the coordinate for any temperature profile of the column are related by the equations:

$$t_R^{TC} = \frac{a_z \tau_0}{4} \int_{z_S}^{z_A} \frac{u_m(T_z)}{Q(T_z)} e^{\frac{\varepsilon_d}{k_B T_z}} dz \tag{4.6}$$

and

$$t_g^{TC} = \int_{z_A}^{z_S} \frac{dz}{w(T_z)} \tag{4.7}$$

Here z_S is the coordinate of the starting point with the temperature T_S and z_A is the position of the zone with the temperature T_A. In the ideal conditions, the latter is identical to T_A^{id} in Fig. 4.1.

Equation 4.6 gives:

$$t_R^{TC} = \frac{a_z T_0}{Q_0} \sqrt{\frac{k_B}{2\pi m}} \, \tau_0 \int_{z_S}^{z_A} \frac{e^{\frac{\varepsilon_d}{k_B T_z}}}{\sqrt{T_z}} dz \tag{4.8}$$

Suppose that the TC column temperature profile is linear, that is:

$$T_z = T_S - gz \quad \text{and} \quad dz = -dT_z/g \tag{4.9}$$

Then the net retention time equals:

$$t_R^{TC} = \frac{a_z T_0}{g Q_0} \sqrt{\frac{k_B}{2\pi m}} \, \tau_0 \int_{T_A}^{T_S} \frac{1}{T_z^{1/2}} e^{\frac{\varepsilon_d}{k_B T_z}} dT_z \quad T_z = T_S - gz \tag{4.10}$$

In the case of an exponential temperature profile with

$$T_z = T_S e^{-\gamma z} \quad \text{and} \quad dz = -dT/\gamma T \qquad (4.11)$$

by analogy with Eq. 4.10, we come to:

$$t_R^{TC} = \frac{a_z T_0}{\gamma Q_0} \sqrt{\frac{k_B}{2\pi m}} \tau_0 \int_{T_A}^{T_S} \frac{1}{T_z^{3/2}} e^{\frac{\varepsilon_d}{k_B T}} dT_z \quad T_z = T_S e^{-\gamma z} \qquad (4.12)$$

The pertinent indefinite integrals and approximating series were listed in Sect. 2.1.6. In practice, the value of the definite integral for T_S is usually much smaller than that for T_A and can be neglected. Taking only two terms in the asymptotic series, we obtain for the case of a *constant temperature gradient* that:

$$t_R^{TC} = \frac{a_z T_0}{g Q_0} \sqrt{\frac{k_B T_A}{2\pi m}} \tau_0 \frac{k_B T_A}{\varepsilon_d} e^{\frac{\varepsilon_d}{k_B T_A}} \left(1 + \frac{3}{2} \frac{k_B T_A}{\varepsilon_d}\right) \quad T_A = T_S - g z_A \qquad (4.13)$$

In literature, the temperature dependence has been occasionally simplified. For example, the mean molecular speed was taken constant equal to that at T_A over the entire range (T_S, T_A). Then:

$$t_R^{TC} = \frac{a_z T_0}{g Q_0} \sqrt{\frac{k_B T_A}{2\pi m}} \tau_0 \frac{k_B T_A}{\varepsilon_d} e^{\frac{\varepsilon_d}{k_B T_A}} \left(1 + \frac{k_B T_A}{\varepsilon_d}\right) \quad T_A = T_S - g z_A \qquad (4.14)$$

The gas hold-up time in a cylindrical column is

$$t_g^{TC,0} = \frac{\pi d_c^2 T_0}{4 g Q_0} \int_{T_S}^{T_A} \frac{dT_z}{T_z} = \frac{\pi d_c^2 T_0}{4 g Q_0} \ln \frac{T_S}{T_A} \quad ; \quad T_A = T_S - g z_A \qquad (4.15)$$

For the *exponential temperature profile* we have

$$t_R^{TC} = \frac{a_z T_0}{\gamma Q_0} \sqrt{\frac{k_B T_A}{2\pi m}} \tau_0 \frac{k_B}{\varepsilon_d} e^{\frac{\varepsilon_d}{k_B T_A}} \left(1 + \frac{1}{2} \frac{k_B T_A}{\varepsilon_d}\right) \quad T_A = T_S e^{-\gamma z_A} \qquad (4.16)$$

and in the simplified case:

$$t_R^{TC} = \frac{a_z T_0}{\gamma Q_0} \sqrt{\frac{k_B T_A}{2\pi m}} \tau_0 \frac{k_B}{\varepsilon_d} e^{\frac{\varepsilon_d}{k_B T_A}} \quad T_A = T_S e^{-\gamma z_A} \qquad (4.17)$$

The hold-up time is now:

$$t_g^{TC,0} = \frac{\pi d_c^2 T_0}{4 \gamma Q_0} \cdot \frac{T_S - T_A}{T_S T_A} \quad T_A = T_S e^{-\gamma z_A} \qquad (4.18)$$

4.2.3 Shapes of Chromatographic Peaks

In the experiments on accelerators the required short-lived products of nuclear reactions are converted into volatile compounds and separated by gas–solid chromatography techniques in a continuous regime. Open columns of a meter in length and a few millimeters in diameter are used. The linear velocity of the carrier gas has varied from centimeters to meters per second. To optimize the separation process, it is important to understand how the experimental conditions, the properties of the separated species and other factors affect the shape and position of the resulting adsorption zone.

In general the term *zone profile* – $\rho^{IC}(z)$ or $\rho^{TC}(z)$ – used below means the (normalized) distribution of the adsorbate as a function of the column longitudinal coordinate after EOB — either found experimentally or expected from theory. With small number of atoms and short lifetimes, the meaning of the term might need to be refined. The experimental chromatogram is now usually a histogram of the number of decay events observed within short sections of the column. The events may be identified in the course of the run or after EOB.

For long-lived γ-active nuclides obtained with sufficient yield, which is the case for lighter homologs of TAEs, the profiles can be revealed by spectrometric measurements from outside the column. The profiles of short-lived α-activities can be measured only when the column is a channel formed by surfaces of electronic particle detectors. Such measurements are done in real time; of course, they also work well for s.f. nuclides. The latter can also be registered in open columns made of the materials which can serve as solid-state track detectors of fission fragments (fused silica, mica); see Sect. 1.2.2.

The shape of the zone, at least a characteristic of its width, is of major concern in all chromatographic techniques. Jonsson [4] and Giddings [5] seem to discuss the case and peculiarities of open columns in more details than do other general monographs on chromatography.

Necessarily, the zone profiles within the column and the more common elution curves do not have the same statistical parameters (see below about the statistical moments). Indeed, even if the internal zone profile, which broadens with time, is completely symmetrical, the elution curve must be skewed because the later fractions of the elution peak spend a longer time in the column [4] than the earlier ones. It is preferable to discuss the internal zone profiles because they provide a common basis for considering the regularities of IC and TC.

The shapes of chromatographic peaks often resemble the Gaussian distribution, though distorted to a variable degree. Theory seldom finds shape functions for chromatographic peaks in explicit form. More frequently, only the Laplace transform of the appropriate differential equation can be solved. Then, even if the solution cannot be inversed, it allows evaluation of the statistical moments of the distribution [4].

Some moments and normalized cumulants of distributions are:

$\tilde{m}_n$ is the n'th moment;
$\bar{m}_n$ is the n'th central moment;

$z_A \equiv \tilde{m}_1$ is the center of gravity of the peak (first moment);
$\sigma_z^2 \equiv \bar{m}_2 = \tilde{m}_2 - z_A^2$ is the variance of distribution — second central moment;
$\Sigma_z \equiv \frac{\bar{m}_3}{\sigma_z^3}$ is the skewness – third normalized cumulant; for Gaussians, $\Sigma_z = 0$; a tailing peak has $S_z < 0$;
$\varepsilon_z \equiv \frac{\bar{m}_4 - 3\sigma_z^4}{\sigma_z^4}$ is the kurtosis excess — fourth normalized cumulant; for Gaussians, $\varepsilon_z = 0$; a flatter peak has $\varepsilon_z < 0$.

From the point of view of mathematics, the zone profiles encountered in chromatography belong to the class of distributions, which can be described in terms of their moments and cumulants [6] by the Gram–Charlier series

$$\rho(z) = \frac{1}{\sigma_z \sqrt{2}} e^{-\frac{u^2}{2}} \left[1 + \frac{\Sigma_z}{6}\left(u^3 - 3u\right) + \frac{\varepsilon_z}{24}\left(u^4 - 6u^2 + 3\right) + \ldots \right] \quad (4.19)$$

where $u \equiv \frac{z - z_A}{\sigma_z}$, and in round parentheses are the Hermite polynomials $H_3(u)$ and $H_4(u)$.

The similar Edgeworth–Cramér asymptotic expansion may offer a better approximation [4]; usually it is truncated like

$$\rho(z) = \frac{1}{\sigma_z \sqrt{2}} e^{-\frac{u^2}{2}} \left[1 + \frac{\Sigma_z}{6} H_3(u) + \frac{\varepsilon_z}{24} H_4(u) + \frac{10 \Sigma_z^2}{144} H_6(u) \right] \quad (4.20)$$

where $H_6(u) = u^6 - 15u^4 + 45u^2 - 15$.

When treating experimental data, calculation of the moments is straightforward. Hence, the above formulae offer a convenient way to fitting the shapes.

According to the extensive review by Marco and Bombi [7], almost 100 dedicated empirical or semiempirical formulae have been proposed for the chromatographic peak shapes. Some of them may reasonably well fit experimental data, even when physicochemical rationale for them is lacking.

The "exponentially modified Gaussian" [8] is widely used

$$\rho(z) = \frac{1}{2\eta_e} \exp\left(\frac{\sigma_G^2}{2\eta_e^2} - \frac{z - \bar{z}_G}{\eta_e} \right) \times \mathrm{erfc}\left[\frac{1}{\sqrt{2}} \left(\frac{\sigma_G}{\eta_e} - \frac{z - \bar{z}_G}{\sigma_G} \right) \right] \quad (4.21)$$

where η_e is the constant of the precursor exponential distribution, σ_G is the standard deviation and $\bar{z}_G$ is the mean coordinate of the precursor Gaussian distribution. The moments of this $\rho(z)$ are:

$$z_A = \bar{z}_G + \eta_e, \quad \sigma_z^2 = \sigma_G^2 + \eta_e^2, \quad \Sigma_z = \frac{2\eta_e^3}{(\sigma_G^2 + \eta_e^2)^{3/2}}, \quad \varepsilon_z = \frac{6\eta_e^4}{(\sigma_G^2 + \eta_e^2)^2} \quad (4.22)$$

Figure 4.2 presents a computer-generated profile of this type and its best fit by the series 4.19. [9] On this occasion the approximation is better than that provided by 4.20. As could be expected, the Gram–Charlier series experiences problems in fitting the exponential tail of the profile.

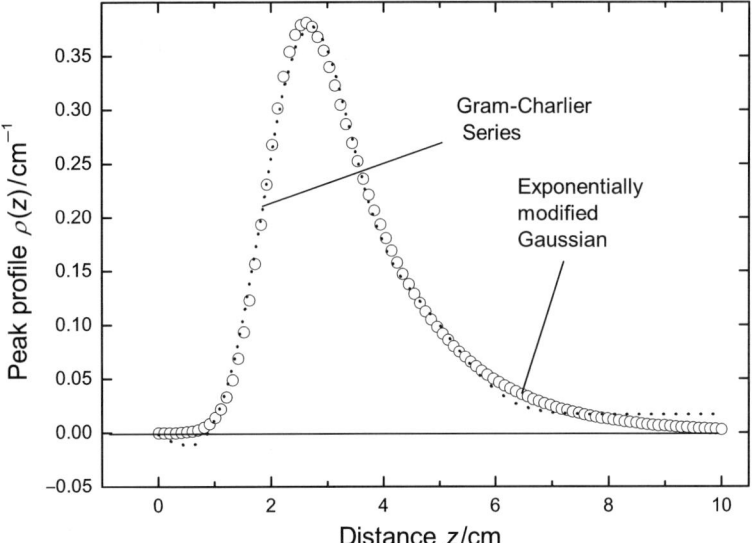

Fig. 4.2 Computer-generated exponentially modified Gaussian function (see Eq. 4.21) with $\bar{z}_G = 3.0$; $\sigma_G = 0.5$, and $\eta_e = 1.5$ as fitted by the series 4.19 [9].

Lan and Jorgensen [8] gave a detailed discussion of the "exponential–Gaussian hybrid function," which describes asymmetric peaks better than the exponentially modified Gaussian. The proposal does not seem to have advantages when applied to the situations considered in this book.

4.2.3.1 Isothermal Chromatography

We start with the case of $\delta(z)$ injection profile into an isothermal column under laminar gas flow regime. The factors which cause broadening of the initial profile are longitudinal diffusion and radial diffusion across streamlines. These continuously change the concentration distribution and gradients. In the absence of any adsorption sojourn time of molecules on the wall (i.e.: negligible adsorption–desorption energy), and only in this case, the two above diffusion processes yield:

$$\frac{\sigma_z^2}{z} = \frac{2}{w}\left[D + \frac{d_c^2 w^2}{192D}\right] \qquad \varepsilon_d = 0 \qquad (4.23)$$

Here and below $D \equiv D_{1,2}$. In terms of flow rate:

$$\frac{\sigma_z^2}{z} = 2\left[\frac{\pi d_c^2}{4Q}D + \frac{d_c^2 Q}{48\pi D}\right] \qquad \varepsilon_d = 0 \qquad (4.24)$$

Equation 4.23 is the well-known expression for the dispersion of a solute in flow of fluid, which was first derived by Taylor [10, 11] and Aris [12]. Its appearance is modified here to emphasize that, besides the true gaseous diffusion coefficient, the laminar velocity profile gives another effective diffusion-like contribution to the longitudinal dispersion of the spike. Notice that $d_c w/2D$ is the diffusional Peclet number (Pe), so that the second term in parentheses in Eq. 4.23 becomes $Pe^2/48$.

It should be emphasized that this broadening cannot help any separation. Still, Eq. 4.23 does find an application — it is the measurement of $D_{1,2}$ for two non-adsorbable gases. In a chromatography-like experiment, the flow of a gas is spiked with a little of the other; the elution curve is recorded, and $D_{1,2}$ is evaluated from the curve width. Equation 4.23 then serves to account for the nonzero flow velocity [13].

Chromatographic separations are possible only when the analytes are adsorbed for an appreciable time upon the tube surface. Then they migrate down the column more slowly than the carrier gas and the migration speed depends on ε_d. For the laminar flow in open cylindrical isothermal columns [5, 13]:

$$\frac{\sigma_z^2}{z} = \frac{2}{w}\left[D + \frac{\left(11 - 16\mathfrak{R} + 6\mathfrak{R}^2\right)d_c^2 w^2}{96D}\right] \qquad \varepsilon_d > 0 \qquad (4.25)$$

or

$$\frac{\sigma_z^2}{z} = 2\left[\frac{\pi d_c^2}{4Q}D + \left(11 - 16\mathfrak{R} + 6\mathfrak{R}^2\right)\frac{Q}{48\pi D}\right] \qquad \varepsilon_d > 0 \qquad (4.26)$$

where $\mathfrak{R} \leq 1$ is the velocity of the center of gravity of the adsorption zone, divided by the mean velocity of mobile phase (see Sect. 4.3.1 for evaluation). Again, the second term in the square brackets is the effective diffusion coefficient. Now it also depends on the thermodynamic characteristics of adsorption. The function of $\mathfrak{R}$ in parentheses reaches 11 with $\mathfrak{R}$ approaching zero. Its graph is given in Fig. 4.3, below, alongside a simpler reasonable approximation, which works well enough in applications.

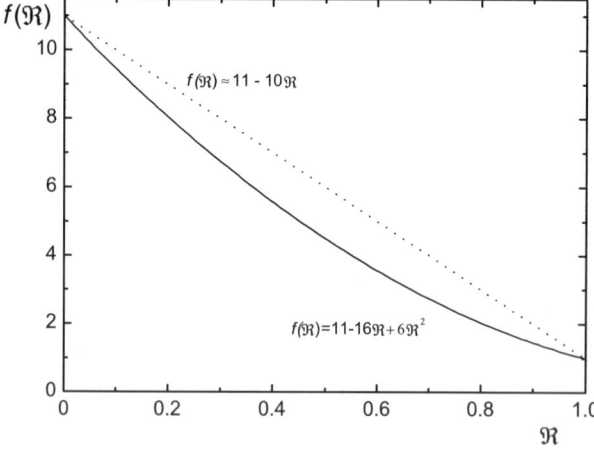

Fig. 4.3 Graph of the factor $\left(11 - 16\mathfrak{R} + 6\mathfrak{R}^2\right)$ and its approximation.

4.2 Theory

The dispersion given by Eq. 4.25 can be considered as two terms. One is due to the longitudinal diffusion and the other originates from the laminar flow patterns and from the adsorption retention. In practice, one of the terms is relatively small and may be neglected. In the conditions of present interest, it is mostly the case with the longitudinal diffusion. This omission will be implied if not stated otherwise.

The first approximation to the peak shape is a Gaussian:

$$\rho^{IC}(z) = \frac{1}{\sqrt{2\pi}\sigma_z} \exp\left[-\frac{(z-z_A)^2}{2\sigma_z^2}\right] \qquad (4.27)$$

Tunitskii, et al. [14] also gave the third central moment of the distribution 4.25; it yields the skewness:

$$\Sigma_z = \frac{wd_c^2}{80D\sigma_z} \qquad (4.28)$$

Then a better approximation is (cf. the series 4.19 and 4.20, above):

$$\rho^{IC}(z) = \frac{1}{\sqrt{2\pi\sigma_{\mathfrak{R}}^2}} e^{-\frac{(z-z_A)^2}{2\sigma_{\mathfrak{R}}^2}} \left\{ 1 + \frac{wd_c^2}{480D\sigma_z} \left[\left(\frac{z-z_A}{\sigma_z}\right)^3 - 3\frac{z-z_A}{\sigma_z} \right] \right\} \qquad (4.29)$$

4.2.3.2 Thermochromatography

When separating mixtures by chromatography, the major concern is the width of zones. Novgorodov and Kolatchkowski [15, 16] seem to be the first to consider the dispersion of TC zone profiles analytically. They assumed that the factors which determine the dispersion of TC zones are the same as those in chromathermography. The latter method [17, 18] is based on placing a long "cold" column into a considerably shorter *movable thermochromatographic furnace*. At the start of processing, the furnace heats the head segment of the column; then, simultaneously with the spike of the analyzed mixture, it begins to move down the column – more slowly than the carrier gas. Volatile analytes first move faster than the furnace, towards lower temperature, until each of them reaches its own characteristic temperature, at which the zone proceeds right with the speed of the furnace. Such equilibration takes appreciable time and column length, but now the analytes "feel" like being isothermally chromatographed at the mean zone temperatures. The difference with IC is that the zone spreading by diffusion processes is reduced by a relative deceleration of the front and an acceleration of the tail of the zone because of the temperature gradient. As a result, the zone dispersion achieves a steady value.

In thermochromatography the zone continuously moves to even lower temperature, though with decreasing linear speed. Then the temperature-dependent expanding and compression "forces" may not be in equilibrium at each point, for example, when the carrier gas flows very fast and the temperature profile is steep. Nevertheless, the authors of Refs. [15, 16] almost straightforwardly applied the formulae derived for the width of the chromathermographic zone in filled columns to the TC zones in open columns. To obtain a formula for the steady dispersion σ_z, they equated the absolute values of the time derivatives $d\sigma_z/dt$ for the two competing processes. The compression term was derived like in Ref. [18]:

$$\left(\frac{d\sigma_z}{dt}\right)_{cpr} = -g\sigma_z \frac{\varepsilon_d}{k_B T_A^2} \Re w \tag{4.30}$$

It essentially implies using the approximation

$$\exp\left(-\frac{\varepsilon_d}{k_B(T_A + g\sigma_z)}\right) - \exp\left(-\frac{\varepsilon_d}{k_B T_A}\right) \approx \frac{g\sigma_z \varepsilon_d}{k_B T_A^2} \exp\left(-\frac{\varepsilon_d}{k_B T_A}\right) \tag{4.31}$$

which is accurate enough for the purpose, only if $g\sigma_z/T$ is less than 0.01. Meanwhile, the real thermochromatograms are mostly much broader, and such approximation may be too rough.

To apply this approach to thermochromatography, the expansion term should have been obtained by differentiation of Eq. 4.25, which results in:

$$\left(\frac{d\sigma_z}{dt}\right)_{xpn} = \frac{1}{\sigma_z}\left[D + \frac{\left(11 - 16\Re + 6\Re^2\right) d_c^2 w^2}{192 D}\right] \tag{4.32}$$

Now, equating the two competing terms would give:

$$\sigma_z^2 = \frac{k_B T_A^2}{g\varepsilon_d}\left(\frac{D}{\Re w} + \frac{\left(11 - 16\Re + 6\Re^2\right) d_c^2 w}{192 D}\right) \tag{4.33}$$

Thus, the dispersion does not explicitly depend on time, but we remember that T_A depends on the retention time.

The authors of Refs. [15, 16] erroneously evaluated the expanding term 4.32 by differentiating Eq. 4.23. They also used this equation to find the optimal gas flow rate (which gives the smallest height of the theoretical plate) for the columns with a constant temperature gradient and came to:

$$Q_0^{opt} = \pi D_0 d_c \sqrt{\frac{6(T_S + T_A)}{T_0}} \quad \text{or} \quad w_0^{opt} = \frac{4 D_0}{d_c}\sqrt{\frac{6(T_S + T_A)}{T_0}} \tag{4.34}$$

Because Eq. 4.23 does not describe chromatographic processing, its use may result in misleading conclusions. Suffice to say that the $\Re$ values at maximums of TC zones are usually very small; see Sect. 4.3.1 for the formulae. Hence, a value of σ_z calculated from Eq. 4.25 is nearly 10 times larger than that from Eq. 4.23. This flaw limits the usefulness of this otherwise interesting pioneering attempt, which also considered influence of the temporal injection profile (δ-function, short plug, exponential) on the zone shapes.

The approach as such may be prospective, provided that one uses correct starting formulae, which would be valid for fast separations and broad peaks. It means starting with Eq. 4.25 and introducing a more accurate compression term than that given by Eq. 4.30.

4.2 Theory

The problem of the TC *peak shape* was first tackled by Zhuikov [19]. Attempting to evaluate analytical formulae, he used the differential equations appropriate to the molecular kinetic picture. Surface concentration of an adsorbate changes in time as the difference between the rates of adsorption and desorption

$$\frac{\partial c_a}{\partial t} = \frac{u_m a_z}{4} c_g - v_0 e^{-\frac{\varepsilon_d}{k_B T}} c_a \tag{4.35}$$

while the equation of balance, assuming constant temperature gradient, is:

$$\frac{\partial c_g}{\partial T} = \frac{1}{gQ} \frac{\partial c_a}{\partial t} \tag{4.36}$$

Then he made a number of assumptions and approximations. First, by differentiating Eq. 4.35, excluding $\partial c_g / \partial T$, and omitting some small terms he came to:

$$\frac{\partial^2 c_a}{\partial t \partial T} = \frac{u_m a_z}{4gQ} \frac{\partial c_a}{\partial t} - v_0 e^{-\frac{\varepsilon_d}{k_B T}} \frac{\partial c_a}{\partial T} - \frac{\varepsilon_d}{k_B T^2} v_0 e^{-\frac{\varepsilon_d}{k_B T}} c_a \tag{4.37}$$

To obtain an analytical solution to this equation, he made the principal assumption that c_g at any given T_z is determined only by desorption of adsorbate from a *short* adjacent upstream section with the length Δ_T in K (that is Δ_T/g in cm). Then, from Eqs. 4.35 and 4.36, he found that, in the absence of any income of the adsorbate into the short section, the value of c_a at T_z would be reduced with time like

$$\frac{c_a(t)}{c_a(0)} = \exp\left[-bt \exp\left(-\frac{\varepsilon_d}{k_B T}\right)\right] \tag{4.38}$$

where

$$b \approx \frac{4gQv_0}{u_m a_z \Delta_T} \tag{4.39}$$

He considered the case of practically momentous injection of the analyte and postulated that, in the region of the thermochromatographic peak, the entire amount of the adsorbate above T_z is effectively concentrated within the Δ_T segment. Combining this postulate with Eq. 4.38 gave him:

$$\frac{1}{g} \int_{T_S}^{T_z} c_a dT_z = \exp\left[-bt \exp\left(-\frac{\varepsilon_d}{k_B T_z}\right)\right] \tag{4.40}$$

Taking the temperature derivative of the above, neglecting the dependence of b on T, and passing on to the linear coordinate resulted in the principal formula:

$$\rho^{TC}(z) = bt_R^{TC} \frac{\varepsilon_d}{k_B T_z^2} \exp\left[-bt_R^{TC} \exp\left(-\frac{\varepsilon_d}{k_B T_z}\right)\right] \times \exp\left(-\frac{\varepsilon_d}{k_B T_z}\right). \tag{4.41}$$

His value of Δ_T, consistent with the formula for the maximum of the zone, is given by:

$$\Delta_T \approx \frac{k_B T_A^2}{\varepsilon_d} \approx 0.04 T_m \tag{4.42}$$

He also presented the formulae for the frontal regime of injection. This TC peak profile, on its high temperature side, grows roughly as $\exp(\varepsilon_d/k_B T_z)$, that is proportionally to the adsorption isobar; it is expected for frontal processing from first principles; cf. Fig. 4.1. However, the low temperature side of the theoretical peak is not described correctly. Indeed, Eq. 4.41 yields a tail which is determined mainly by the reciprocal of the Boltzmann factor, and by the local temperature gradient. Its extent is independent of D and slightly affected by Q. The absence of D in any of the equations is a principal flaw of the work. The proposed formulae are evidently applicable only to very small flow rates of the carrier gas. Otherwise, a non-equilibrated fast flow carries substantial portions of the analyte beyond T_{id}, where it is mostly subject to irreversible diffusional deposition (see Sect. 4.3.3). Then, unlike in the Zhuikov's treatment, the resulting peak tail is almost independent of the column temperature profile and of the desorption energy.

The published graphs [20] of the Zhuikov's theoretical profiles show nearly symmetric curves. The conditions were: $t_R^{TC} = 30$ min, $Q_0 = 0.33 \text{ cm}^3\text{s}^{-1}$, constant $g = 40 \text{ K cm}^{-1}$ and $d_c = 0.4$ cm. For the values of $\varepsilon_d N_A$ equal to 200, 160, 120 and 80 kJ mol^{-1}, the temperatures T_A were 870, 700, 520 and 380 K, and the FWHMs were 2.1, 1.7, 1.25 and 0.82 cm, respectively.

This interesting and stimulating approach deserves thorough analysis of the involved assumptions and simplifications. It is important to find the limits of application of the formulae in terms of experimental conditions and properties of adsorbates. To our knowledge, the work [19] is the only attempt to evaluate formulae for the profiles of TC zones.

4.3 Mathematical Modeling of Gas–Solid Chromatography

As was evidenced above, the attempts to derive the zone profiles by solution of the appropriate differential equations meet severe difficulties. In such situations, it seems reasonable to trace the influence of various variables, as well as of the experimental conditions, by computer simulation of the histories of single molecular entities migrating down the column.

A tracer molecule moves downstream as a result of numerous displacements between consecutive encounters with the wall. Evidently, a just-desorbed molecule erratically diffuses through the column volume by random free flights between the collisions with the molecules of carrier gas, until it again strikes the surface and adsorbs to it. Mostly, after only a few free flights, the molecule will be back to the wall near the point of origin because, so close to the wall, the convective transportation is very slow. Hence, the consecutive adsorption events mostly occur in rapid series at near coordinates. As the random flights in gas are three-dimensional and unequal, the short displacements should happen both up and down the column — not exactly symmetrically owing to the convection, whatever slow. Very seldom the molecule does diffuse far enough from the wall to be carried by the flow over a large downstream distance before the next encounter with the wall. Let $\rho(\eta^*)$ be the

4.3 Mathematical Modeling of Gas–Solid Chromatography

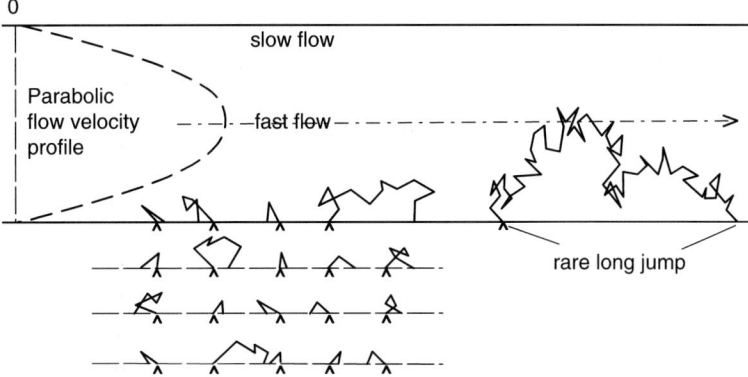

Fig. 4.4 Schematic of the random migration of molecules down the column [21]. The free flights and the column diameter are not to scale. More simulated events are shown on separate lines.

Reproduced from Journal of Radioanalytical and Nuclear Chemistry Articles, 204(1), Zvara I, Problems in thermochromatographic separation of radioelements, 123–134, © 1996, with permission from Springer.

probability density function (pdf) of η^* – the projection of a random displacement on the z-axis. Now, we can expect $\rho(\eta^*)$ to have the shape shown schematically in Fig. 4.4, with small, but nonzero, chance of long "jumps."

4.3.1 Monte Carlo Simulation of Individual Molecular Histories

It would take too much computer time to simulate all the elementary erratic migrations like those depicted in Fig. 4.4. It is desirable to describe the migration by a smaller number of effective displacements. To that end, Zvara [22] proposed a Monte Carlo approach suitable for simulation of chromatography in open columns with moderate to large laminar flow velocities, with the longitudinal diffusion neglected. The technique allows reducing the mean number of steps in each individual history by three orders of magnitude or so. A few sound approximations allow simulating the microscopic migration histories by a series of random, exponentially distributed downstream jumps, which are interspersed with a large random number of adsorption–desorption events occurring in sequence and proceeding without change in the coordinate. The jumps are responsible for the migration, while the collisions with the wall followed by the adsorption sojourns make the total net retention time. A contribution also comes from the gas hold-up time.

When looking for a reasonable approximation of $\rho(\eta^*)$, one can consult the equations of Sect. 2.2. They describe diffusional deposition in channels under various flow regimes, when the tracer is evenly distributed over the inlet area. The solutions for both hydrodynamically and diffusionally developing flow directly apply to the problem of the first jump down the column. It is the common experimental situation

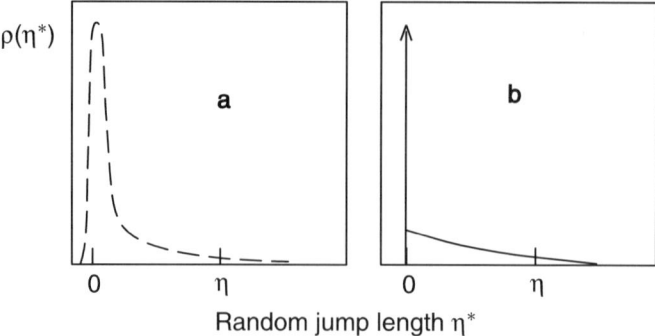

Fig. 4.5 Probability density distributions of elementary displacements $\rho(\eta^*)$ between two encounters of a molecule with column surface. a – character of the real distribution; b – the accepted approximation [22].

Reproduced from Radiochimica Acta, 38(2), Zvara I, Simulation of thermochromatographic processes by the Monte Carlo method, 95–101, © 1985, with permission from Oldenbourg Wissenschaftsverlag.

of the gas being fed from a large reservoir into a tube. However, the case of major interest here is different. We consider a molecule, which starts making the jumps, being near the surface. Mathematically, it is injection into the tube as a point or ring at this very small distance from the wall. Even if the solution were available, it would not accurately describe the results of the actual three-dimensional random walks, which also result in longitudinal diffusion. Still, we can expect that the later exponential terms in the pertinent equations of Sect. 2.2.2 have much higher relative weight (i.e., a_i values) for the ring input.

In the meantime, the above semi-quantitative microscopic picture of the migrations suggests that $\rho(\eta^*)$ has a shape like in Fig. 4.5a, and that a reasonable two-term approximation to it, convenient for simulations, would be Fig. 4.5b. The first term considers that the mean of the short displacements must be close to zero; hence, they can be all put equal zero through $\delta(\eta^*)$. The other term accounts for the occasional intermittent, rather long, downstream jumps; it does it through only one exponential distribution. Thus:

$$\rho(\eta^*) = (1 - a_\eta)\delta(\eta^*) + \frac{a_\eta}{\eta} e^{-\frac{\eta^*}{\eta}} \tag{4.43}$$

The quantity a_η is evidently the reciprocal of the average number of collisions of a molecule with a wall in a column section of the length η.

The problem is to find the effective value of the parameter η. As it follows from Eq. 2.9, the number of collisions per unit length of smooth open cylindrical column is $\pi d_c u_m / 4Q$, so that:

$$a_\eta = \frac{4Q}{\pi d_c u_m \eta} \tag{4.44}$$

The variance of the distribution 4.43 is:

$$\sigma_\eta^2 = a_\eta \eta^2 (2 - a_\eta) \approx 2 a_\eta \eta^2 \tag{4.45}$$

4.3 Mathematical Modeling of Gas–Solid Chromatography

On the other hand, in the simulated IC, the variance of the zone profile σ_z^2, after the zone has migrated the average distance z_A, may be written as:

$$\sigma_z^2 = \frac{\pi d_c u_m}{4Q} z_A \sigma_\eta^2 \tag{4.46}$$

Then, taking into account Eq. 4.45, we obtain:

$$\frac{\sigma_z^2}{z_A} = 2\eta \tag{4.47}$$

We have already seen the solution for the variance of the zone profile in open isothermal columns – cf. Eq. 4.26. Usually, its right-hand side term which accounts for longitudinal molecular diffusion can be neglected. Equating of the right-hand sides of Eqs. 4.26 and 4.27 provides the effective value of η to serve the Monte Carlo simulations:

$$\eta = \frac{(11 - 16\Re + 6\Re^2)Q}{48\pi D} \tag{4.48}$$

The value of $\Re$ is the ratio of the times spent by nonadsorbable and adsorbable species to pass an isothermal column. The net retention time is Z_z (from Eq. 2.9) multiplied by τ_a. In a smooth open cylindrical column of the length $w \times 1$ s, we have $Z_z = \pi d_c w u_m / 4Q = u_m / d_c$. It leads to:

$$\Re = \frac{1}{1 + n_{az}\tau_a} = \frac{d_c}{d_c + u_m \tau_a} \tag{4.49}$$

If $\Re$ approaches zero, Eq. 4.48 reduces to:

$$\lim_{\Re \to 0} \eta \approx \frac{11Q}{48\pi D} = \frac{Q}{4.36\,\pi D} \tag{4.50}$$

The limit is somewhat shorter than the longest component of the diffusional deposition profile from laminar flow given by Eq. 2.45, for which the numerical coefficient in the denominator is 3.65; cf. the discussion in Sect. 2.2.2. From the formula 4.50 it follows that the limiting η depends on the temperature as $\eta \sim 1/\sqrt{T}$, but *does not* depend on pressure or on the column diameter.

The next step is to evaluate the adsorption residence time resulting from a series of encounters. Evidently, the number of short displacements, and also the adsorption events in a sequence, has the discrete geometric probability distribution

$$P(N_a^*) = \frac{(N_a - 1)^{N_a^*-1}}{N_a^{N_a^*}} \tag{4.51}$$

with the average value $N_a = 1/a_\eta$. The residence time τ_N, the sum of N_a^* exponentially distributed values of τ_a has the pdf:

$$p(\tau_N^*) = \left[\frac{(\tau_N^*)^{N_a^*-1}}{\tau_N^{N_a^*}(N_a^* - 1)!} \right] \exp\left(-\frac{\tau_N^*}{\tau_a}\right) \tag{4.52}$$

Finally, the random residence time in the series of nearby adsorptions τ^*_{Ns} is distributed like:

$$\rho(\tau^*_{Ns}) = \sum_{N^*_a=1}^{\infty} P(N^*_a) \cdot \rho(\tau^*_N) = \frac{1}{N_a \tau_a} \exp\left(-\frac{\tau^*_{Ns}}{N_a \tau_a}\right) \quad (4.53)$$

4.3.2 Calculational Procedure

The pdfs for displacements and retention times determined by Eqs. 4.48, 4.43 and 4.53 form the basis for Monte Carlo simulations of the individual migration histories in both isothermal and thermochromatographic columns. In the TC case, a long jump may cover a considerable range of temperature. Hence, each time one would have to select a reasonable effective temperature to calculate the temperature-dependent value of Q/D, which affects the value of η. Concrete simulations showed [9] that substituting the temperature at the starting coordinate of the jump negligibly affects the results, compared to other choices within the jump length.

A vision of the "actual path" is schematically shown in Fig. 4.6 with two levels of detailing, because the number of microscopic displacements is of the order of

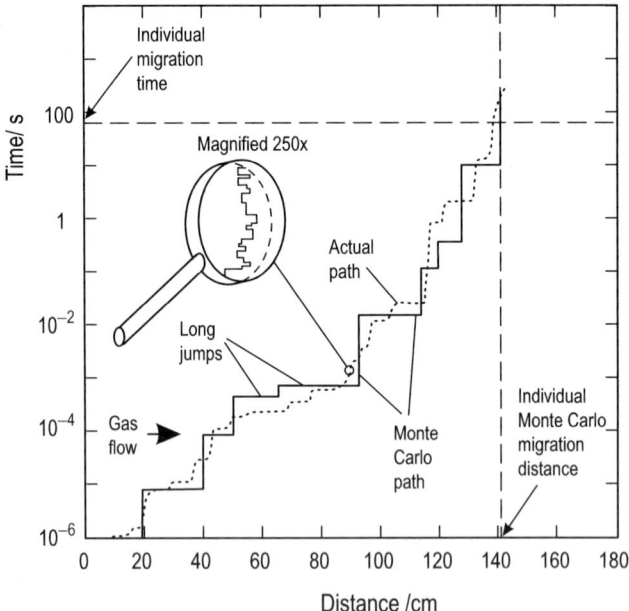

Fig. 4.6 Actual migration path of a molecule in space and time compared with a Monte Carlo simulation [21].
Reproduced from Journal of Radioanalytical and Nuclear Chemistry Articles, 204(1), Zvara I, Problems in thermochromatographic separation of radioelements, 123–134, © 1996, with permission from Springer.

hundreds per centimeter. The Monte Carlo migration distance is fixed when the total retention time becomes longer than the stochastic individual migration time.

4.3.2.1 Random Processing Times of Individual Events

When simulating the molecular histories, especially for the experiments with short-lived nuclides, it is necessary to note that the actual processing time is a random variable with an individual value for each molecule. It is so not only because of the individual lifetimes of the involved nuclei, but also due to some external factors. These are, for example, the nominal t_R^{TC} and the temporal profile of injection. However, equally important is the temporal regime of detection of the decay events — whether they are registered in the course of the run or after its end.

A concrete example comes from the experimental works with the radionuclides characterized by t_λ, much shorter than t_R^{TC}. A long continuous on-line experiment formally looks like the frontal chromatography of the products. However, when the decay events of such nuclides are registered immediately by the material of the column, the resulting internal chromatogram is the same as it would be in the elution regime. The random processing time t_f^* equals the random individual lifetime of the nucleus.

More situations are considered in Ref. [22].

4.3.2.2 Flowchart of Simulations

Migration of a molecule down the column is traced by computer simulation according to the flowchart in Fig. 4.7. After taking an individual processing time t_f^*, starting at $z_{i=0} = 0$, a random long jump η_j^* is picked up. The latter is distributed according Eq. 4.48, while η is calculated from Eq. 4.3.1 using the preset $T_z = T(z)$. The coordinate of the molecule becomes $z_i = z_{i-1} + \eta_j^* = \sum_{j=1}^{i} \eta_j^*$ and the temperature dependent $\tau_a(z_i)$ is calculated from Eq. 2.23. A random τ_{Ns}^* is picked up from the pdf by Eq. 4.53. Next, the adsorption residence times are summed over all jumps and the total is compared with t_f^*. If $\sum_{j=1}^{i} \tau_{sN,j}^* \geq t_f^*$, then the last coordinate z_i, which is the migration distance in Fig. 4.5, is stored and the next molecular event is simulated. If the mean residence time of molecules in the mobile phase is comparable to t_λ of the nuclide, the condition

$$\sum_{j=1}^{i} \tau_{Ns}^* + \sum_{j=1}^{i} \eta_j^*/w(T_{z,j}) \geq t_f^* \tag{4.54}$$

approximately accounts for the time spent in the gaseous phase.

The program can be designed to calculate and sum up the contributions of each event to the zero-point moments $\tilde{m}_n$ of the profile and sort the events for subsequent

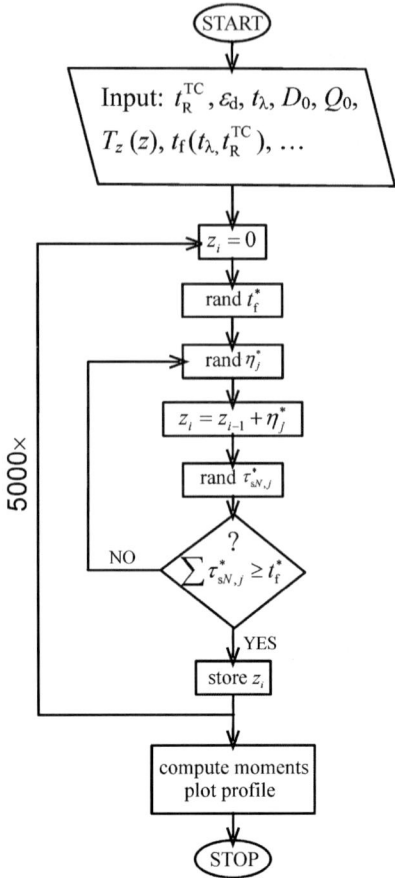

Fig. 4.7 Flowchart of computer program for simulation of GC peak profiles [22].

Reproduced from Radiochimica Acta, 38(2), Zvara I, Simulation of thermochromatographic processes by the Monte Carlo method, 95–101, © 1985, with permission from Oldenbourg Wissenschaftsverlag.

construction of a histogram of the zone. It allows evaluating various statistical characteristics of the profile, fitting the simulated peak by appropriate pdfs and other actions.

4.3.3 Sample Results of Simulations

Most of the results mentioned below come from Ref. [22], where 5,000 molecular histories were followed to obtain each zone profile. The experimental conditions and properties of the processed species were broadly varied. Listed in that paper are various characteristics of the sampled zones: z_{id} and T_{id}, z_A and T_A, coordinate and temperature of the zone maximum, as well as variance, skewness and kurtosis excess

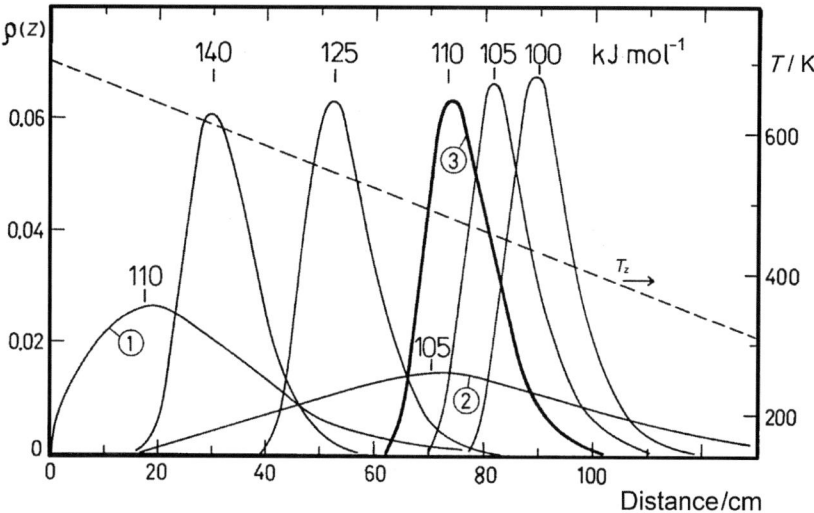

Fig. 4.8 Simulated zone profiles at various $\varepsilon_d N_A$ [22]. All curves: elution regime; $t_R^{IC} = t_R^{TC} = 2000$ s; $Q_0 = 10 \text{ cm}^3 \text{ s}^{-1}$, $d_c = 0.3$ cm, $p = 1$ bar, $D_0 = 0.1 \text{ cm}^2 \text{ s}^{-1}$, $\tau_0 = 1 \cdot 10^{-12}$ s, $M = 220 \text{ g mol}^{-1}$. Curves 1 and 2: IC at $T_c = 500$ K. Other curves: TC with $T_z = T_S - 3z$ (see also graph of T_z).

Reproduced from Radiochimica Acta, 38(2), Zvara I, Simulation of thermochromatographic processes by the Monte Carlo method, 95–101, © 1985, with permission from Oldenbourg Wissenschaftsverlag.

of the profiles. Figure 4.8 compares some elution IC and TC zones for several values of ε_d. The mean lifetimes were much longer than t_R^{TC}. Profile no. 3 was chosen as a standard; as such it is also shown in Fig. 4.9.

Figure 4.9 presents zones of short-lived isotopes of certain elements with different half-lives in conditions similar to those in Fig. 4.8, except that the nominal duration of the experiment t_R^{TC} was much longer than any of the t_λ values. Comparison of the standard profile with that for an isotope with $t_\lambda = 2{,}000$ seconds demonstrates the extent of broadening due to the random lifetimes of the radioactive atoms. The seeming "isotopic separation" is not real — the picture shows the points at which the nuclei ceased existence. Since the longer the lifetime, the greater the mean migrated distance, the actual mean processing time has nothing to do with the nominal t_R^{TC}. Such a situation frequently occurs in TAEs studies.

It is interesting to compare the statistical parameters of the MC simulated zones and the parameters of the best fits to them by the exponentially modified Gaussian distribution. It was done [9] for the zones resulting from three dissimilar thermochromatographic regimes, which were related by some common experimental conditions. These conditions, as well as the zone profiles, are shown in Fig. 4.10; the statistical parameters are listed in Table 4.1. The first regime was the elution thermochromatography of a nonradioactive nuclide, when all individual processing times equal the duration of the run (top of Fig. 4.10). Then the ideal

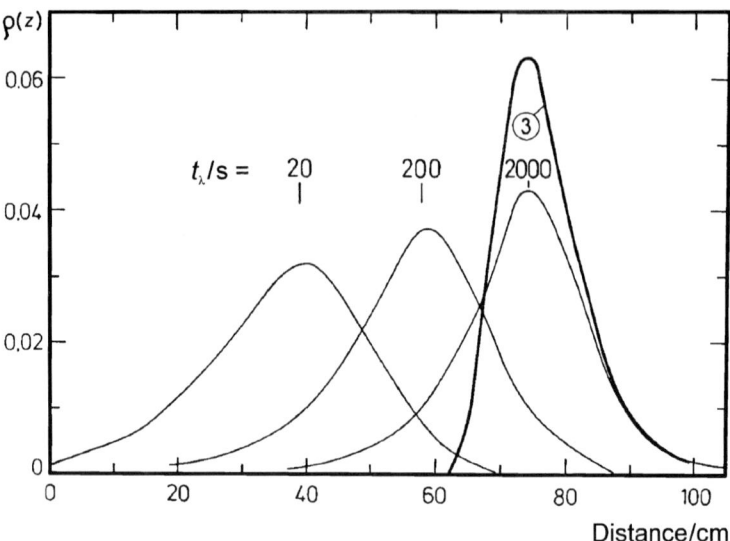

Fig. 4.9 Thermochromatography of short-lived isotopes of an element having different mean lifetimes [22]. Experimental parameters are the same as in Fig. 4.8. Peak No. 3 is identical to that in Fig. 4.8.

Reproduced from Radiochimica Acta, 38(2), Zvara I, Simulation of thermochromatographic processes by the Monte Carlo method, 95–101, © 1985, with permission from Oldenbourg Wissenschaftsverlag.

thermochromatogram is merely $\delta(z_A)$, and the dispersion of the real one originates from the statistics of the long jumps and of the adsorption sojourn time. The bottom of Fig. 4.10 shows the corresponding frontal thermochromatogram. Now the ideal zone has a finite width; the real zone is broader than that in the top of the figure, and its maximum is at a higher temperature.

Experimenters are very interested in simulating long-lasting continuous experiments with much shorter-lived nuclides, when the coordinates of the decay events are recorded in real time. Then the "chromatogram" of decay events does not depend on the time profile of injection, including simultaneous pulse injection of the molecules at any time. The specific feature is the exponential probability distribution of the individual processing times, which are identical with the individual lifetimes. It makes the effective processing regime intermediate between elution and frontal. In the middle of Fig. 4.10 are the distributions of decay events of a nuclide if its average lifetime of 2,000 seconds is much shorter than t_R^{TC}. The ideal zone profile is determined solely by the scatter of the lifetimes. Both the ideal and the real zones are somewhat similar in appearance to their frontal counterparts, but are broader and shifted towards lower temperature because of the larger span of the individual processing times.

It was found that all three simulated TC peaks in Fig. 4.10 can be well fitted with the exponentially modified normal distribution, which was discussed in Sect. 4.2.4 and is described by Eq. 4.21. The success is shown in Fig. 4.10. Important details

4.3 Mathematical Modeling of Gas–Solid Chromatography

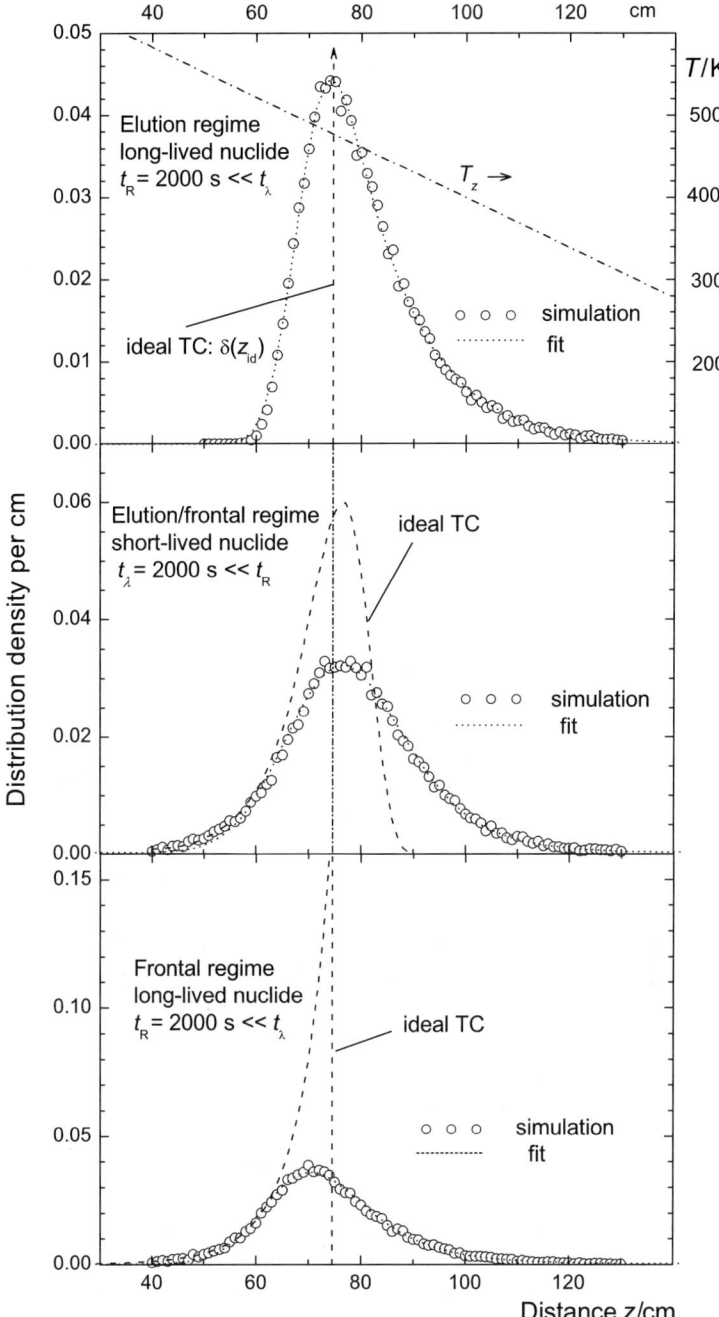

Fig. 4.10 Fits of various simulated thermochromatograms by the exponentially modified Gaussian [9]. The values of parameters were $\varepsilon_d N_A = 110000\,\text{J mol}^{-1}$, $Q_0 \equiv 10\,\text{cm}^3\,\text{s}^{-1}$, $d_c = 0.3\,\text{cm}$, $p = 1\,\text{bar}$, $D_0 = 0.05\,\text{cm}^2\,\text{s}^{-1}$, $\tau_0 = 1\cdot 10^{-12}\,\text{s}$, $M = 220\,\text{g mol}^{-1}$. $T_z = T_S - 3z$ (see also graph of T_z).

Table 4.1 Statistical Parameters of the Zone Profiles Simulated by MC Technique and of the MC Profiles Fitted by the Exponentially Modified Gaussian

Quantity	Evaluation procedure	TC regime			
		Elution	Short-lived	Frontal	Short-lived ideal
η/cm	MC[a]	11.6	11.6	11.4	—
η_e/cm	fit	12.2	9.2	10.1	0.13
z_G/cm	fit	68.7	69.9	64.0	74.8
σ_G/cm	fit	4.7	9.3	7.5	—
z_{av}/cm	MC	80.8	78.8	73.8	72.3
$(z_G + \eta_e)$/cm	fit	80.9	79.1	74.1	—
σ_z/cm	MC	12.4	15.0	14.2	6.2
$\sqrt{\sigma_G^2 + \eta_e^2}$/cm	fit	13.1	13.0	12.6	—
Σ_z	MC	1.5	0.45	0.6	1.7
$\dfrac{2\eta_e^3}{(\sigma_G^2 + \eta_e^2)^{3/2}}$	fit	1.6	0.7	1.0	—
ε_z	MC	3.9	2.2	2.5	5.6
$\dfrac{6\eta_e^4}{(\sigma_G^2 + \eta_e^2)^2}$	fit	4.9	1.5	2.5	—

[a] the parameter was predetermined (see Sect. 4.3.1)

are provided in Table 4.1. Here the statistical parameters of the simulated zone are compared to the parameters evaluated from the best fit using the relationships 4.22.

Let us consider some quantities which affect and characterize the simulated peak profiles.

1. The effective mean jump length in MC simulations from Eq. 4.50 is close to the limit of the mean deposition length from Eq. 2.45. Thus, it is in general accordance with the assumed physical picture of molecular migration down the column. Even if $\Re$ is not *much* smaller than unity, this picture holds unchanged – now the zone "runs after" the molecules carried by flow and its variance corresponds effectively to shorter steps calculated from 4.48. It is evident that the simulated profile on its low temperature tail approaches an exponential curve with η as the characteristic length (if we neglect its temperature dependence).
2. An analysis of Eq. 4.26 shows [22] that the longitudinal diffusion contribution to η is to be taken into account when $Q/Dd_c < 2$. In this case, one should add a random diffusional displacement, positive or negative, to the length of jump each time.
3. It was emphasized above that η is chosen to correctly reproduce the variance of the isothermal peak. Indeed, the characteristics of the two IC profiles in Fig. 4.8 satisfy Eq. 4.47 within statistical uncertainties. Such η also yields a reasonable skewness $\Sigma_k = 3\sqrt{\eta/2z}$ for the pdf by Eq. 4.43 at $a \ll 1$, while the exact solution for a zone in laminar flow [14] yields the numerical coefficient 2.4 rather than three.
4. The simulated profiles of isothermal chromatograms have something in common with the results of Refs. [5, 23, 24]. In these studies, the elution chromatography was formally characterized by the rate constants of adsorption and

4.3 Mathematical Modeling of Gas–Solid Chromatography

desorption, and stochastic summation of residence times in the stationary and mobile phases yielded formulae for the elution curve (time-dependent concentration at the exit). Further developments in these lines established equivalence between the microscopic (molecular) model of chromatography and the macroscopic lumped kinetic model [24, 25].

5. When using the described MC simulations, we suppose that the column surface is absolutely smooth and homogeneous with respect to τ_0 and ε_d, and that the sticking coefficient is unity. Meanwhile, we tacitly ignore the possibility of lateral surface diffusion. All this may be rather far from being true. Yet, the above quantities characterize a certain model of chromatography. It cannot be rigorous because of the complex structure of real surfaces, which will be discussed in Chapter 5.
6. The simulations, according to the proposed scheme, automatically account for the compression action of the temperature gradient on the zone width, which was of concern in Sect. 4.2.3.1.
7. With a slight modification, the flowchart in Fig. 4.7 can serve to simulate isothermal chromatography of short-lived nuclides when we detect the nuclei which survived at the column exit; see Fig. 1.7.
8. Some gas-chromatographic mechanisms differ from the mere adsorption–desorption events of unaltered molecules in that they involve chemical reaction(s). The resulting chromatograms can be simulated as well if, when formulating the individual processing time, we allow for the time of the chemical reaction (in gas or on surface) and for the difference in adsorption (and diffusion) characteristics of the initial and final species. See Sect. 5.7 for more details.

For some 20 years the above-outlined MC simulations have been widely utilized for evaluation and interpretation of thermodynamic characteristics of tracer compounds, especially in TAE chemistry studies. Sometimes, this resulted in too far-reaching conclusions and statements, which could be avoided if one paid attention to the warnings already issued in the original publication [22]. It should be stressed that the basic formulae serving for simulation were deduced after making several rather fundamental assumptions and simplifications. One must be aware of them, especially when trying to upgrade the approach and the corresponding calculational programs.

Important advantages of the MC simulations are in the versatility of taking into account numerous circumstances of real experiments. For instance, it is easy to allow for any column temperature profile and temporal regime of injection, as well as for the decay chains of the involved nuclides. The temperature along the column can be described merely by a numerical table and a procedure of interpolation. The simulations allow finding the "best" values of τ_0 and ε_d which would correctly fit the experimental zone profiles.

The versatility of the approach, in combination with the ever increasing available computational power, obeys good prospects for future studies. It will be seen in later chapters that it is necessary to introduce a more complex picture of the adsorption processes. Presently, we simulate only what is mobile adsorption on a homogeneous surface at near-zero coverage. As the next step, it is desirable to account for the heterogeneity of real surfaces and for the surface diffusion.

The proposed approach may also be useful in simulating thermochromatography in vacuum columns, chromathermography and other separation techniques in open columns. Moreover, repeated Monte Carlo "experiments" with small number of molecules serve to visualize the uncertainties imposed by poor statistics. They are very helpful in evaluating Bayesian confidence intervals for the parameters measured in the experiment performed in non-ideal conditions, when any attempt to obtain an analytical solution fails completely. This will be discussed and illustrated in Sect. 6.2.

4.4 Vacuum Thermochromatography

Thermochromatography in evacuated columns differs in some aspects from the gas flow-driven chromatographic techniques. Let us consider columns similar in diameter and length to those in gas chromatography and suggest that the analyte is initially placed into the higher temperature end, which is closed. The term vacuum thermochromatography (VTC) implies that the molecules do not experience collisions when in the column volume. This is the "Knudsen regime" — the total gas pressure is so low that the mean free path λ_m is much larger than d_c and, strictly speaking, even longer than the column. For the column diameters in millimeter range, it means pressures much less than 0.001 mmHg. In prevalent conditions, the molecular entities striking the wall are "diffuse reflected" rather than elastically scattered. It means that, even if the sojourn time is practically negligible, the molecule gets attached for a time necessary to loose memory about its vector momentum before the strike. As a result, it moves along the column owing to uncorrelated flights of random length and direction between two encounters with the wall.

In an *isothermal* vacuum tube, with the tracer initially concentrated at its closed end as $\delta(z)$, the random flights are reflected at $z = 0$ and must produce [26] a zone with a shape of the half of a symmetrical distribution (not necessarily normal). This distribution will ever broaden with time, and there cannot emerge any maximum away from the zero coordinate. In VTC such a peak does appear. The obvious reason is that the flight frequency of the molecules rapidly decreases with lower temperature, which produces a seeming flow towards the cold end of the column.

4.4.1 Retention Time

Because of the principal absence of convective flow in VTC, it is not easy to derive the basic relations by analogy with the treatment of gas–solid chromatography processes. Help is offered by a concept of vacuum physics related to Knudsen regime – the conductance C_{VC} of a tube with the length l_c. The quantity has dimension of the flow rate:

4.4 Vacuum Thermochromatography

$$C_{VC} = \frac{4d_c}{3l_c}\frac{u_m}{4}\frac{\pi d_c^2}{4} = \frac{\pi}{12}\frac{u_m d_c^3}{l_c} \text{ cm}^3 \text{ s}^{-1} \qquad (4.55)$$

This applies whenever a reservoir of rarified gas is connected by the tube (strictly speaking — straight and very long) to a completely evacuated reservoir, and if both volumes are so large that the two pressure values are not changed by the amount flowing through the tube. Then C_{VC} is the volume of the "pressurized" gas penetrating the tube per time unit. It does not depend on the inlet pressure because each molecule moves independently from the others; of course, the mass flow of the gas is proportional to the pressure. The tube volume divided by the conductance is the mean time spent by molecules in the tube. Notice that the number of molecules which cross the opening of the channel per second is constant, and that the molecules already inside the column can also get back to the pressurized volume. In the closed evacuated column with the diffusion started and reflected at $z = 0$, the mean time spent in the tube by molecules of a nonadsorbing gas is two times shorter [27].

Gaggeler, et al. [28] took the z coordinate divided by the spent time as an effective flow velocity for the given column segment

$$w^{VTC} = 2z \frac{\pi u_m d_c^3}{12z} \frac{4}{\pi d_c^2 z} = \frac{2}{3} \frac{u_m d_c}{z} \qquad (4.56)$$

and assumed that t_R^{VTC} is incremented over dz, like in the ideal linear IC. It gives:

$$\frac{dt_R^{VTC}}{dz} = \frac{k_i(T)}{w_T^{VTC}} + \frac{1}{w_T^{VTC}} \qquad (4.57)$$

where the partition coefficient k_i equals:

$$k_i = \frac{u_m}{d_c} \tau_0 e^{\frac{\varepsilon_d}{k_B T_z}} \qquad (4.58)$$

Eichler and Schädel [29] alternatively used $\tau_0 = h/k_B T$. It is justified in view of the real uncertainty of the two quantities. The v_0 values for metals have been estimated through several physical approaches. Ref. [30] contains a useful compilation and comparison of the derived values. By integrating Eq. 4.57, the authors of Ref. [28] evaluated the adsorption characteristic as a function of $t_\Re^{VCT}$ and other experimental parameters. The resulting formula is quite cumbersome, because the two terms in Eq. 4.47 depend on temperature as $\frac{T_S-T}{\sqrt{T}} e^{\frac{\varepsilon_d}{k_B T}}$ and $\frac{T_S-T}{\sqrt{T}}$, respectively. After a number of approximations, a version of the relationship became:

$$t_R^{VTC} = \frac{3}{2g^2 d_c^2} \frac{k_B T_A}{\varepsilon_d} e^{\frac{\varepsilon_d}{k_B T_A}} \frac{h}{k_B} T_S \qquad T_z = T_S - gz \qquad (4.59)$$

The approach is also briefly outlined in [31].

4.4.2 Description by Random Flights

The actual random flight mechanism behind VTC deserves more detailed consideration. During a typical experiment a molecule must experience some 10^4 free flights. It enables direct simulation of each flight and of the sojourn time by a Monte Carlo technique. Modern computer workstations make it easy. The advantage is that both the position and shape of the peak are obtained in a consistent straightforward way. Besides assuming the cosine law of reflection (see below), there is another substantial assumption which we will examine in Sect. 5.4.1 – the absence of any lateral diffusion. In other words, we imply localized adsorption.

Random free flights were considered as starting from a point at the inner surface of a cylinder [9]. The x-axis was taken perpendicular to the surface and so to the z–y plane, while the cylinder extended along the z-axis. In the spherical coordinates, θ^* denoted the random polar angle of the flight vector to the x-axis, and φ^* was the random angle between the projection of the vector on the z–y plane and the z-axis. Then the random values of the flight length ς and its projection on z-axis ς_z are:

$$\frac{\varsigma^*}{d_c} = \frac{\cos\theta^*}{1 - \sin^2\theta^* \sin^2\varphi^*} \quad \text{and} \quad \frac{\varsigma_z^*}{d_c} = \frac{\cos\theta^* \sin\theta^* \sin\varphi^*}{1 - \sin^2\theta^* \sin^2\varphi^*} \tag{4.60}$$

The angular distribution of the desorbed molecules is not isotropic, but follows the cosine law — the probability of emission into the elementary solid angle is $d\varphi d\theta \cos\theta$. Now, by integration of the above expressions, accounting for the cosine law, the mean values are found to be:

$$\varsigma = \sqrt{\frac{6}{\pi}} d_c = 1.386 d_c \quad |\varsigma_z| = \frac{d_c}{2} \quad \varsigma_z^2 = \frac{2}{3} d_c^2 \tag{4.61}$$

Suppose that the VTC zone results from symmetrical random flights in the presence of a reflecting barrier at $z = 0$, and that the flight length characteristics are those from Eq. 4.61. If the sojourn time can be neglected, the free flights obviously happen with the frequency u_m/ς. Otherwise, in the chromatographic regime, the frequency is $u_m/(\varsigma + u_m \tau_d)$. From Eq. 2.21, we obtain for the effective linear diffusion coefficient D_{Kn} of nonadsorbing gases in cylindrical tubes in Knudsen regime

$$D_{Kn} = \frac{1}{2} \frac{u_m}{\varsigma} \varsigma_z^2 = \frac{1}{2} \sqrt{\frac{\pi}{6}} \frac{u_m}{d_c} \frac{2}{3} d_c^2 = \frac{d_c u_m}{4.16} \tag{4.62}$$

and in the chromatographic conditions:

$$D_{Kn} = \frac{u_m d_c}{4.16} \frac{\varsigma}{(\varsigma + u_m \tau_d)} = \frac{u_m d_c^2}{3(1.386 d_c + u_m \tau_d)} \tag{4.63}$$

Pollard and Present [32], on the basis of molecular kinetics and dynamics, rigorously evaluated the effective diffusion coefficient of nonadsorbable gases in capillaries of finite length to find that it increases with higher l_c and its limiting value is:

4.4 Vacuum Thermochromatography

$$\lim_{l_c/d_c \to \infty} D_{Kn} = \frac{u_m d_c}{3} \tag{4.64}$$

This limit differs from Eq. 4.62 only in the numerical coefficient. Other treatments of the problem, including the first one by Knudsen [33], also gave slightly different values of the coefficient.

The random flight frequency of nonadsorbable molecules is some 10^5 s^{-1}. Their diffusivity, according to Eq. 4.62, is of the order of 10^3–10^4 cm^2 s^{-1}; consequently, the mean diffusional displacement in the first second goes up to 100 cm!

In the conditions of longitudinally decreasing temperature and much longer sojourn time, D^{VC} drastically lessens (see Eq. 4.63), giving rise to a peak at a distance from zero. Thus, the random flight model and its diffusion equivalent make an alternative approach to evaluating t_R^{VTC} possible, without seeking an effective convective flow. To reach a mean square displacement of z^2, the number of walks must be the square of the distance measured in flights, $(z/\varsigma_z)^2 = 4z^2/d_c^2$; cf. Eq. 4.61. Then, in an evacuated isothermal tube, the necessary time t_R^{Kn} is the number of displacements multiplied by the mean sojourn time, plus the total time spent by the molecule in flight on its way to point z:

$$t_R^{Kn} = \tau_d \frac{4z^2}{d_c^2} + \sqrt{\frac{6}{\pi} \frac{d_c}{u_m} \frac{4z^2}{d_c^2}} \tag{4.65}$$

Now the proposed procedure for evaluation of t_R^{VTC} is to differentiate the above t_R^{Kn} by time and then integrate the derivative along the temperature profile of the column. In shorthand:

$$t_R^{VTC} = \int_{T_S}^{T_A} \left(\frac{d}{dT_z} t_R^{Kn} \right) dT_z \tag{4.66}$$

The second term in Eq. 4.65 is usually negligible, compared with the first one, and we can omit it already when differentiating. In the case of the constant temperature gradient we come to:

$$\frac{d}{dT_z} t_R^{Kn} = \frac{4}{d_c^2} \left(2z \frac{dz}{dT} \tau + z^2 \frac{d\tau}{dT} \right)$$

$$= \frac{4\tau_0}{d_c^2 g^2} \left[-2(T_S - T_z) e^{\frac{\varepsilon_d}{k_B T_z}} - (T_S - T_z)^2 \frac{\varepsilon_d}{k_B T_z^2} e^{\frac{\varepsilon_d}{k_B T_z}} \right] \tag{4.67}$$

Then the integral is:

$$t_R^{VTC} = \frac{4\tau_0}{d_c^2 g^2} \left[-2 \int_{T_A}^{T_S} T_z e^{\frac{\varepsilon_d}{k_B T}} dT_z + \left(2T_S + \frac{\varepsilon_d}{k_B} \right) \int_{T_A}^{T_S} e^{\frac{\varepsilon_d}{k_B T_z}} dT_z \right.$$

$$\left. -2T_S \frac{\varepsilon_d}{k_B} \int_{T_A}^{T_S} \frac{e^{\frac{\varepsilon_d}{k_B T_z}}}{T_z} dT_z + T_S^2 \frac{\varepsilon_d}{k_B} \int_{T_A}^{T_S} \frac{e^{\frac{\varepsilon_d}{k_B T_z}}}{T_z^2} dT_z \right] \tag{4.68}$$

Assuming that the values of the integral at T_S can be abandoned compared with those for T_A, and taking into account only the first member of the asymptotic series, one comes to:

$$t_R^{VTC} = \frac{4\tau_0 e^{\frac{\varepsilon_d}{k_B T_A}}}{d_c^2 g^2} \left[2\frac{k_B T_A}{\varepsilon_d}(T_S - T_A)T_A + (T_S - T_A)^2 \right] \quad (4.69)$$

and

$$t_R^{VTC} \approx \frac{8}{g^2 d_c^2} \frac{k_B T_A}{\varepsilon_d} \tau_0 e^{\frac{\varepsilon_d}{k_B T_A}} (T_S - T_A)T_A \quad (4.70)$$

In the meantime, substitution of $\tau_0 = h/k_B T_A$ into Eq. 4.59 gives:

$$t_R^{VTC} = \frac{3}{2g^2 d_c^2} \frac{k_B T_A}{\varepsilon_d} \tau_0 e^{\frac{\varepsilon_d}{k_B T_A}} T_A T_S \quad (4.71)$$

The two last formulae are in agreement within the uncertainties brought in by the approximations. Actually, we are interested in ε_d, so the difference is negligible.

4.4.3 Monte Carlo Simulation

The procedure for Monte Carlo simulation of the cosine law is simple. It can be derived by following the general scheme for continuous distributions [34]. This way we come to

$$\theta^* = \arcsin\sqrt{\xi^*} \quad \text{and} \quad \varphi^* = \frac{\pi}{2}\xi^* \quad (4.72)$$

where ξ^* is a random number with the standard uniform distribution and the two ξ^*'s are picked up independently. Then the random angles are substituted into Eq. 4.60 to calculate the migration distance increment.

Some Monte Carlo simulations of vacuum thermochromatography were published in Refs. [29, 35]; sample data are shown in Fig. 4.11. Also simulated was diffusion in an isothermal column [35] to find the time, after which 50 percent of a short-lived activity emerges from the column outlet, as a function of T_c and ε_d. Only later, these authors presented [31] their prescriptions for the simulation of random flights. Their Monte Carlo procedure for the cosine law was faulty, so their probability density distribution and the mean of displacements are in error. It would be interesting to make similar simulations on the basis of Eq. 4.60; the rigorous pdf of the longitudinal projections of the random flights seems to have non-trivial statistical characteristics. The same is true for isotropically directed free flights; they also have to be examined because the cosine law may not be valid for very rough surfaces.

From the contents of Chapter 5 it follows that the original evaluation of the VTC data in terms of the energy of desorption presented in Refs. [29, 31, 35] should be revised. The value of ε_d was deduced implying the model of mobile adsorption. Meanwhile, if the model were valid, the unhindered movement of the adsorbed molecules across the surface would result in the column, independently of

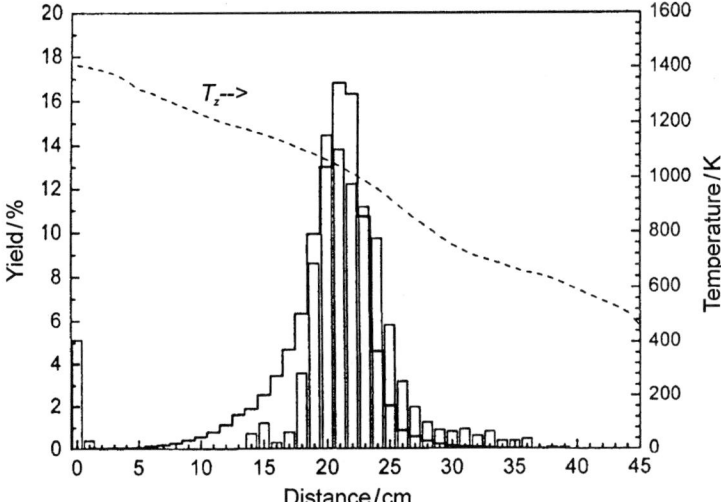

Fig. 4.11 Vacuum thermochromatography of gold in quartz column [35]. Black bars — distribution of ^{192}Au, solid histogram — the result of simulation. Conditions: $d_c = 0.3$ cm, $l_c = 100$ cm, $t_R^{VTC} = 1$–2 hours, column temperature profile – see the broken line.

Reproduced from Radiochimica Acta, 92(8), Hohn A, Eichler R, Eichler B, Investigations on adsorption and transport behavior of carrier free Ag, Au and Pt in quartz columns under vacuum conditions, 513–516, © 2004, with permission from Oldenbourg Wissenschaftsverlag.

its temperature profile, to be momentously emptied through the open end. With both ends closed, the molecules would be practically evenly distributed over the column length. The actual localized character of adsorption can be accounted for if one takes effective v_0 values, which are by several orders of magnitude larger; see Sect. 5.6.4. This will yield a higher value of ε_d.

References

1. Giddings JC (1962), J Chem Educ 39:569
2. Rudolph J, Bächmann K (1979) J Chromatogr 178:459
3. Zvara I (2003) Phys Atomic Nuclei 66:1161
4. Jonsson JA (1987) Dispersion and peak shapes in chromatography. In: Jonsson JA (ed) Chromatographic science series vol 38, Chromatographic theory and basic principles. Dekker, New York
5. Giddings JC (1965) Dynamics of chromatography, Part I, Principles and theory. Dekker, New York
6. Grushka E (1972), J Phys. Chem 76:2586
7. Di Marco WB, Bombi GG (2001) J Chromatogr A 930:1
8. Lan K, Jorgenson JW (2001) J Chromatogr A 915:1
9. Zvara I Unpublished results
10. Taylor GI (1953) Proc Roy Soc, A 219:186
11. Taylor GI (1954) Proc Roy Soc, A 225:473

12. Aris R (1956) Proc Roy Soc, A 235:67
13. Grushka E, Manyard V (1972) J Chem Educ 49:565
14. Tunitskii NN, Kaminskii V.A., Timashev SF (1972) Metody fizokokhimicheskoi kinetiki (Methods of physicochemical kinetiks). Khimiya, Moskva
15. Novgorodov AF, Kolatchkowski (1979) Report P-6-12457. JINR, Dubna
16. Novgorodov AF, Rösch F, Korolev NA (2003) Radiochemical separations by thermochromatography. In: Vértes A, Nagy S, Klencsár (eds) Handbook of nuclear chemistry, vol 5. Kluwer, Dordrecht, p 227
17. Zhukhovitskii AA, Zolotareva OV, Sokolov VA, Terkeltaub NM (1951) (in Russian) Dokl Akad Nauk SSSR 77:453
18. Ohline RW, DeFord DD (1963) Anal Chem 35:27
19. Zhuikov BL (1982) PhD thesis. JINR, Dubna
20. Zvara I, Eichler B, Domanov VP, Zhuikov BL, Kim SCh, Timokhin, SN (1983) (in Russian). In: Internat school seminar on heavy ion physics, Proceedings. Report D7-83-644. JINR, Dubna, p76
21. Zvara I (1996) J Radioanal Nucl Chem Articles 204:123
22. Zvara I (1985) Radiochim Acta 38:95
23. Giddings JC, Eyring HJ (1955) J Phys Chem 59:416
24. Felinger A, Cavazzini A, Dondi F (2004) J Chromatogr, A 1043:149
25. Dondi F, Cavazzini A., Pasti L (2006) J Chromatogr, A 1126:257
26. Feller W (1970) An introduction to probability theory and its applications. Wiley, New York
27. Grover JR (1969) J Inorg Nucl Chem 31:3697
28. Gäggeler H, Eichler B, Greulich N, Herrmann G, Trautmann N (1986) Radiochim Acta 40:137
29. Eichler R, Schädel M (2002) J Phys Chem B 106:5413
30. Eichler B, Kratz JV (2000) Radiochim Acta 88:475
31. Eichler B, Eichler R (2003) Gas-phase adsorption chromatographic determination of thermochemical data. In: Schädel M (ed) The chemistry of superheavy elements. Kluwer, Dordrecht, p 205
32. Pollard WG, Present RD (1948) Phys Rev 73:762
33. Knudsen M (1909) Ann Physik 28:75
34. Yermakov SM, Mikhailov GA (1982) Statisticheskoye modelirovaniye (Statistical modeling). Nauka, Moskva
35. Hohn A, Eichler R, Eichler B (2004) Radiochim Acta 92:513

Chapter 5
Evaluation and Interpretation of the Experimental Data

Abstract Of primary interest in gas-phase radiochemistry are the energies (enthalpies) of desorption of molecular entities from various surfaces. For a new element only single chromatographic measurement of the adsorption equilibrium constant becomes available, and the van't Hoff equation cannot be applied. The only way is to calculate the entropy change from the first principles. Usually, one has assumed smooth homogeneous surface and mobile adsorption. For known compounds their desorption energies from silica correlate with the vaporization (or sublimation) energies; moreover, the values differ little. The finding is difficult to rationalize for smooth bare surfaces and stimulates considering real surfaces. Those of silica and metals are inherently rough and heterogeneous. In the chromatographic experiments with halides, they become covered (modified) by bonded fragments of the reagent molecules. Then a molecule adsorbed at the possible sites in the form of nanoscale wells (pockets) needs similar energies to escape as to leave its own condensed phase. However, accounting for the mostly localized adsorption in the entropy calculations further raises the above "experimental" desorption energy. Meanwhile, heterogeneity makes the latter be an effective value, necessarily less than the maximum in the spectrum. Hence, the highest desorption energy may be considerably larger than the sublimation energy. Possibly the source of it, as well as of the considerable scatter of the data, is incomplete and poorly reproducible modification of the surface. Gas-solid chromatography, besides mere adsorption–desorption of unaltered molecules, may be based on chemical mechanisms; some of them are evidenced and characterized.

Studies of gas-phase radiochemistry of the heaviest elements have focused primarily on physical adsorption of molecular compounds on various surfaces. Generally, physisorption is characterized by relatively low adsorption energies. In the 1930s the founders of the adsorption theories used to give a certain upper limit for this energy, usually a few tens of kJ mol^{-1}, implying that higher values indicate chemisorption mechanisms. It is not to be misleading — they discussed only adsorption of nitrogen and other light gases. Actually, a fundamental characteristic of the physical adsorption is that it stems exclusively from the van der Waals interactions between adsorbates and adsorbents, that is from the electrostatic attraction due to the permanent, induced and instantaneous dipoles.

The London dispersion forces are of general importance because they are omnipresent. Consistently we assume that the chemical nature of the adsorbing surface or of the particular adsorption site is not changed by an adsorption–desorption event, and that perturbation of the electronic structures of the adsorbent and adsorbate, if any, is very small. There do not emerge chemical bonds due to the formation of molecular orbitals or because of the analogous interactions between the electrons of a metallic adatom and those of metallic substrate. Such considerations do not allow for any well-substantiated upper limit on the energy of physical adsorption of molecular adsorptives. Indeed, suppose that the adsorbent is a molecular crystal (rather than an ionic one); then, why couldn't the desorption energy of the particular adsorbate amount up to a considerable fraction of its energy of sublimation, however large the latter might be? Physisorption in different adsorbate–adsorbent systems has some common features: the adsorption and subsequent desorption usually do not require activation energy, and an adsorbent is seldom very selective for some particular compound(s).

Fewer radiochemical works have exploited chemical adsorption, which is characterized by forming true chemical or metallic bonds. In the latter case the valence electrons of the adatom are delocalized and shared with the electron cloud of the foreign metal lattice. Changes in the electronic state may yield chemisorption energies as high as several hundreds of $kJ\ mol^{-1}$. Strong chemisorption can be very selective. It may also require some activation energy for forming and then disrupting the bonds, and may even irreversibly alter the adsorption site.

An example of exploiting chemisorption in radiochemistry of heavy elements was given by the adsorption studies of the atoms of heaviest actinoids on refractory metals to judge the metallic valence of trace elements; see Sect. 1.5.2. Another case of interest was the high adsorption energy of some molecular halides on the surface of alkali halides. It takes place owing to the formation of complexes like K_2ZrCl_6, which are well known as bulk phases; see Sect. 5.1.2 below.

In principle situations intermediate between physisorption and chemisorption can also occur; they can be discussed only on an individual basis.

5.1 Adsorption Enthalpy on Homogeneous Surface

Adsorption (desorption) energies or enthalpies of molecules and atoms on various surfaces are of primary and major interest in the experimental gas-phase radiochemical studies of the heaviest elements. In practice, pertinent data can be obtained almost exclusively in the experiments based on chromatographic principles. In the pioneering works [1–3] the required values were derived using the simplest description of the processes in columns in terms of molecular kinetics (see Sect. 4.2). Later [4] the task of finding the adsorption enthalpies was examined using a thermodynamic approach. It revealed that the molecular-kinetic treatment

5.1.1 Thermodynamic Approach

This approach is based on classical thermodynamics and statistical mechanics; the latter makes the link to the microscopic picture of the adsorption processes. From the point of view of thermodynamics, adsorption can be treated like a chemical reaction. It will be shown that adsorption can proceed with or without a change of the number of gaseous molecules, so that the thermodynamic equilibrium constant expressed in partial pressures may or may not equal the constant expressed in concentrations. Notice that the standard values of some quantities accepted here when deriving formulae for the adsorption characteristics are not the standard states commonly used in chemical thermodynamics. In particular, it concerns the concentrations.

5.1.1.1 Equilibrium Constants

Let c_a in cm^{-2} be the equilibrium number concentration of the tracer entities in the adsorption layer, and c_g in cm^{-3} the concentration in the gaseous phase. Experimental studies have shown that a range of small c_a values (in the sense of surface coverage much less than unity) always exists, over which the dimensional ratio of the concentrations k_a, the distribution coefficient

$$k_a = \frac{c_a}{c_g} \tag{5.1}$$

is constant at a given temperature (a sort of Henry's law). The primary data obtained in the isothermal elution chromatography is the migration distance as a function of the processing time, column temperature, column surface area and flow rate of the carrier gas. In columns with a microscopically uniform surface, k_a can be found from the evident relationship

$$z_A = \frac{t_R^{IC} v_z w c_g}{a_z c_a} = \frac{Q t_R^{IC}}{a_z k_a} \tag{5.2}$$

which gives:

$$k_a = \frac{Q t_R^{IC}}{z_A a_z} \tag{5.3}$$

To analyze the thermodynamics of adsorption, it is necessary to know some basic characteristics of the microscopic profile of the adsorption energy across the surface, as well as the characteristics of the state of the adsorbate. Otherwise, we must make some a priori assumptions concerning these factors. Suppose that the surface is ideally uniform, that is, without even minor patches or single sites with different adsorption properties, and that the adsorption potential across the surface is ideally flat. Then the adsorbate must behave like two-dimensional gas. Its surface pressure is $c_a k_B T$ – similar to the pressure of three-dimensional gas (see Sect. 2.1). Thus, the ratio of concentrations equals that of pressures and must be constant even at large concentrations; the dimension of k_a is cm. The model is called mobile adsorption.

Another idealized model is the localized adsorption, a rationale for the Langmuir isotherm. Now the surface is supposed to consist solely of identical distinct adsorption sites with the number concentration C_a in cm^{-2}; the latter has a definite value for each particular adsorbent–adsorbate system. Let the fractional surface coverage be $\theta \equiv c_a/C_a$. In such a model, proportional to c_g is $c_a/C_a(1-\theta)$, the surface concentration divided by the concentration of free adsorption sites. Hence, strictly speaking, the corresponding dimensional constant k_a^{lc}, now in cm^3, is:

$$k_a^{lc} = \frac{c_a}{(1-\theta)C_a} \frac{1}{c_g} = \frac{k_a}{(1-\theta)C_a} \tag{5.4}$$

At small degrees of coverage, the constant can be approximated by

$$k_a^{lc} \approx \frac{c_a}{C_a c_g} = \frac{k_a}{C_a} \text{cm}^3 \quad \text{if } \theta \ll 1 \tag{5.5}$$

and the model, which deals with the number of sites, rather than with the area per unit length of the column, yields the same migration distance:

$$z_A = \frac{t_R^{IC} v_z w}{a_z C_a} \frac{C_a c_g}{c_a} = \frac{Q t_R^{IC}}{a_z k_a} \tag{5.6}$$

Thus, the initial straight part of the adsorption isotherm does not contain any information about the nature of the adsorbed state — the experiments allow obtaining only k_a evaluated from the primary data through Eq. 5.3.

To obtain the adsorption enthalpy, we need the dimensionless equilibrium constants for partial pressures, or for concentrations K_p or K_c, which would obey the fundamental relations

$$\ln K_p = -\frac{\Delta_{ads} H}{RT} + \frac{\Delta_{ads} S}{R} \tag{5.7}$$

$$\ln K_c = -\frac{\Delta_{ads} U}{RT} + \frac{\Delta_{ads} S}{R} \tag{5.8}$$

where $\Delta_{ads} H$, $\Delta_{ads} U$, and $\Delta_{ads} S$ are enthalpy, internal energy and entropy changes in the process of adsorption. It requires accepting some standard values of concentrations or pressures. In the case of the mobile adsorption, more convenient for the purpose is to set the standard molar volume V and the standard molar area A [4]; their ratio is the reciprocal to the ratio of the standard concentrations or pressures. The choice of the standards is arbitrary; it will be discussed in more detail later.

5.1 Adsorption Enthalpy on Homogeneous Surface

There is only one natural limit for each of them: A should be taken larger than the molar area of the closely packed monolayer of the adsorbate, and V should be larger than the molar volume of the condensed adsorptive. Because we are going to deal only with small degrees of coverage, it seems reasonable to also comply with it when choosing the standard values, and take V and A much larger than their indicated natural limits. This will be implied below, even though formally the only eligible standard is the ratio A/V. Indeed, the thermodynamic constants for the mobile adsorption K_c^{mb} and K_p^{mb} are related by:

$$K_c^{mb} = K_p^{mb} = \frac{A}{V} k_a \qquad (5.9)$$

In the case of the localized adsorption model, we must take into account Eq. 5.3 and also accept an eligible standard value for the fractional coverage $\theta°$. The standard coverage can obviously be interchanged with the ratio of the minimum possible molar area, N_A/C_a, to the standard molar area:

$$\theta° = \frac{N_A}{A C_a} \qquad (5.10)$$

Now, in principle, V and A must be specified separately rather than only by their ratio. The corresponding constants are denoted K_c^{lc} and K_p^{lc}. We obtain:

$$K_c^{lc} = \frac{c_a}{c_g(1-\theta)C_a} \frac{AC_a}{V}(1-\theta°) = \frac{(1-\theta°)}{(1-\theta)} K_c^{mb} \qquad (5.11)$$

In the gas-phase radiochemistry experiments, the surface coverage by the tracer is much less than unity. Then it is reasonable (though not necessary) to also take the standard coverage that small. If so, successive approximations to Eq. 5.11 give:

$$K_c^{lc} \approx K_c^{mb} \quad \text{if} \quad \theta \ll 1 \quad \text{and} \quad \theta° \ll 1 \qquad (5.12)$$

On the other hand, the concentration equilibrium constant does not equal the partial pressure constant because now in the "reaction of adsorption" one mole of gas is lost; it results in:

$$K_p^{lc} = \frac{K_c^{lc}}{RT} \approx \frac{K_c^{mb}}{RT} \qquad (5.13)$$

5.1.1.2 Adsorption Enthalpy by van't Hoff's Equation

Notice that both the adsorption enthalpy and the entropy in the adsorbent–adsorbate systems are negative. Because these parameters relatively weakly depend on temperature and their experimental values are known with only moderate accuracy, we shall neglect this dependence.

In principle the adsorption equilibrium data derived from the experimental data at different column temperatures allow us to evaluate $\Delta_{ads} H$ by pursuing the standard thermodynamic "second law" procedure. Differentiation of Eq. 5.7 gives the enthalpy change in the case of the mobile adsorption model:

$$\frac{-\Delta_{ads} H}{R} = \frac{d \ln K_p^{mb}}{d(1/T_c)} \quad (5.14)$$

Thus, the required value of $\Delta_{ads} H$ can be obtained from the slope of the appropriate linear regression though, in principle, from but two experimental values of the constant. The same result is also provided by simply differentiating the net retention time or the retention volume, because Eq. 5.6 shows that $k_a \sim Q t_R^{IC}$. For example,

$$\frac{-\Delta_{ads} H}{R} = \frac{d \ln t_R^{IC}}{d(1/T_c)} \quad (5.15)$$

Assuming the localized adsorption and consulting Eq. 5.13, we find that:

$$\frac{-\Delta_{ads} H}{R} = \frac{d \ln \left(K_p^{mb}/RT \right)}{d(1/T)} = \frac{d \ln K_p^{mb}}{d(1/T)} - \frac{d \ln T}{d(1/T)} = \frac{d \ln K_p^{mb}}{d(1/T)} + T \quad (5.16)$$

Thus treatment of the data with this alternative adsorption model yields a more negative enthalpy change by RT than the mobile model.

Isothermal chromatographic experiments with the detection of elution curves at several column temperatures are feasible in the separation practice of numerous fields of chemistry, and Eq. 5.15 has been widely exploited. In the radiochemical studies of our concern, such an approach could be realized only if long-lived nuclides with high enough activity were available; this, unfortunately, is not the case.

In the meantime, a reasonably fast radioactive decay makes a different technique for obtaining data on t_R^{IC} feasible, even with rather low activities. One can measure the fraction r_λ^{IC} of the nuclei introduced into the IC column which survives at its exit. This principle can be used in on-line experiments with the nuclides which have mean lifetimes much less than the nominal duration of the run; in practice, it means the range from seconds to hours. For a nuclide with the particular t_λ, two or more measurements at different temperatures must be done. At least one at a temperature when $t_R^{IC} \ll t_\lambda$ to find the production rate of the detectable activity, as well as one when t_R^{IC} is of the order of t_λ and so the surviving fraction is in the range $0 < r_\lambda^{IC} < 1$. From the point of view of the statistics, most desirable is to aim at r_λ^{IC} near 0.5. Obviously, $r = \exp\left(-t_R^{IC}/t_\lambda\right)$, hence:

$$\ln t_R^{IC} = \ln \left[\ln \left(1/r_\lambda^{IC} \right) \right] + \ln t_\lambda \quad (5.17)$$

Substitution of this into Eq. 5.15 gives:

$$\frac{-\Delta_{ads} H}{R} = \frac{d \ln \left[\ln(1/r_\lambda^{IC}) \right]}{d(1/T)} \quad (5.18)$$

Measurements of r_λ^{IC} were pioneered for systematic studies of short-lived transactinoids in Ref. [5]. Sample experimental data are illustrated by the figures in Sect. 1.3.

5.1 Adsorption Enthalpy on Homogeneous Surface

The data obtained by the temperature-programmed chromatography [6] could also provide the sought values of the thermodynamical quantities. In this method, the temperature of a longitudinally isothermal column is steadily raised from certain original T_0 with the rate γ_t (in K s^{-1}). One measures the "retention temperature" T_A, at which the compound emerges at the column exit. Sample chromatograms can be seen in Sect. 1.5.2. The basic equation serving to evaluate the adsorption enthalpy and entropy is:

$$\frac{RT_0 a_z z_A \gamma_t e^{\Delta_{ads}S/R}}{Q_0 p_0 \cdot (\Delta_{ads}H/R)} = \frac{e^{-\Delta_{ads}H/RT_0}}{(1+\Delta_{ads}H/RT_0)^2} - \frac{e^{-\Delta_{ads}H/RT_A}}{(1+\Delta_{ads}H/RT_A)^2} \quad (5.19)$$

It cannot be solved for the required values in closed form. Therefore, the T_A values were plotted versus γ_t/w_0, and the graph was numerically fitted by the equation to find the enthalpy and entropy change [6]. The technique made it possible to study several compounds simultaneously (like by thermochromatography), and so to evaluate their relative behavior under identical processing conditions. The precision of determination is much less than in IC studies because of technical reasons – larger uncertainty in the temperature corresponding to elution of the zone center of gravity.

The thermodynamic formulae for thermochromatography will be discussed in more detail later in this chapter. The one essential for the purpose is [4]

$$t_R^{TC} = \frac{a_z T_0}{g Q_0} \frac{V}{A} \exp\left(\frac{\Delta_{ads}S}{R}\right) \left[\text{Ei}\left(-\frac{\Delta_{ads}H}{RT_A}\right) - \text{Ei}\left(-\frac{\Delta_{ads}H}{RT_S}\right)\right] \quad (5.20)$$

cf. Sect. 2.1.6 for the properties of the integral exponential function Ei(x). In this case, t_R^{TC} is the net duration of the experiment and T_A is the temperature of the zone in the internal thermochromatogram. One can measure T_A as a function of t_R^{TC} and again apply a best-fit procedure to the above equation to obtain the thermodynamic characteristics of adsorption. To our knowledge, this possibility has not yet been exploited. The reason, again, is low precision of T_A values because of the large width of zones and of the poor statistics of counts.

There seems to be one more approach to the thermodynamic evaluation of TC data which can be applied to the lighter congeners of TAEs. It is illustrated by some once-published [7, 8] data on simultaneous thermochromatography of several short-lived isotopes of hafnium in the form of tetrachloride and tetrabromide. The nuclides were produced by bombarding a Sm target with ^{22}Ne ions. Except for the 16-h ^{170}Hf, their TC zones were measured indirectly after EOB – through the radiation of their long-lived descendants, isotopes of the lanthanoid elements. The latter, in the atmosphere of gaseous $SOCl_2 + TiCl_4$ or $Br_2 + BBr_3$, were obviously present as nonvolatile trichlorides or tribromides. Thus, the profiles of the deposition zones of the (grand-) daughters closely followed the profiles of the mother hafnium activities. In Fig. 5.1 the zones are plotted in an "integral form," that is the ordinate is the total percentage of the short-lived Hf nuclei which have decayed ahead of the given abscissa. The mean duration of the experiment equals the lifetime of the particular short-lived activity, so that the migration distance grows with longer t_λ. Using a smoothed integral representation, the coordinate and so temperature at, e.g., the

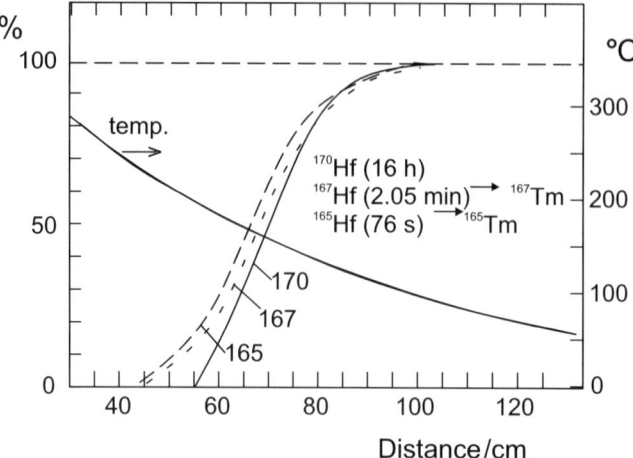

Fig. 5.1 Integral of the TC zones of short-lived neutron-deficient isotopes of hafnium measured by gamma radiation of long-lived members of their decay chains [7, 8].
Reproduced (adapted) from Radiokhimiya, 18(3), Zvara I, Belov VZ, Domanov VP, Shalaevski MR, Chemical separation of nielsbohrium, 371–377, © 1976, with permission from Academizdat "Nauka" Publishers.

medians of the zone can be obtained with better precision. It is important for the subsequent fitting Eq. 5.20 to evaluate the sought parameters. It has not yet been done.

Finally we pay attention to the ideal frontal TC (cf. Fig. 4.1). The high temperature front of the zone profile is obviously proportional to the adsorption isobar and so, at least for the localized adsorption model, to the adsorption constant. As such, it would obey Eq. 5.14. It holds for the activities which do not appreciably decay in the course of run. As for the shorter-lived nuclides, both the elution and the formally frontal TC result in non-ideal frontal chromatograms. Their shapes are close to what would arise from ideal processing during $\approx t_\lambda$ but they are smeared due to the random lifetimes of nuclei. Still the initial part of the thermochromatogram might be useful for evaluation of the required quantity, provided that the statistics of detected decay events is good.

5.1.2 Experimental Values from Second Law

Unfortunately, in the continuous on-line experiments with transactinoid elements, the application of the above Second Law and related procedures has not been possible because of the very poor statistics of detected decays. Measurements of this type have been done only with long-lived lighter homologs of the transactinoids in batch tests. Packed columns and low flow rates of the carrier gas were employed to achieve narrow peaks. Major contribution came from Bächmann and co-workers. In Refs. [6, 9–11] they measured the adsorption enthalpies and entropies on solid

5.1 Adsorption Enthalpy on Homogeneous Surface

Table 5.1 Selected Data on Adsorption Enthalpies, $\Delta_{ads}H$ in kJ mol^{-1}, from IC and the Programmed Temperature Chromatography Experiments Obtained by the Second Law Approach [6]

Group	Compd	Adsorbent			Group	Compd	Adsorbent		
		SiO$_2$	NaCl	CsCl			SiO$_2$	NaCl	CsCl
4	ZrCl$_4$	−97*	−99	–	5	NbCl$_5$	−65*	−76*	−86*
4	HfCl$_4$	–	−99	−104	5	TaCl$_5$	−66	−96	–
7	TcCl$_4$	−85*	–	−75*	(5)	PaCl$_5$	–	−126	−126
8	OsCl$_4$	–	−101	–	6	MoCl$_5$	−57*	−72*	−76*
14	SnCl$_4$	−54	−29	−31	7	ReCl$_5$	–	−64	–
16	TeCl$_4$	−57*	−70	−42	15	SbCl$_5$	−34	−36	−39
16	PoCl$_4$	−69	−88	–					
7	HTcO$_4$	−52*							
		−47							
7	HReO$_4$	−69*							
		−42							
8	RuO$_4$	−10.2* ± 1.7							
8	OsO$_4$	−9.8* ± 2.0							

* obtained by IC, others by programmed temperature chromatography

non-porous adsorbents (quartz, alkali and alkaline earth chlorides, graphite) at high temperatures (650 to 1,100 K) for a number of inorganic chlorides. They utilized the "classical" IC technique [10], as well as temperature-programmed gas chromatography [6]; sample data are shown in Table 5.1. To evaluate the thermodynamic quantities, they assumed the mobile adsorption and accepted the standard states suggested by de Boer [12]: the molar volume at STP for V while $A = 6.7 \times 10^{10}$ cm^2, which is the area with the same average separation between molecules like in the bulk gas (3.34×10^{-7} cm). Thus, the data in Refs. [6, 9–11] were evaluated using $A/V = 3.26 \times 10^6$ cm and $\ln(A/V/\text{cm}) = 15.0$. The average error of $\Delta_{ads}H$ values from IC experiments was stated as ≈ 5 kJ mol^{-1}, that in temperature-programmed chromatography – about 10 to 20 kJ mol^{-1}. In the latter case, the errors came mostly from the uncertainty in the retention temperature.

Bächmann and co-workers interpreted their data for molecules halides in columns loaded with alkaline halides as evidence for the formation of complexes between the adsorbate and the crystalline halide. Such complexes were considered for application in radiochemical separations earlier by Zvara and co-workers in [3,13,14], who estimated the adsorption enthalpies from the rough adsorption isobar obtained in thermochromatographic experiments using Eq. 5.14. The adsorption enthalpy for NbCl$_5$ on KCl [14] is intermediate between the data of Table 5.1 for the adsorbents NaCl and CsCl, so that the results of the two groups are consistent.

The numerous values obtained in [6,9–11] would deserve more analysis, discussion and comparison with later data. There are some unexpected trends and deviations in the $\Delta_{ads}H$ values; possibly, they originate from ambiguous chemical states of the particular elements. The experimental data on $\Delta_{ads}S$ are also of fundamental interest. As will be seen later in this chapter, evaluation of the experiments with TAEs is based on calculation of the adsorption entropy from the first principles. The studies [6, 9–11] reported observation of a correlation between the experimental values of $\Delta_{ads}H$ and $\Delta_{ads}S$, which is not predicted by such calculations. It is well-known that empirical correlations of the two quantities have been reported many times in the studies of the equilibria and kinetics of a variety of chemical processes. On the other hand, it proved difficult to find a rationale for such correlations. It became a standing problem because of its general importance. A number of dedicated works [15–19] concluded that the correlations are phantom phenomena, which emerge because the conventional statistical treatment does not evaluate the enthalpy and entropy changes independently.

Steffen and Bächmann [20, 21] also measured the adsorption characteristics on quartz for the volatile oxides of Tc, Re, Ru, Os and Ir, as well as for the hydroxides of Tc and Re. The techniques were isothermal chromatography and thermochromatography; the prescriptions were outlined in Sect. 5.1.1. Some of their data are also included in Table 5.1.

Measurement of the adsorption energies and entropies on the basis of the Second Law of thermodynamics is not feasible in the radiochemistry of TAEs. In general, the difficulties in the production of the attainable nuclides and their radioactive properties impose strong limitations on the allowable experimental conditions. It is unfortunate because the method guarantees much better accuracy (veracity) of the numerical values than that expected with the calculated entropies.

5.1.3 Quasi Third Law Approach – Entropy from Statistical Mechanics

Typically the experiment with a transactinoid element lasts days or even weeks, and results in a single experimental value of the adsorption constant. The only possibility to obtain an estimate of the adsorption enthalpy based on such a result is to calculate the entropy change from the first principles [4] and substitute it into Eq. 5.7. The values of $\Delta_{ads}S$ are calculated from the formulae of statistical mechanics for the particular model of the adsorbed state. The evaluation starts with the partition function of single molecule q_m and with the molar partition function Z to calculate the absolute molar entropy from the general equation:

$$\frac{S}{R} = \ln Z + T\frac{\partial \ln Z}{\partial T} \qquad (5.21)$$

The available literature on the entropy of adsorption of gases on solid surfaces is almost exclusively devoted to the experimental measurement of $\Delta_{ads}S$ in real systems to obtain information about the adsorbed state and about the structure of the surface. Only scarce attempts have been made to calculate or simulate (by Monte Carlo techniques) the entropy change for more or less real situations, which differ so much from the two idealized models of adsorption described in Sect. 5.1.1.1.

To illustrate the real values or their orders of magnitude obtained from the formulae, we will substitute them into typical values of the quantities which characterize the tracers and experimental conditions. These values are listed in Table 5.2.

5.1 Adsorption Enthalpy on Homogeneous Surface

Table 5.2 Common Values of Some Molecular Quantities at $T = 500$ K for $M_1 = 300$ g mol^{-1}

m_1	$u_m/4 = \sqrt{k_B T/2\pi m_1}$	$\Lambda_m = h/\sqrt{2\pi m_1 k_B T}$	$k_B T$[a]	$h/k_B T$	C_a	ν_0
5×10^{-22} g	4500 cm s^{-1}	3×10^{-10} cm	6.9×10^{-14} erg	9.6×10^{-14} s	10^{14} cm^{-2}	$10^{12} - 10^{13}$ s^{-1}

[a] $k_B T = 0.043$ eV; $N_A k_B T = 4.16 \times 10^{10}$ erg mol^{-1} = 4.16 kJ mol^{-1}

5.1.3.1 Partition Functions of Gaseous and Adsorbed Molecules

Evaluation of the entropy change in adsorption by the statistical mechanics approach can be found in numerous sources. Making use of Refs. [22, 23], we will outline what is required for the present purpose. The partition function for a gaseous molecule q_m is the product of the translational component q_{tr} and the internal components: rotational q_{rot}; vibrational q_{vib} and electronic q_{el}.

For a molecular entity freely moving in the potential box of the size a_m the translational partition function is:

$$q_{tr3} = \left[\frac{a_m \sqrt{2\pi m k_B T}}{h}\right]^3 = \left(\frac{a_m}{\Lambda_m}\right)^3 \quad (5.22)$$

Here Λ_m is the "thermal de Broglie wavelength"

$$\Lambda_m \equiv \frac{h}{\sqrt{2\pi m k_B T}} = \frac{u_m}{4}\frac{h}{k_B T} \quad (5.23)$$

which is roughly the average de Broglie wavelength of a particle in the ideal gas at the specified temperature. The typical Λ_m equals 3×10^{-10} cm or so (cf. Table 5.2). It is about 1,000 times shorter than the distance between gaseous molecules at STP and about 100 times shorter than the distance between surface atoms of the adsorbents. The translational partition functions for two dimensional and linear boxes are the appropriate roots of q_{tr3}. In particular:

$$q_{tr2} = \left(\frac{a_m}{\Lambda_m}\right)^2 \quad (5.24)$$

For the molar partition function Z_{tr} we obtain

$$Z_{tr} = \frac{q_{tr}^{N_A}}{N_A!} \quad (5.25)$$

which, using the Stirling approximation, yields:

$$Z_{tr} = \left(\frac{q_{tr} e}{N_A}\right)^{N_A} \quad (5.26)$$

Then the molar translational entropy of a three-dimensional gas is (Sackur–Tetrode equation):

$$\frac{S_{tr3}}{R} = \ln\left[\frac{V}{N_A}\frac{e^{5/2}}{\Lambda_m^3}\right] = \ln\left[\frac{V}{N_A}\frac{64e^{5/2}}{u_m^3}\left(\frac{k_B T}{h}\right)^3\right] \qquad (5.27)$$

It, for example, gives $20.1\,R$ for the entropy of xenon gas at STP.

The translational entropy of two-dimensional gas is:

$$\frac{S_{tr2}}{R} = \ln\left(\frac{A}{N_A}\frac{e^3}{\Lambda_m^2}\right) \qquad (5.28)$$

If we accept the de Boer's standard area $A = 6.7 \times 10^{10}\,\text{cm}^2$ mentioned in Sect. 5.1.2, which reasonably corresponds to the (by STP born) $V = 22{,}414\,\text{cm}^3$, the above equation yields an entropy of about $13\,R$ for two-dimensional gaseous xenon.

At moderate temperatures the internal degrees of freedom make different contribution to the entropy of free molecules. That due to the excited electronic states is mostly zero. For one vibrational degree of freedom

$$q_{vib} = \frac{e^{h\nu_0/2k_B T}}{e^{h\nu_0/k_B T} - 1} \quad \text{so that} \quad \frac{S_{vib}}{R} = \ln\left(\frac{k_B T}{h\nu_b}\right) \qquad (5.29)$$

where ν_b is the vibrational frequency of the bond. The vibrational energy microstates are usually well separated. Thus, they make negligible contribution to the entropy of light small molecules, but non-negligible in the case of large and heavy molecules. The rotational partition function for a molecule is

$$q_{rot} = \frac{1}{\pi\,\sigma_{rot}}\left(\frac{8\pi^3 I k_B T}{h^2}\right)^{n_{rot}/2} \qquad (5.30)$$

so that

$$\frac{S_{rot}}{R} = \ln\left[\frac{e^{3/2}}{\pi\,\sigma_{rot}}\left(\frac{8\pi^3 I k_B T}{h^2}\right)^{n_{rot}/2}\right] \qquad (5.31)$$

where σ is the rotation number, n_{rot} is the number of rotational degrees and I is the geometrical mean of the appropriate moments of inertia. With the exception of atoms, this contribution to entropy is very significant because molecules possess numerous low-lying, tightly spaced rotational levels.

5.1.3.2 Mobile Adsorption Model

When evaluating the entropy change in the mobile adsorption it is assumed that the rotational, vibrational and electronic degrees of freedom of the molecules are preserved. It allows considering only the translational entropy of the surface gas and the entropy of vibrations of its molecules perpendicular to the surface. These oscillations are induced by the characteristic vibration frequency of the adsorbent

5.1 Adsorption Enthalpy on Homogeneous Surface

lattice ν_0, which was discussed in Sect. 2.1.5. The entropy of such oscillations is also calculated from Eq. 5.29.

The two-dimensional gas model assumes no mutual interaction of the adsorbed molecules. It is believed that the adsorbent creates a constant (across the surface) adsorption potential. Thus, in the framework of statistical thermodynamics, the model describes adsorption as the transition of a gas with three translational degrees of freedom into an adsorbed state with one vibrational and two translational degrees. Assuming ideal behavior and using molar quantities, one obtains the standard entropy in the adsorbed phase as the sum of the translational and vibrational entropies from Eqs. 5.28 and 5.29:

$$\frac{S_a^{mb}}{R} = \ln\left(\frac{A}{N_A}\frac{e^3 2\pi m k_B T}{h^2}\right) + \ln\left(\frac{k_B T}{h\nu_0}\right) \quad (5.32)$$

Because $k_B T/h\nu_0$ is of the order of unity, the contribution of the second term is small.

It gives for the standard adsorption entropy:

$$\frac{\Delta_{ads} S^{mb}}{R} = \frac{S_a^{mb} - S_{tr3}}{R} = \frac{S_{tr2} + S_{vib} - S_{tr3}}{R}$$

$$\frac{\Delta_{ads} S^{mb}}{R} = \ln\left[\frac{A}{V}\frac{e^{1/2}}{\nu_0}\left(\frac{k_B T}{2\pi m}\right)^{\frac{1}{2}}\right] = \ln\left[\frac{A}{V}\frac{e^{1/2}}{4}\frac{u_m}{\nu_0}\right] \quad (5.33)$$

Because the choice of the standard states is arbitrary, it seems convenient to define them only by their ratio, namely by $A/V = 1$ cm. Of course A cannot be taken less than the molar area of the closely packed monolayer of the adsorbate molecules. The latter is $N_A d_m^2 \sqrt{3}/2$, usually of the order of 10^9 cm^2, which gives the maximum monolayer capacity of the order of 10^{14} cm^{-2}. Because a one-centimeter-high stack contains as many as some 10^7 monolayers, taking equal numerical values for A and V means that the accepted standard volume always exceeds its physically possible minimal value by many orders of magnitude. In addition, when choosing the standard values to derive formulae for application to tracers, it seems reasonable to accept an A much larger than the possible minimum to keep with the real very low coverage. Ref. [4] also discusses other reasonable standard states. These have nothing in common with the "STP" state, as is the case with the state defined just above.

If one observes the above limitations for the standard values, a working formula which explicitly shows some dependencies of the entropy change in adsorption looks like

$$\frac{\Delta_{ads} S^{mb}}{R} = \ln\left(\frac{A}{V}\frac{u_m}{\nu_0}\right) - 0.9 \quad (5.34)$$

and with $A/V = 1$ cm, the standard entropy change $\Delta_{ads} S^{\circ,mb}$ is:

$$\frac{\Delta_{ads} S^{\circ,mb}}{R} = \ln\left[\frac{1}{\nu_0}\left(\frac{T}{M}\right)^{1/2}\right] + 8.7 \quad (5.35)$$

Remember that the two formulae above do not take into account possible changes in the internal entropy.

5.1.3.3 Localized Adsorption Model

According to this model the adsorbed molecules are randomly distributed over the available adsorption sites. They possess three vibrational degrees of freedom, the frequency of which is close to the characteristic frequency of the adsorbent v_0. Then the molar standard entropy of the adsorbate is the sum of the molar vibrational entropy and the configurational entropy S_{cf}. The latter arises from the fact that the molecules can be distributed over the adsorption sites in different ways. The molecules are indistinguishable, but the sites are always distinguishable because of each one's unique coordinates. Let N_A particles be randomly distributed over N_A/θ sites, where θ is the fractional coverage. The partition function equals the number of possible configurations, and the formulae of combinatorics give the molar value as:

$$Z_{cf} = \frac{(N_A/\theta)!}{N_A!(N_A/\theta - N_A)!} \tag{5.36}$$

Substitution of this into Eq. 5.21 and application of the Stirling's approximation yield the integral molar configurational entropy as:

$$\frac{S_{cf}}{R} = -\left[\ln\theta + \frac{1-\theta}{\theta}\ln(1-\theta)\right] \tag{5.37}$$

The partial (differential) molar configurational entropy $\bar{S}_{cf}$ is [24]:

$$\frac{\bar{S}_{cf}}{R} = -\ln\frac{\theta}{1-\theta} \approx \ln\frac{1}{\theta} \tag{5.38}$$

The two above equations show that, at small values of θ, the molar configurational entropies (both the integral and the partial) increase as $\ln(1/\theta)$. It is illustrated by Fig. 5.2. Meanwhile, S_{cf} is larger by a term, which decreases with lower θ approaching unity, and thus becomes less important. It can be seen from Eq. 5.37 that the integral entropy goes to zero at the complete coverage. It is because indistinguishable molecules can completely fill a monolayer only in one way. Notice that the configurational entropy is principally independent of temperature.

A peculiar property of the configurational entropy (not of the molar one) shows up when we consider the entropy-versus-coverage curve for a fixed number of the adsorption sites. The curve is completely symmetrical; it goes to zero both at $\theta \to 0$ and $\theta \to 1$, reaches one maximum (which is a function of the number of sites), and, at a given $\theta < 0.5$, the entropy is proportional to the number of sites. When calculating the standard $\Delta_{ads}S$ equations of the statistical mechanics, we face the necessity to introduce standard configuration entropy. This can be done through the standard value of the coverage $\theta°$, which we first met in Sect. 5.1.1.1. The number

5.1 Adsorption Enthalpy on Homogeneous Surface

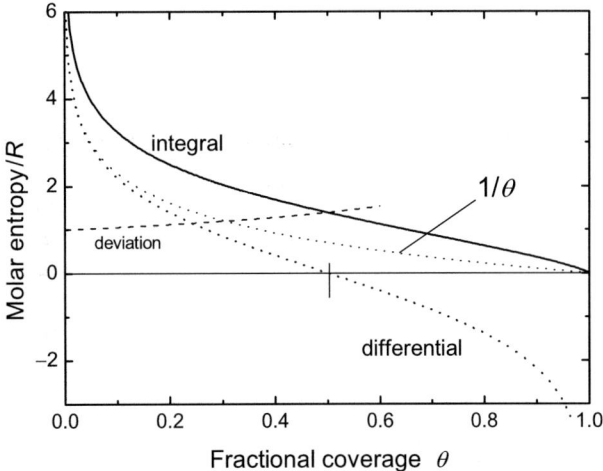

Fig. 5.2 Molar configurational entropies, differential and integral, as the function of the fractional coverage. Also shown are the $1/\theta$ approximation of the two entropies and the accurate difference between them.

of adsorption sites per unit area in localized adsorption is not necessarily the same as in the array of closely packed molecules, which we referred to when considering mobile adsorption. Equally close packing on a surface consisting of deep, regularly arranged, distinct potential wells may be impossible simply for geometrical reasons. As a result, the smallest allowable standard molar area N_A/C_a is most likely somewhat larger (not more than several times) than such an area on a structureless surface. Later, we shall have to consider adsorption states intermediate between the ideal models. Therefore, it is desirable that the values of A and V chosen for mobile adsorption be compatible with $\theta°$ for the localized model. It is a conceptual problem. It can be solved if we assume that the complete coverage value is the same in both models, so that C_a *does equal* the capacity of the close-packed monolayer. The factual difference cannot bring larger uncertainties to the final results than those contributed by other inevitable assumptions, simplifications and approximations. If we take the standard fractional surface coverage $\theta°$ much less than unity, it merely means that the chosen A must be much larger than N_A/C_a. Below, when numerically evaluating the formulae and examples, we will accept the following standards:

$$V/\text{cm}^3 = A/\text{cm}^2 = N_A \quad \text{so that} \quad \theta° = 1/C_a \tag{5.39}$$

We shall not discuss other possible alternatives, because the one above makes the formulae much simpler.

The entropy change in localized adsorption is interpreted as the loss of three translational degrees of freedom and the gain of three vibrational degrees of freedom. Strictly speaking, the vibration frequencies ν_x, ν_y, ν_0 are not exactly equal to each other, but for the purpose we put $\nu \equiv (\nu_x \nu_y \nu_0)^{1/3}$. From Eq. 5.29 we obtain for the vibrational entropy of the localized state

$$\frac{S_{vb}^{lc}}{R} = \ln\left(\frac{k_B T}{h\nu}\right)^3 \tag{5.40}$$

which, when combined with the configurational entropy from Eq. (5.38), yields:

$$\frac{S^{lc}}{R} = \ln\frac{1}{\theta} + \ln\left(\frac{k_B T}{h\nu}\right)^3 \tag{5.41}$$

Therefore,

$$\frac{\Delta_{ads} S^{lc}}{R} = \frac{S^{lc} - S_{tr3}}{R} = \ln\left(\frac{k_B T}{h\nu}\right)^3 - \ln\frac{N_A}{AC_a} - \ln\left[\frac{V}{N_A}\frac{64 e^{5/2}}{u_m^3}\left(\frac{k_B T}{h}\right)^3\right]$$

$$= \ln\left[\frac{AC_a}{V}\frac{u_m^3}{64 e^{5/2}\nu^3}\right] \tag{5.42}$$

For the standard states given by Eq. 5.39 we have:

$$\frac{\Delta_{ads} S^{o,lc}}{R} = \ln\left[\frac{C_a}{\nu^3}\left(\frac{T}{M}\right)^{3/2}\right] + 22.1 \tag{5.43}$$

It is interesting to compare the entropy change for the two models of adsorption. Their difference evaluated from Eqs. 5.34 and 5.42 is:

$$\frac{\Delta_{ads} S^{lc} - \Delta_{ads} S^{mb}}{R} = \ln\left[\frac{AC_a}{V}\frac{u_m^3}{64 e^{5/2}\nu^3}\right] - \ln\left[\frac{A}{V}\frac{e^{1/2}}{4}\frac{u_m}{\nu_0}\right] = \ln\left(\frac{C_a}{16 e^3}\frac{u_m^2}{\nu_{xy}^2}\right) \tag{5.44}$$

The character of the dependence on variables is better seen from the formula:

$$\frac{\Delta_{ads} S^{o,lc} - \Delta_{ads} S^{o,mb}}{R} = \ln\left(\frac{C_a}{\nu_{xy}^2}\frac{T}{M}\right) + 13.4 \tag{5.45}$$

For homogeneous surfaces, the difference of the entropies is around $-10R$. It is getting more negative with lower monolayer capacity. If we substitute typical values of M and T then:

$$\frac{\Delta_{ads} S^{o,lc} - \Delta_{ads} S^{o,mb}}{R} \approx \ln\left(\frac{10^6 C_a}{\nu_{xy}^2}\right) \tag{5.46}$$

Commonly, ν_0 is in the range 10^{12} to 10^{13} s^{-1} (cf. Sect. 2.17), but ν_x and ν_y are somewhat less than 10^{12} s^{-1} [22]; the value of C_a cannot be larger than 10^{14}–10^{15} cm^{-2}. Hence, the value in parentheses in the above formula is probably always less than unity and the entropy change in localized adsorption at low coverage is even more negative than in the case of mobile adsorption. More entropy may be lost also because of hindrance to the internal degrees of freedom of the molecules upon adsorption; it seems to concern mostly the rotational entropy.

Now we can use the formulae for entropy to estimate the adsorption enthalpy on the basis of the experimental constant and the Third Law-like procedure.

In the case of IC experiments, assuming mobile adsorption, from Eqs. 5.2, 5.9 and 5.33 we obtain

$$\frac{-\Delta_{ads}H^{mb}}{RT_c} = \frac{-\Delta_{ads}U^{mb}}{RT_c} = \ln K_p^{mb} + \frac{-\Delta_{ads}S^{mb}}{R} = \ln\left(\frac{A \, t_R^{IC} Q}{V \, a_z z_A}\right) + \ln\left(\frac{V}{A}\frac{4v_0}{\sqrt{eu_m}}\right) \quad (5.47)$$

or explicitly:

$$\frac{-\Delta_{ads}H^{mb}}{RT_c} = \ln\left(\frac{4}{e^{1/2}} \frac{t_R^{IC} Q v_0}{a_z z_A u_m}\right) = \ln\left(v_0 \frac{t_R^{IC} Q_0}{a_z z_A} \sqrt{T_c M}\right) - 14.3 \quad (5.48)$$

Thus, the result does not depend on the accepted ratio V/A.

Analogous expressions for the localized adsorption, which follow from Eqs. 5.42 and 5.12, are:

$$\frac{-\Delta_{ads}H^{lc} - RT}{RT} = \frac{-\Delta_{ads}U^{lc}}{RT} = \ln K_c^{lc} + \frac{-\Delta_{ads}S^{lc}}{R}$$
$$= \ln\left(\frac{A \, t_R^{IC} Q}{V \, a_z z_A}\right) + \ln\left[\frac{V}{AC_a} \frac{64 e^{5/2} v^3}{u_m^3}\right] = \ln\left(\frac{64 e^{5/2} t_R^{IC} Q}{C_a a_z z_A} \frac{v^3}{u_m^3}\right) \quad (5.49)$$

Again, the form showing various dependencies is:

$$\frac{-\Delta_{ads}H^{lc} - RT}{RT} = \frac{-\Delta_{ads}U^{lc}}{RT} = \ln\left(v^3 \frac{t_R^{IC} Q_0}{C_a a_z z_A} \frac{M^{3/2}}{T^{1/2}}\right) - 27.7 \quad (5.50)$$

It can be checked that the difference in the values provided by Eqs. 5.48 and 5.49 is the same as between the reduced entropy values, except for the sign.

5.2 Adsorption Enthalpy from Thermochromatographic Experiments

In Chapter 4 we considered a naïve model of IC chromatography based on a molecular-kinetic description of the competing process of adsorption and desorption. The rate of adsorption was the number of molecules hitting the unit surface per unit time, and that of desorption was governed by the Boltzmann factor with the desorption energy in the exponent. It resulted in Eq. 4.1, which can be rewritten like:

$$\frac{\varepsilon_d}{k_B T_c} = \ln \frac{4 t_R^{IC} Q v_0}{a_z u_m z_A} \quad (5.51)$$

Comparing the latter with Eq. 5.48 we see that the values $-\Delta_{ads}H^{mb}$ and $N_A\varepsilon_d$ coincide, except for the small term $RT_c/2$. The latter is the difference between the mean kinetic energy of the molecules in three-dimensional and two-dimensional gas. Thus, the molecular kinetic approach is equivalent to the model of mobile adsorption.

5.2.1 Basic Equations

Making use of Eqs. 5.5, 5.7, 5.9 and 5.33, we may write:

$$t_R^{IC} = \frac{z_A a_z}{Q} e^{-\frac{\Delta_{ads}S}{R}} e^{-\frac{\Delta_{ads}H}{RT_c}} \tag{5.52}$$

Evaluation of the thermodynamic equations for thermochromatographic retention times can be done by analogy with the procedures outlined in Sect. 4.2.2. Namely, we start with the derivative dt_R^{TC}/dT and proceed with integration from the starting high temperature T_S down to the final zone deposition temperature T_A. If we assume that the enthalpy and entropy changes are independent of temperature, in the case of the constant temperature gradient, the integration yields:

$$\text{Ei}\left(\frac{-\Delta_{ads}H}{RT_A}\right) - \text{Ei}\left(\frac{-\Delta_{ads}H}{RT_S}\right) = \frac{V}{A}\frac{t_R^{TC}Q_0g}{a_zT_0}\exp\left(\frac{-\Delta_{ads}S}{R}\right) \tag{5.53}$$

In practice, the second term on the left side of this and similar equations is almost always much smaller than the first one and can be neglected. Then, we obtain for the here accepted standard state that:

$$\text{Ei}\left(\frac{-\Delta_{ads}H^\circ}{RT_A}\right) = \frac{t_R^{TC}Q_0g}{a_zT_0}\exp\left(\frac{-\Delta_{ads}S^\circ}{R}\right) \quad T_z = T_S - gz \tag{5.54}$$

If the column temperature profile is exponential, then:

$$\frac{RT_A}{-\Delta_{ads}H^\circ}\exp\left(\frac{-\Delta_{ads}H^\circ}{RT_A}\right) = \frac{t_R^{TC}Q_0\gamma}{a_zT_0}\exp\left(\frac{-\Delta_{ads}S^\circ}{R}\right) \quad T_z = T_S e^{-\gamma z} \tag{5.55}$$

The analogous expression for IC from 5.52 is:

$$\exp\left(\frac{-\Delta_{ads}H^\circ}{RT_A}\right) = \frac{t_R^{IC}Q_0T_c}{a_zT_0z_A}\exp\left(\frac{-\Delta_{ads}S^\circ}{R}\right) \quad T_z \equiv T_c \tag{5.56}$$

Notice that in the three above equations the values of g, γ and T_c/z_A, respectively, have the dimension K cm^{-1} and play an analogous role.

The above equations, especially Eq. 5.54 (and so the mobile adsorption model) obtained wide use in radiochemistry of TAEs. The adsorption entropy was calculated from Eq. 5.33 accepting $A/V = 1$. Several authors proposed approximate

5.2 Adsorption Enthalpy from Thermochromatographic Experiments

formulae to facilitate rapid estimates of the adsorption enthalpy. They used the asymptotic series, Eq. 2.23, for Ei(x) because the common enthalpy values obey the condition $-\Delta_{ads} H^{\circ,mb}/RT_A \gg 1$. Possible contribution of T_S was already neglected in Eq. 5.54. Refs. [25, 26] used a simple, but rather crude estimate:

$$\frac{-\Delta_{ads} H^\circ}{RT_A} \approx \ln\left(\frac{t_R^{TC} Q_0 g}{12.55 a_z}\right) + \frac{-\Delta_{ads} S^\circ}{R} \quad (5.57)$$

Jin and Zvara [27] solved Eq. 5.54 numerically for the values of its right side in the range 10^4–10^{14}. They found that the formula

$$\frac{-\Delta_{ads} H^\circ}{RT_A} \approx 1.047 \left(\ln \frac{7.85 \, t_R^{TC} Q_0 g}{a_z T_0} + \frac{-\Delta_{ads} S^\circ}{R} \right) \quad (5.58)$$

is accurate enough for any practical purpose. They checked the results against the solution of the "rigorous" Eq. 5.53. It turned out that if $-\Delta_{ads} H^\circ/RT_A \approx 20$ and $T_S > 1.1 T_A$, neglecting the term Ei($-\Delta_{ads} H^\circ/RT_S$) produces less than 1 percent error in the enthalpy change. In addition, when the temperature decreases down the column monotonously, though not strictly linearly, a sufficiently accurate result is obtained by substituting the local value of g at z_A into Eq. 5.58. This is obviously equivalent to replacing the temperature profile by the tangent to the temperature profile at the abscissa z_A. The authors also traced the slight changes of the results with various functional dependence of the entropy change on temperature. They noted that a factor of 10 in the value of the right side of Eq. 5.54 changes the calculated $\Delta_{ads} H^\circ/RT_A$ by about 12 percent. The possible uncertainties in a_z and $\exp(\Delta_{ads} S^\circ/R)$ will be discussed in detail later in this chapter.

Qualitative observations made in the pioneering experiments in heavy metal halide radiochemistry evidenced some correlation between the vaporization and the adsorption energies of the compounds. This is important when judging bulk properties of new elements and compounds from their adsorption parameters. So every attempt was directed toward quantitative evaluation of the characteristics like adsorption energy or enthalpy.

5.2.2 Third Law-based Results for Halides

From the very beginning of the studies of gas-phase radiochemistry, the experimental adsorption enthalpies were examined for a possible relationship with the macroscopic characteristics of volatility. Shortly after it was reported [2] that a few investigated chlorides showed heats of adsorption (evaluated somewhat differently than today) equal to about two-thirds of their heats of vaporization. This roughly linear correlation took place for both molecular and ionic compounds in the range of the vaporization heat from 40 to 200 kJ mol^{-1}. Since then, the adsorption enthalpies have been reported in the literature for many more chlorides and oxychlorides. The values which were deduced from the experimental data in a consistent way — in particular, using the mobile adsorption model — could be correlated with the boiling points of the compounds [28] by:

$$-\Delta_{ads}H/\text{kJ mol}^{-1} = (0.087 \pm 0.005)T_b + 32.9 \pm 6.4 \quad (5.59)$$

In the meantime, according to Trouton's rule, there is a simple relationship between the enthalpy of vaporization and the normal boiling temperature, $T_b \approx \Delta_{vap}H/10.5R$. If we substitute it into the above correlation, then:

$$-\Delta_{ads}H/\text{kJ mol}^{-1} \approx \Delta_{vap}H/\text{kJ mol}^{-1} + 33 \quad (5.60)$$

The error bars are omitted because of the approximate character of the rule. Notice that it looks like there is a fundamental contribution from the vaporization enthalpy. Figure 5.3 shows a plot of the $\Delta_{ads}H$ data versus the vaporization enthalpy.

Correlating $\Delta_{ads}H$ with $\Delta_{sub}H$ might seem preferable because the accurate data on sublimation are more abundant than those on $\Delta_{vap}H$. Meanwhile, $\Delta_{sub}H$ is the sum of $\Delta_{vap}H$, and the enthalpy of melting $\Delta_{fus}H$ (strictly speaking, only at the triple point). The corresponding correlation [30] is:

$$-\Delta_{ads}H = (0.600 \pm 0.025)\Delta_{sub}H + (21.6 \pm 5.2) \quad (5.61)$$

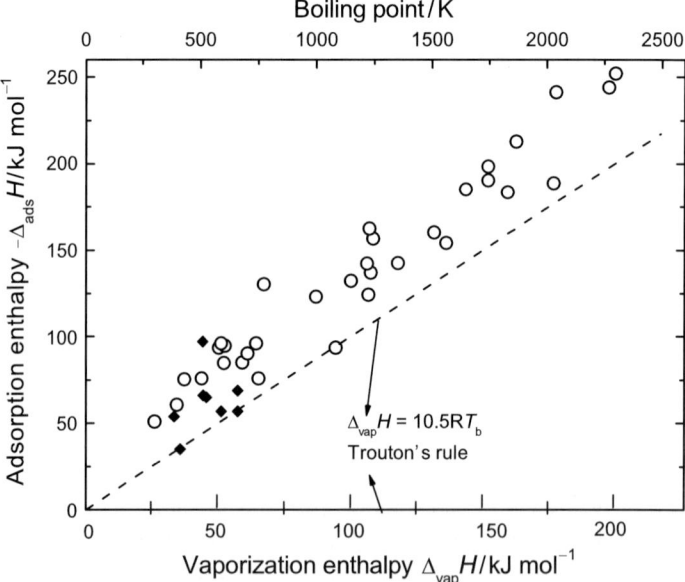

Fig. 5.3 The enthalpy change of adsorption of chlorides on fused silica (in the presence of chlorinating agents) versus enthalpy change of vaporization of the compounds.

The author is grateful to B. Eichler for the numerical data on adsorption enthalpies (circles). Rhombs are some data by Bächmann, et al. from Table 5.1; these were not included in calculation of the regression parameters in Eq. 5.59.

Thus, the adsorption enthalpies evaluated by the procedures which have been in use to date do roughly correlate with the vaporization or sublimation parameters of bulk compounds.

5.3 Real Structure of Column Surfaces

The mere observation of the above correlations has been widely used to judge the parameters of macro volatility of the (oxy)halides of TAEs. Not enough attention has been paid to the fact that the evaluated $-\Delta_{ads}H$ values fall between the corresponding $\Delta_{vap}H$ and $\Delta_{sub}H$. It is not at all obvious that it must be so be if we imagine a molecule interacting with bare flat surface of fused silica and compare the situation with that of a molecule in the condensed phase, where it is tightly surrounded by a number of its own replicas; see Fig. 5.4.

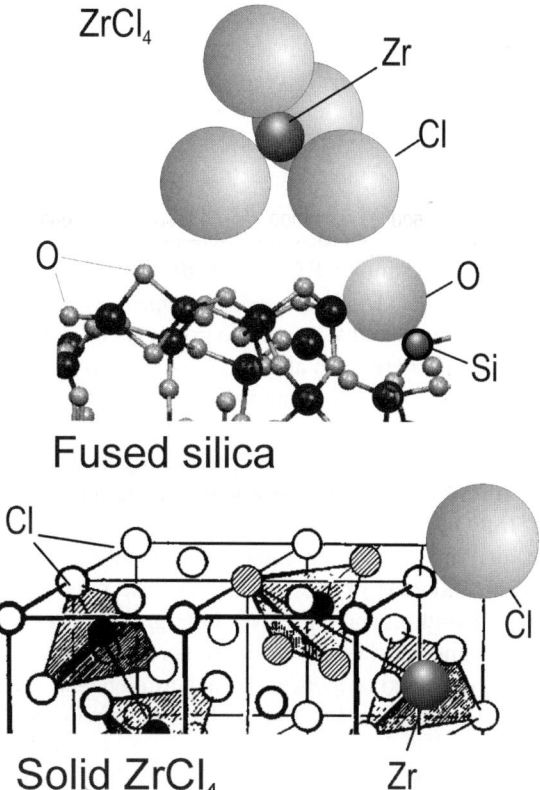

Fig. 5.4 Comparison between the imaginary situations of a $ZrCl_4$ molecule near the bare silica surface, and the molecules in their own molecular crystal. Relative sizes of the 3-D presentation of Zr, Si, Cl and O are in accordance with the model of touching hard spheres. Most of the Cl and O atoms in solids are diminished to make visible the geometry of the lattices.

Table 5.3 Sizes of Ions (Atoms) in the Compounds of Interest in this Book for Particular Oxidation and Coordination Numbers. {Numbers in italics are estimates (intra– or extrapolated) [29]}

Element	ON	CN	Radius/Å	Element	ON	CN	Radius/Å
F	−1	1	*1.14*	B	+3	3	0.15
Cl	−1	1	*1.62*	Si	+4	4	0.4
Br	−1	1	*1.74*	Ti	+4	4	0.56
I	−1	1	*1.97*	Zr, Hf	+4	4	0.72
O	−2	1, 2	*1.2*	Nb, Ta	+5	5	0.69
S	−2	1	*1.62*	W	+6	4	0.55
				W	+6	5	0.64
				Re	+7	4	0.52
				Os	+8	4	0.53

Knowledge of the relative sizes of atomic particles, which are shown in the figure, helps us to obtain a better feeling about the microscopic pictures of surface phenomena. The sizes are essentially Shannon's [31] additive "crystal" radii, which, in his opinion, better correspond to the real sizes of the ions than do the traditional "ionic" radii. The crystal radii depend on the oxidation number (ON) as well as on the coordination number (CN); their summation yields quite accurate values of the interatomic distances in crystals. Moreover, they also do it for the molecules of volatile halides [29]. The only difference is that CN of the central ion is usually smaller than in crystals, and the radius is lacking in Shannon's table. However, the missing values can be easily obtained by extrapolation or interpolation of the tabulated data; no arbitrary judgment is necessary, and the uncertainty of the estimates is about ± 0.02 Å. The values for the ions of present interest are listed in Table 5.3. The radii of homologous transactinoids Rf, Db, Sg, Bh and Hs should not significantly differ from the values for Hf, Ta, W, Re and Os, respectively.

Looking at Fig. 5.4, we see that one can hardly expect equal energetics of desorption and sublimation. There were also some indications [32] that the adsorption behavior of some halides on such dissimilar column material like silica, glass or nickel is the same! Figure 5.5 depicts a concrete situation.

These are the most important findings. The linearity of the correlations is not of major concern — it becomes questionable when taking the narrower region of the $\Delta_{ads} H$ values of truly molecular liquids.

This uneasy situation, with such important quantities, calls for a search of the physicochemical rationale of the regularities. The poor understanding of the absolute values of adsorption enthalpy gives little ground to any far-going conclusions from the experimental data. It concerns, for example, the manifestation of relativistic effects in chemical properties of the new elements, which is believed to be evidenced if a TAE compound is more volatile (less adsorbable) then that of a lighter homolog.

The clue to the similarity of micro and bulk volatility parameters seems to be in the true structure of the column surface. There is overwhelming evidence that real surfaces of solids are, almost in all respects, heterogeneous rather than homogeneous. The starting point is the common roughness, which increases the real surface area, compared with the smooth case. Most in-depth studies evidence a broad

5.3 Real Structure of Column Surfaces

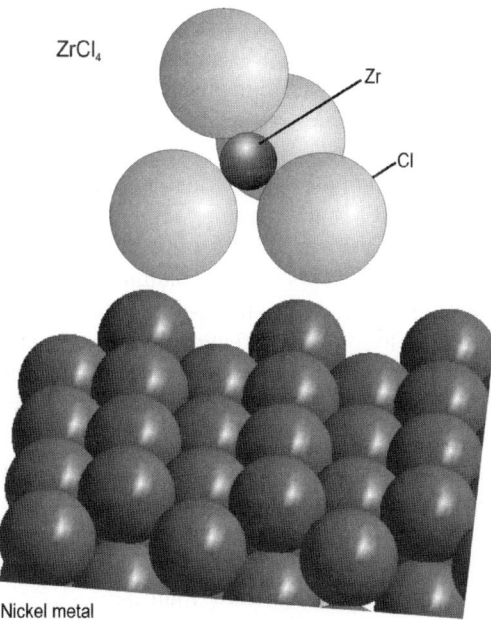

Fig. 5.5 Imaginary picture of a ZrCl$_4$ molecule near the surface of nickel metal. Relative sizes of Zr, Cl and Ni are in accordance with the model of touching hard spheres.

continuous spectrum of the adsorption enthalpies. The adsorption potential minima – various in origin, size, and depth – are necessarily separated, but are linked with relatively low potential barriers. These are the conditions for the existence of activated surface diffusion of the adsorbate molecules.

The mobile adsorption state seems to seldom occur in reality. De Boer [12] and other authors present the adsorption of krypton on the surface of liquid mercury as the only good example; they do not mention any case of adsorption on solids. The conditions for mobile adsorption can hardly take place in the adsorption of heavy element halides on silica or metallic columns. Doubts can also be cast on the simplest picture of the ideal "localized" adsorption. An ideal crystal face does show ordered, equally deep potential wells on a map of the adsorption energy; moreover, cutting of the crystal by certain planes (perpendicular to the surface) produces sections, which show one-dimensional adsorption wells separated by barriers reaching up to the zero adsorption potential. However, most of the possible sections show barriers, which do not reach the zero potential energy. As a consequence, a molecule can visit many neighboring sites before it is desorbed from the surface.

5.3.1 Geometrical and Chemical Structure of Fused Silica Surface

Roughness of surfaces can be defined as the geometric morphology at suprananometer scale. It can be quantitatively characterized by setting a mean surface level with subsequent evaluation of the root-mean-square (rms) of the surface height deviations from it. For certain purposes, the peak-to-valley rms roughness may characterize the surface better. It can be evaluated in an analogous way, now with respect to a "ground" level; the numerical values are necessarily larger than those for the height variations around the mean.

Modern instrumental techniques like scanning electron microscopy (SEM), atomic force microscopy (AFM) and scanning tunneling microscopy (STM) allow one to see details of surfaces down to the atomic level. Unfortunately, at least at present, scanning of a few percents of the column surface to obtain significant statistics of the defects or become sure that they are absent is hardly possible. The findings about the morphology of imperfections presented below should warn the researchers against overrating the observations made only on a few small spots.

A great general interest in the sorption technologies with silica adsorbents, as well as their widespread use, have motivated numerous experimental and theoretical studies of the structure and chemistry of these materials. In particular, it is true for the investigations of surfaces of amorphous silica, both its bulk and high specific area varieties. Several books and monographs are fully or partially concerned with this topic [33–36]. Below, we shall selectively cite some more recent journal publications; they contain references to more sources.

5.3.1.1 Roughness

Surfaces of real commercial fused-silica chromatographic columns and wafers have recently been repeatedly visualized by SEM, AFM, and STM techniques. The imaging showed that the surfaces commonly feature scratches, digs, pits, nodules, foreign inclusions, eventual marks from polishing and machining, as well as attached particles of dust.

The inner surface of the fused silica tubes (typically of 3 mm i.d.) which were used in the TC experiments with TAEs was inspected [37] by SEM. The tubes came from different manufacturers and batches. Sample results are presented in Fig. 5.6. Some of the observed morphology on the micrometer scale was unexpected and even exotic (shots numbers 1 to 3). The statistics of such peculiarities were not quantitatively measured, but most parts of the surfaces seemed to be smoother (shots numbers 4 to 6), with roughness of a few tens of nanometers; the flatness on this scale could be significantly improved by chemical etching and subsequent annealing of the tubes at high temperatures.

Bonvent, et al. [38] obtained AFM pictures of the bare surfaces of real fused-silica capillaries, which they used for electrophoretic separations; the capillaries are considerably more narrow than the columns used in the gas-phase radiochemistry. Their results for certain specimen presented in Fig. 5.7 clearly show nanometer roughness.

Knaupp, et al. [39] reported systematic examination of the surfaces of 75 μm i.d. capillaries for electrophoresis by SEM. They found that etching with 10 percent hydrofluoric acid reduced the roughness. Gupta, et al. [40] studied the surfaces of

5.3 Real Structure of Column Surfaces

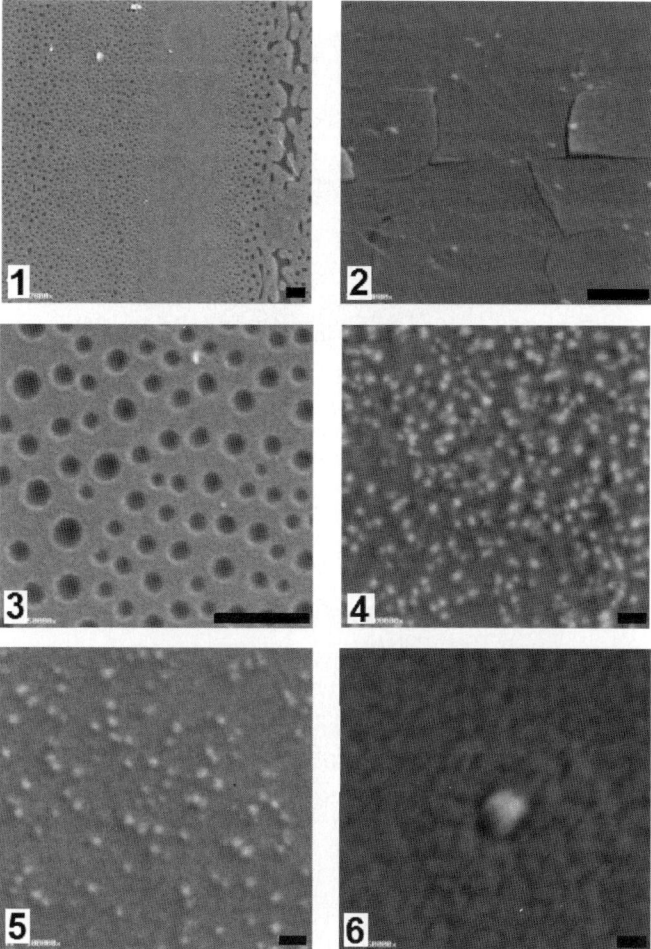

Fig. 5.6 Scanning electron microscopy [37] of the surface of the thermochromatographic columns, which were used in the thermochromatographic experiments at FLNR Dubna and PSI Villigen. The black bars in the left corners indicate 1 μm in shots one to three, and 100 nm in shots four to six.

fused silica (as well as of a multicomponent glass) with the goal to discover similarities or differences between the surfaces which were produced and investigated in very different conditions. These were, for example, the fracture surfaces produced at room temperature immediately before the measurements and the melt-formed surfaces produced relatively slowly, at high temperatures, and so difficult to be studied in the pristine state. All the measured surfaces exhibited nanoscale roughness, like that in Fig. 5.8. In general, the melt-formed surfaces were smoother than the fracture surfaces, and the fused silica surfaces were smoother than the corresponding glass surfaces. Peak to valley roughness was 1.5 nm for melt-formed silica, 3.1 nm

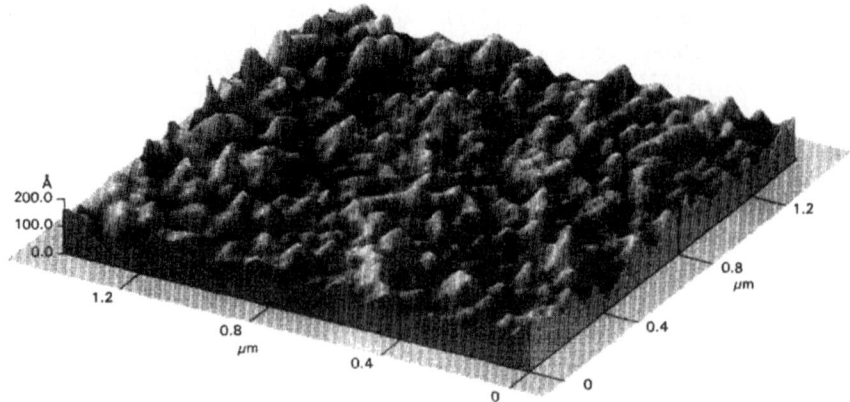

Fig. 5.7 Atomic tunneling microscopy of the surface of a 100 μm i.d. fused-silica capillary [38].

Reprinted from Journal of Chromatography, A756 (1–2), Bonvent JJ, Barberi R, Bartolino R, Capelli L, Righetti PG, Adsorption of proteins to fused-silica capillaries, 233–234, © 1996 with permission from Elsevier.

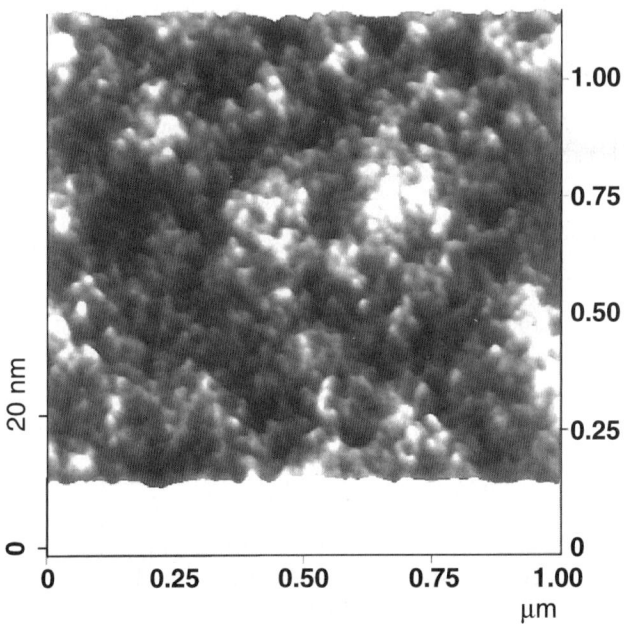

Fig. 5.8 A SEM picture of the surface of a freshly fractured synthetic silica rod [40].

Reprinted from Journal of non-crystalline solids, 262(1–3), Gupta PK, Inniss D, Kurkijan CR, Zhong Qian, Nanoscale roughness of oxide glass surfaces, 200–206, © 2001 with permission from Elsevier.

for fractured synthetic silica and 4.5 nm for annealed fractured silica. The figures for the rms roughness were 0.18 nm, 0.40 nm and 0.62 nm, respectively.

Henke, et al. [41] investigated how chemical cleaning affects the roughness of fused silica wafers to be used for biosensor preparation; such application needs precise values of the surface density of the immobilized molecules. After the wafers were chemically cleaned by washing with basic peroxide and then by acidic peroxide, they still observed numerous nodules, scratches and pits on the surface. The average nodule size was typically 5 to 10 nm (largest reached 50 nm), scratches were 2 to 4 nm in depth and pits – about 6 nm; see Fig. 5.9. The rms roughness was evaluated based on the measurement of the total integrated light scattering or directly from the surface profile measurements by AFM or STM. The chemical treatment reportedly decreased the original rms roughness from 3 nm to 0.7 nm and, consistently, reduced the original 30 percent excess of the true specific area over the geometric one to a mere 10 percent. This resulted mostly from the smaller number and size of the surface nodules, which dominated in roughness, despite that the pits and scratches became somewhat deeper.

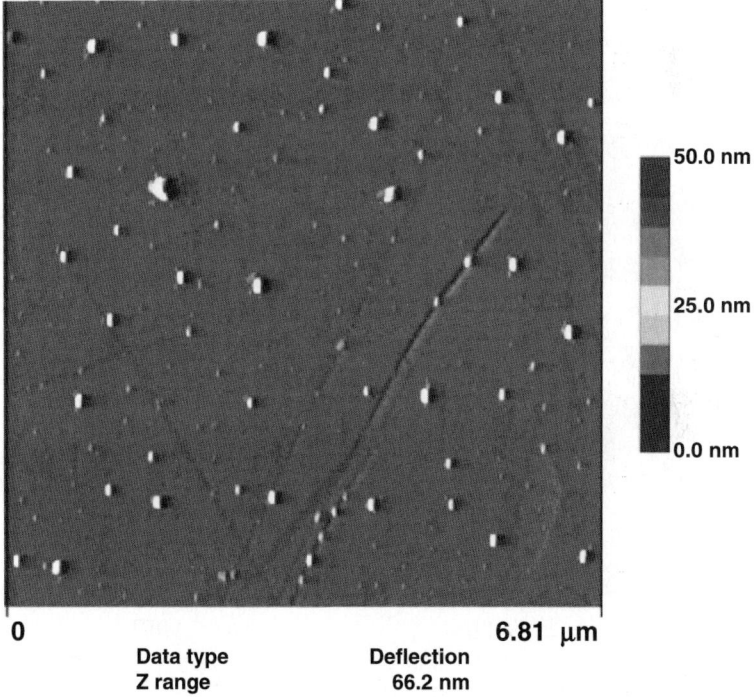

Fig. 5.9 Deflection AFM images of fused-silica wafers. Nodules are clearly evident.

Reprinted from Biosensors and bioelectronics, 17(6–7), Henke L, Nagy N, Krull UJ, An AFM determination of the effects on surface roughness caused by cleaning of fused silica and glass substrates in the process of optical biosensor preparation, 547–551, © 2002, with permission from Elsevier.

Thus, the scale of the non-inherent imperfections in the structure of amorphous silica surfaces is from fractions of a micrometer down to nanometers.

5.3.1.2 Structure on the Atomic Level

The building blocks of bulk silica, both crystalline and amorphous, are SiO_4 tetrahedra, which are linked by sharing corners, while each O is bonded to two Si. However, in the amorphous variety, the three-dimensional network of the tetrahedra lacks the symmetry and periodicity characteristic of the crystalline forms. The Si-O distances are similar to other varieties, about 0.16 nm, but the Si-O-Si angles deviate by as much as 20° from the mean 153° [42]. Naturally, the structures on the surface are even more complex and variable.

Poggemann, et al. [43] managed to obtain high spatial resolution AMF pictures of a fresh fracture surface of the amorphous silica, like that shown in Fig. 5.10. Seen are O–O and Si–O atomic distances, the SiO_4 tetrahedra, rings of such tetrahedra, as

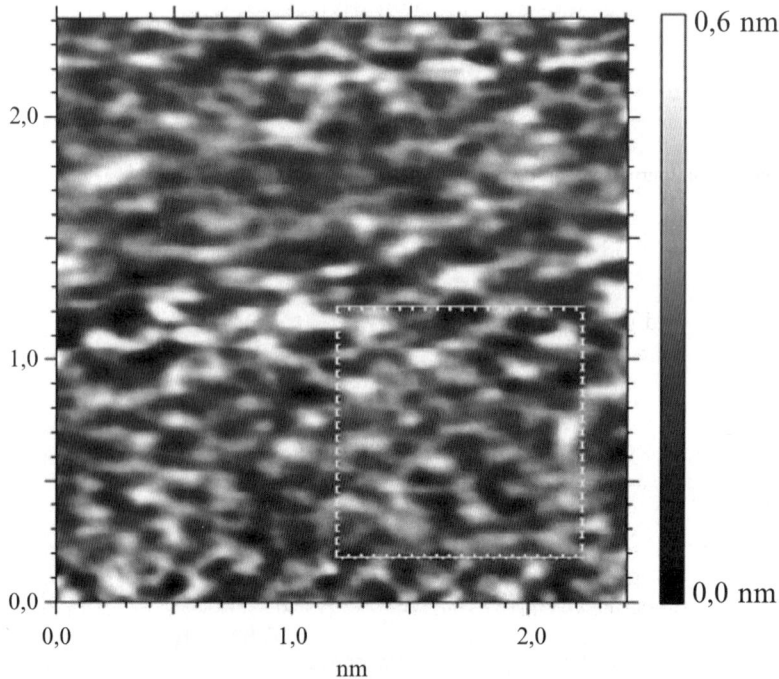

Fig. 5.10 Non-contact AFM image of a silica glass fracture prepared 1×10^{-11} mbar and imaged at 1×10^{-8} mbar [43].

Reprinted from Journal of non-crystalline solids, 281(1–3), Poggemann JF, Goss A, Heide G, Radlein E, Frischat GH, Direct view of the structure of a silica glass fracture surface, 221–226, © 2001, with permission from Elsevier.

5.3 Real Structure of Column Surfaces

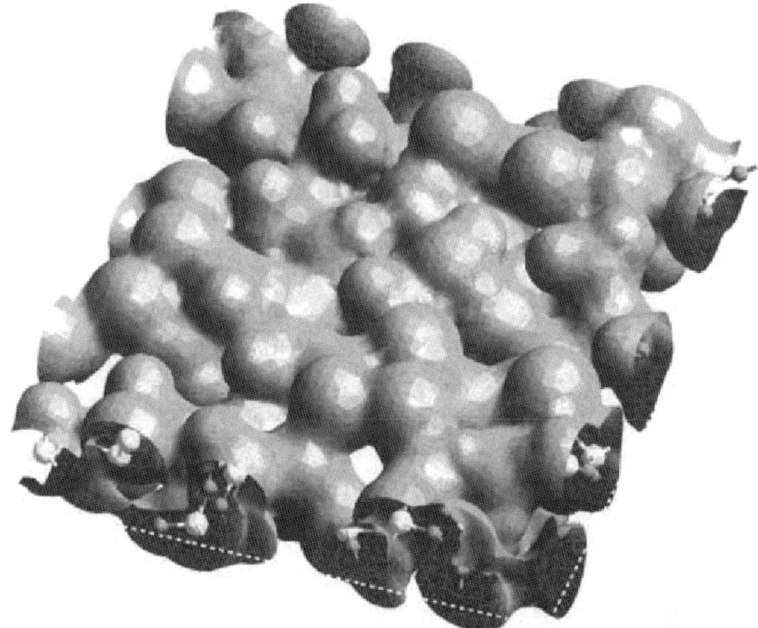

Fig. 5.11 Three-dimensional view on the calculated energy surface of a box with the same size as in the previous (experimental) figure. [43].

Reprinted from Journal of non-crystalline solids, 281(1–3), Poggemann JF, Goss A, Heide G, Radlein E, Frischat GH, Direct view of the structure of a silica glass fracture surface, 221–226, © 2001, with permission from Elsevier.

well as holes and grooves between these structural elements. The structure and relaxation of the surfaces at the atomic scale, as well as some reactions on the surface, have been treated in a growing number of theoretical papers. The approach consists in molecular dynamics simulations based on density functional theory; various two-body and three-body interaction potentials between atoms were used.

The numerical simulations of such a surface, made by the same authors, are illustrated by Fig. 5.11. They yielded pictures similar to the experimental data because of an intrinsic randomness in the structure of bulk amorphous silica and its surface.

Stallons and Inglesia [44] simulated for comparison the ordered surface of a crystal obtained by cutting the bulk material, the unrelaxed cut-off amorphous surface, as well as the latter relaxed. The last case was the random structure created by the Monte Carlo sphere packing method. They calculated the adsorption potential surface for some weakly bound adsorbates (N_2, Ar, CH_4) with the aim of judging the fidelity of the surface models by comparison with the available experimental data on the heats of adsorption and surface diffusivity. The adsorption energy profile in Fig. 5.12 gives an interesting look of the surface from the point of view of the problems discussed in this book; the concrete data will be called for an analysis in later sections.

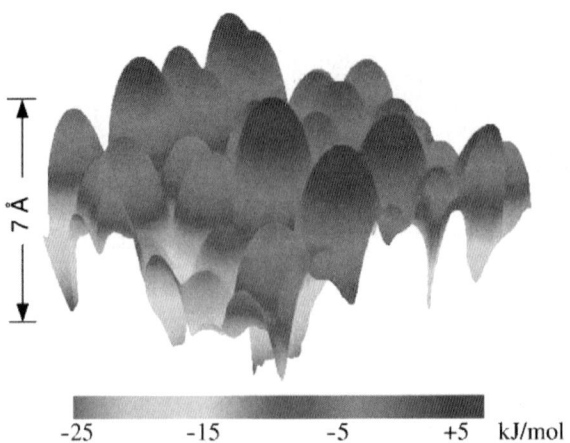

Fig. 5.12 Potential energy surface for N_2 on the simulated relaxed silica surface [44].

Reprinted from Chemical Engineering Science, 56(14), Stallons JM, Inglesia E, Simulations of the structure and properties of amorphous silica surface, 4205–4216, © 2001, with permission from Elsevier.

5.3.2 Silanols and Siloxanes on Silica Surface

The above pictures discussed ideal surfaces consisting only of the silicon and oxygen atoms. Meanwhile, an extremely important and characteristic property of the bare surfaces of real silicas is their readiness to react with water molecules when exposed to ambient atmosphere. The interactions result in the formation of numerous ≡Si-OH groups (silanols) on the surface; Kiselev [45] discovered it some 70 years ago.

To date there has been found only one contrasting exception. Namely, Bakaev and Steele [46] predicted an extreme "hydrophobicity" for surfaces dominated by bridging O atoms in Si–O–Si moieties. Such a surface, strikingly unreactive toward water, could indeed be obtained [47, 48]. It required controlled oxidation of a monoatomic Si film deposited on a Mo (112) template under ultra high vacuum. The obtained ultrathin SiO_2 film was relatively free of defects. No evidence for dissociation of water at the surface was found; even at low coverage, water molecules made three-dimensional clusters. Thus, the water–silica interaction was weaker than water–water hydrogen bonding.

5.3.2.1 Hydroxylation

Numerous experimental studies, made mostly with silica gels and other high specific area silicas exposed to water vapor, reported abundances of the silanol groups as high as 6 per 100 Å^2 [33,49]. Zhuravlev [50] believes that the number of OH groups per unit area, when the surface is hydroxylated to the maximal degree, is a physicochemical constant, equal to (4.6 to 4.9) hydroxyls per 100 Å^2. He obtained this value

5.3 Real Structure of Column Surfaces

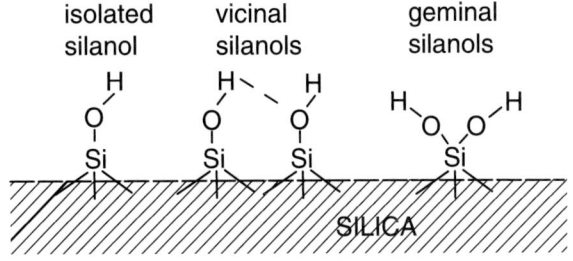

Fig. 5.13 Various types of silanol groups on the surface of silicas.

from the measurements of different varieties of amorphous silica like aerosils, silica gels and porous glasses, with specific areas ranging from 1 to $1,000\,\mathrm{m^2 g^{-1}}$. His value is somewhat smaller than the largest figures reported in literature; possibly, the latter were based on inaccurate estimates of the true surface area. At 5 radicals per $100\,\text{Å}^2$, the silanol groups are on average only some 5 Å apart, and the area per a group is about $20\,\text{Å}^2$. The Zhuravlev's value is in agreement with the 4.6 Si atoms per $100\,\text{Å}^2$ on the octahedral face of β-cristobalite.

According to present knowledge, the silanol groups differ in their attachment to the surface and in their mutual position on it [51]. It is illustrated by Fig. 5.13. Isolated (free) silanol groups occur with the hydroxyl attached to a surface Si atom, which has three other bonds dipped into the bulk structure. Next is a pair of different "vicinal" silanols at a distance which makes bridging the hydroxyls by a hydrogen bond possible. Third are "geminal" silanols — two OH groups attached to one surface Si atom. Leonardelli, et al. [52] found a constant fraction of the geminal hydroxyl silanol sites in all investigations of different samples of amorphous silica. The varieties of silanols are schematically depicted in Fig. 5.13.

Hydroxylation, dehydroxylation, as well as modification of the surface of amorphous silica crucially define the structure and properties of real surfaces of the chromatographic columns which are used in the experiments with metal halides and oxides. We need deeper information about the processes involving the surface hydroxyls. The experimental approaches to the problem have mostly leaned upon methods like infrared spectroscopy, Raman spectroscopy and nuclear magnetic resonance spectroscopy. The techniques make use of the spectral lines, the fingerprints of the particular bonds, to measure abundances of the structural groups and, eventually, to judge the peculiarities in the chemical and structural environment from the shifts in the stretching frequencies. Theoretical developments usually proceed from the hydroxylated main faces of β-cristobalite – (100) and (111). This crystalline phase of silica has the density and refractive index close to those of amorphous silica [35]. Moreover, it proved possible to rationalize the spectroscopic data for the amorphous hydroxylated surface by modeling the structure as alternation of the patches existing on the hydroxylated cristobalite surfaces [53].

Hydroxylation involves dissociation of the molecules of water. It is no surprise that the fresh fracture surfaces, with their chemically active bond defects, vigorously

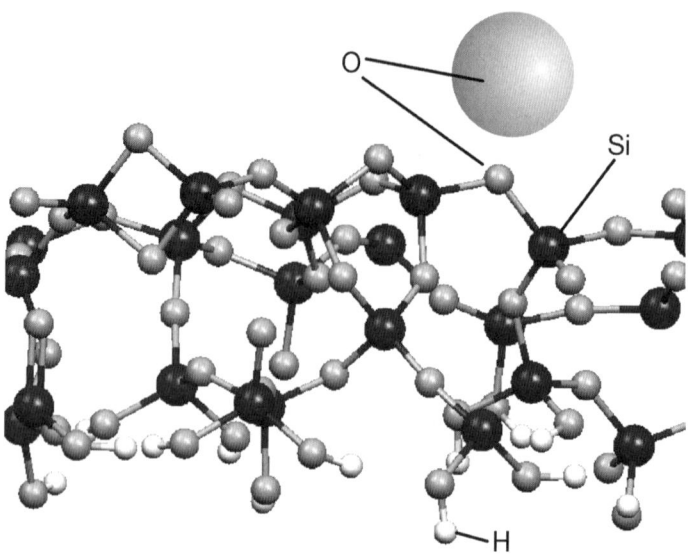

Fig. 5.14 Side view of a slab model of the surface of amorphous silica [55]. The top surface consists only of Si and O atoms; the bottom one is fully hydroxylated. Black, light grey and white spheres are silicon, oxygen and hydrogen atoms, respectively. The large spherical oxygen and all the silicon atoms are to scale with the interatomic distances.

Reproduced from Journal of Physics: Condensed Matter, 14(16): 4133–4144, Masini P, Bernasconi M, *Ab initio* simulations of hydroxylation and dehydroxylation reactions at surfaces: amorphous silica and brucite, © 2002, with permission from IOP Publishing Ltd.

react when exposed to ambient air. For this reason, Poggemann, et al. [43] had to perform their high resolution AFM measurements (vide ante) under ultra-high vacuum conditions. Otherwise, the fracture surfaces immediately took up water from the air and produced a several-nanometers-thick gel layer, which interfered with scanning. Dissociation of water molecules when they react with unsaturated bonds on a fresh fracture is rather straightforward. Souza and Pantano [54] reported an extended study of the hydroxylation and dehydroxylation of fracture surfaces under high vacuum.

Masini and Bernasconi [55] undertook ab initio simulations of the amorphous silica surface consisting only of Si and O atoms, as well as that hydroxylated. Some of their results are presented in Fig. 5.14.

5.3.2.2 Dehydroxylation

It has not yet been mentioned that, at moderate temperatures, the silica surface contains physically adsorbed molecules of water, which are attached to the surface hydroxyls; see Refs. [56, 57] for the theory of the problem. The water is relatively weakly bound, and with increasing temperature, dehydratation of the surface precedes the dehydroxylation. The last monolayer of the physically adsorbed water can

5.3 Real Structure of Column Surfaces

be removed by heating to 180 to 200 °C under vacuum; the hydroxyl cover is still completely retained [58, 59].

The vast amount of available information about thermal dehydroxylation is not completely consistent. Various authors reported that the vicinal hydroxyls are being removed due to splitting out H_2O, starting from 150 to 200 °C, and the process is completed at 450 to 600 °C. The isolated and geminal hydroxyls, which typically make 25 to 30 percent of the initial total, are left behind. They are much harder to remove, requiring temperatures in the range of 600 to 900 °C. Zhuravlev [58] found that 5 percent of the OH groups are still left at 1,000 °C, but less than 2 percent at 1, 100 °C. Sneh and George [60] studied the thermal dehydroxylation on a well defined silica surface, a 5-μm thick SiO_2 layer grown on Si(100), using sophisticated techniques for quantitative measurements. The surface was cleaned and analyzed under high vacuum conditions. Their results on the thermal stability were consistent with earlier measurements of high specific area silica powders. They concluded that, "...the data that may be the least affected by sample preparation do display a fairly consistent picture."

Possible mechanisms of dehydroxylation and the structural features left behind were simulated in some theoretical works. The mechanisms depend on the initial position and attachment of the pairs of hydroxyls. Some of them are obvious, like that upon heating the two vicinal hydroxyls, a "siloxane" group $\equiv$Si–O–Si$\equiv$ is left behind. However, the geminal pairs of OH, which would seem to yield >Si=O groups, were found to persist after drying the specimens at 600 °C [61] because silicon does not readily form >Si=O. Final dehydroxylation probably involves deep local restructuring of the surfaces and also produces the siloxane links.

The surface siloxane groups are worthy of more detailed investigations, like that by Hamann [62]. Models of amorphous silica predominantly contain five-, six-, and seven-membered rings, which should allow easy atomic relaxation of the network. Smaller rings constrain the relaxation; they can be considered like defect centers in amorphous silica and on its surface. These strained rings become sites of extremely high chemical reactivity. Their strain energies were calculated to be 1.23 and 0.25 eV for the two- and three-membered rings, respectively (i.e., $\approx$120 and 24 kJ per mole of the rings); other authors obtained even larger values. The detailed structures of various rings were calculated by Lopez, et al. [63]. Some of them are shown in Fig. 5.15.

Simulations of the surface by Rarivomanantsoa, et al. [64] showed that the most energetically unfavorable small rings are enhanced at the surface, while the large rings show a tendency to disappear.

Hambleton and Hockey [65] first noticed that the inherent roughness of the surface at the nanometer and atomic scale, which was discussed in Sect. 5.3.1, makes novel mutual positions of the hydroxyl radicals possible. For example, two silanol groups found at a small distance may belong to different layers in the structure, and the groups covering the walls of a deep hole can find themselves in an unusual mutual configuration. These situations are illustrated in Fig. 5.16.

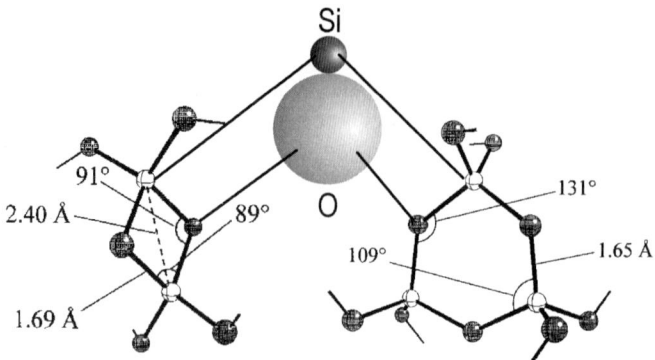

Fig. 5.15 Geometry of two-membered and three-membered rings [63]. Si atoms are shown in white, O atoms – in black. The sizes of Si and O shown by the large spheres (inserted by the present author) correspond to the interatomic distances within the touching hard sphere model.

Reprinted from Journal of non-crystalline solids 271 (1–2), Lopez N, Vitiello M, Illas F, Pacchioni G, Interaction of H_2 with strained rings at the silica surface from ab initio calculations, 56–63, © 2000, with permission from Elsevier.

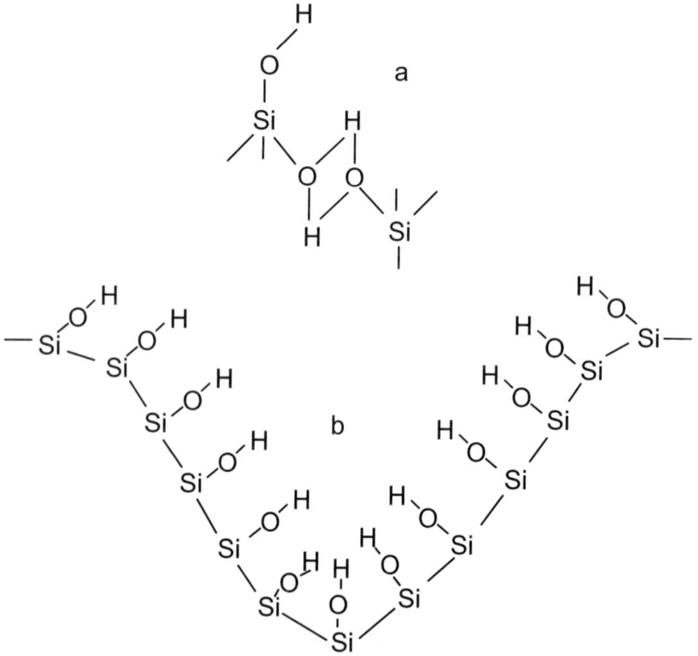

Fig. 5.16 Configurations of silanols on nanometer scale irregularities of the surface structure. (not to scale) [65].

Reproduced from Transactions of Faraday Society, 62, Hambleton FH, Hockey JA, Infra-red spectroscopic investigation of the interaction of BCl_3 with aerosol silicas, 1694–1701, © 1966, with permission from the Royal Society of Chemistry.

5.3.2.3 Rehydroxylation

Lygin [66, 67] considered, step by step, the processes of removal of the silanol groups from the surface, and the subsequent rehydroxylation, to be able to propose reasonable mechanisms. He found that rehydroxylation does not restore the original structure in every detail. It can be seen from his schematics presented in Fig. 5.17 by watching the Si atom in the middle, at steps a through e. His model predicts the

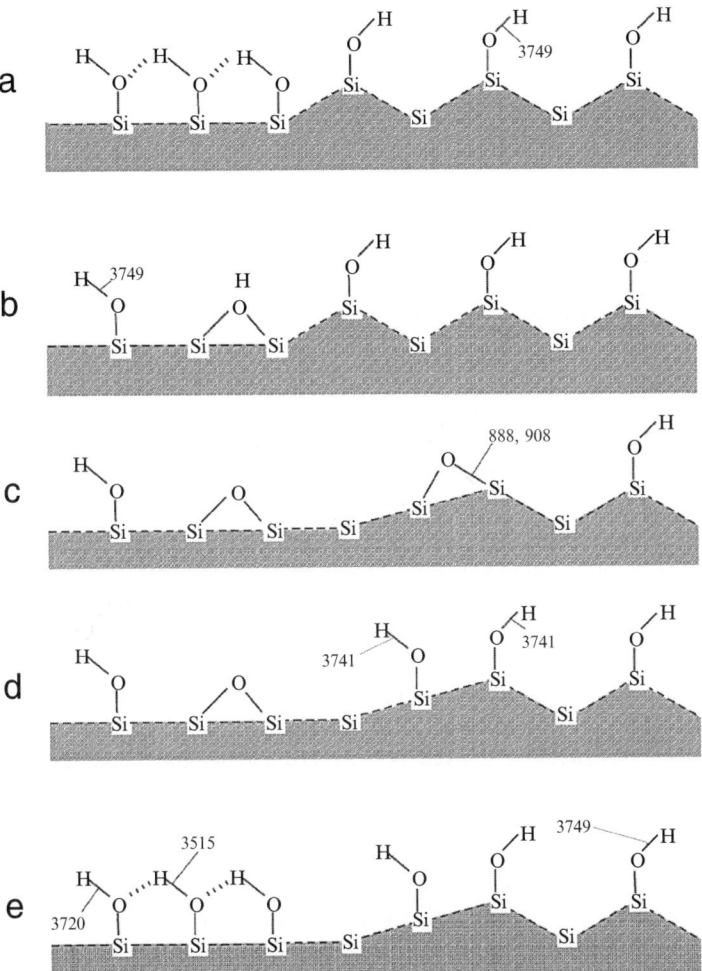

Fig. 5.17 Changes in structure of the hydroxylated surface of silica: a – the original structure; b, c – the structures after dehydroxylation in vacuum at 200 to 400 °C; d, e – the structure after subsequent rehydroxylation by chemisorption of water [66]. The vibration frequencies of bonds are given in cm^{-1}.

Reproduced (adapted) from Zhurnal Rossiiskogo Khimicheskogo Obshchestva, 46(3), Lygin VI, Models of soft and hard surface of silicas, 12–18, © 2002, with permission from the journal Editors.

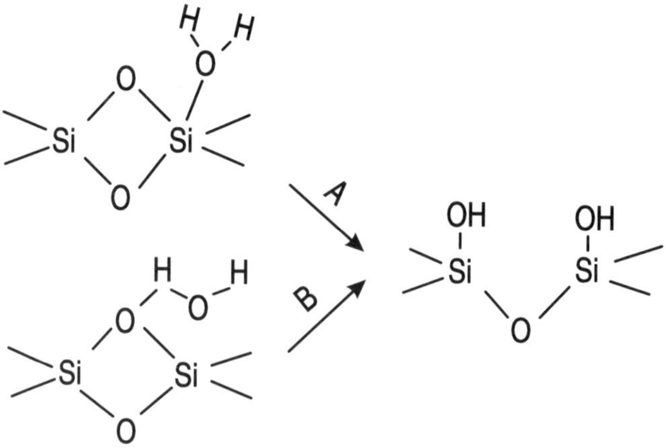

Fig. 5.18 Sketch of the two conceivable reaction paths for the rehydroxylation reaction of the two-membered silicon ring [55].

Reproduced from Journal of Physics: Condensed Matter, 14(16), Masini P, Bernasconi M, *Ab initio* simulations of hydroxylation and dehydroxylation reactions at surfaces: amorphous silica and brucite, 4133–4144, © 2002, with permission from IOP Publishing Ltd.

concentration and distribution of different types of silanol and siloxane groups, and characterizes the energetic heterogeneity of the silica surface as a function of the pre-treatment temperature. Very important are his experimental findings that the adsorption and other surface parameters per unit surface area of silica are identical for the numerous varieties of the amorphous silica.

The strained two-member silicon rings, which are the most peculiar topological defects on the surface of amorphous silica, proved the most reactive [55]. Thus, hydratation occurs preferentially with these rings – they are cleaved through the processes shown schematically in Fig. 5.18. This is in agreement with the most abundant spectroscopic evidence collected in real hydratation and dehydratation experiments and reported in literature.

The data reviewed above showed, both experimentally and theoretically, that fundamental conditions for relatively easy rehydroxylation of the surface of silica exist, even after it was dehydroxylated at a high temperature. Hence, we may expect that it is also true for the fused-silica tubes and (coiled) capillaries used in the gas-phase radiochemistry studies. The capillaries (several tenths of a millimeter in diameter), which are also used for chromatographic separations in many fields of chemistry, are produced by drawing prepared larger tubes heated to 2,000 °C with the speeds up to 10 centimeters per second. It was found that the concentration of silanols on the inner surface of the ready capillary is much higher then the concentration of OH groups in the bulk material. To avoid it, the surface must be prevented from contact with ambient air during manufacturing [68].

When the nascent capillaries or tubes cool down to moderate temperatures, they evidently pass a temperature range favorable for rehydroxylation. Fast cooling may prevent complete rehydroxylation at that moment. However, it leaves behind unrelaxed active structures, which then may react on a longer time scale. These considerations are very important in view of the modification of silica surface by chemicals other than water.

5.3.3 Modification of Silica Surface by Haloginating Reagents

Chapter 1 described many experiments on gas-solid (thermo) chromatography in silica columns, when the carrier gas contained halogenating constituents like CCl_4, $SOCl_2$, $TiCl_4$, HBr, BBr_3 and HCl. Meanwhile, the literature contains rather extensive information about modification of the surface of high specific area silicas by some of the above-mentioned and many other inorganic molecular halides. The works aimed at obtaining information about the reactivity, structure and other properties of the silica surface, first of all because halogenation is an intermediate step in the final modification of the surface by organic and metalorganic agents. It is well-known that the adsorbents produced this way find numerous applications in adsorption, separation, catalysis and other technologies. Several books [34, 69–71] survey the accumulated data on halogenation. In recent years, such studies appeared less frequently. Some concrete studies from this last period, which are mentioned below, provide additional useful references. Because, as we saw earlier, the chemical properties of the silica surface do not critically depend on the specific area of the material, most of the data on halogenation are of interest for researchers in gas-phase radiochemistry.

The experiments were performed at temperatures, partial pressures of the modifying agent and other parameters varying in broad ranges. The effect of chemical and thermal pre-treatment of silica was also investigated. Generally, the halides MX_n of both metals and nonmetals reacted with isolated silanol groups and with vicinal silanols. The concrete case of $TiCl_4$ showed in Fig. 5.19 may serve to visualize the interactions.

Reactions (A) and (B) seem to be of general importance. Reaction (B) is characteristic of two hydroxyls joined by the hydrogen bond. There are indications of possible interactions with three hydroxyls [70]. Even at room temperature, the rate of interaction of $TiCl_4$ with the surface hydroxyls is too rapid to be measured. The consecutive steps (A) and (C) are conceivable, but difficult to be evidenced for $TiCl_4$; analogous mechanism was proved in the case of the modification by $HSiCl_3$. Reaction (D), cleaving the strained siloxane bridge, was postulated in older works. Recently Shrijnemakers, et al. [72] tried to reinvestigate the problem, which was discussed also in earlier Refs. [73, 74]. They explored the effect of pre-treatment temperature, which was mostly $\leq 500\,°C$, on the relative role of different reactions. Below the indicated limit, the contribution of reaction (D) was negligible; for higher temperatures, the conclusions were uncertain. Several chlorinating agents, other than metal chlorides, are able to "directly chlorinate" the surface, which results in chlorine atoms bonded to the surface atoms of silicon. Thionyl chloride is very

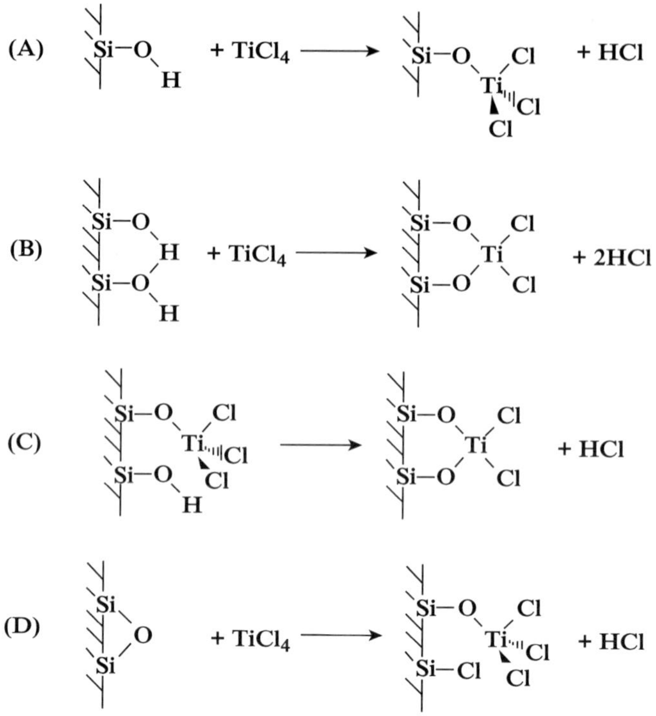

Fig. 5.19 Possible reactions of TiCl$_4$ with a silica surface [72].
Reproduced from Physical Chemistry, Chemical Physics, 1, Schrijnemakers, P. Van Der Voort, E. F. Vansant, Characterization of a TiCl$_4$-modifed silica surface by means of quantitative surface analysis, 2569–2572, © 1999, with permission of the Royal Society of Chemistry.

effective; moreover, its interaction has interesting peculiarities. When the silica is exposed to SOCl$_2$ vapor at elevated temperatures, the result can be described by the reaction:

$$\equiv\text{SiOH} + \text{SOCl}_2 \rightarrow \equiv\text{SiCl} + \text{SO}_2 + \text{HCl},$$

that is, effectively, like substitution of the OH in silanol by an atom of Cl. The reaction (realized in boiling benzene solution to achieve higher working temperatures) is even used to activate completely hydroxylized silica before its modification by organic reagents or metalorganics. Gorlov, et al. [75] firmly established that SOCl$_2$ reacts at moderate temperatures with the silanols by analogy with the chlorides of metals:

$$\equiv\text{SiOH} + \text{SOCl}_2 \rightarrow \equiv\text{Si-O-SO-Cl} + \text{HCl}$$

However, upon heating above some 200 °C, the molecule of SO$_2$ splits off and Cl bonds to the surface Si.

Direct chlorination by CCl$_4$ and Cl$_2$ was evidenced through identification of the infrared frequencies of the $\equiv$SiCl, and even the =SiCl$_2$ species on the surface [76]. Carbon tetrachloride was reported to react with silanols at 400 °C [77] and with both

5.3 Real Structure of Column Surfaces

silanols and siloxanes at 600 to 800 °C to yield Cl attached to Si atoms [78]. Direct chlorination is also possible with S_2Cl_2, $COCl_2$ and CH_3COCl.

Chlorine could be even used for efficient dehydroxylation of silica gels [79] at 800 °C due to the reaction:

$$\equiv SiOH + Cl_2 \rightarrow \equiv SiCl + HCl + 1/2 O_2$$

Haukka, et al. [74] confirmed it. They did not find evidence for HCl cleaving the siloxane bridges.

While hydrogen fluoride readily reacts according to

$$\equiv SiOH + HF \rightarrow \equiv SiF + H_2O$$

the analogous chlorination with HCl seems to be poorly studied. A conceivable mechanism would be the formation of a free Cl atom on a site with energetic oxygen and subsequent replacement of the hydroxyl by the atom [70]. Bromination data are extremely scarce. According to McDaniel [80], gaseous $SOBr_2$ shows less reactivity compared with the earlier studied $SOCl_2$ [78]. It brominates silanols, but not siloxanes. In the presence of CO, both silanol and siloxane groups reacted with bromine above 500 °C; the maximum coverage reached 3.5 Br per 100 $Å^2$.

5.3.4 Morphology of Metal Surfaces

The real surfaces of metals are relatively clean and better understandable compared with the surfaces of oxide materials, which are most frequently used in gas-solid chromatography of molecular inorganic compounds.

At the atomic level, the smoothest observed surfaces of metal crystals still contain structures like those shown in Fig. 5.20.

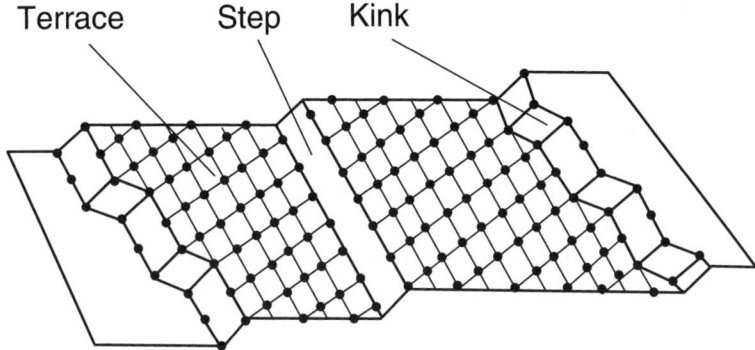

Fig. 5.20 Typical structures of the surface of crystalline metals. It is difficult to obtain terraces extending over macroscopic dimensions.

Such well-ordered surfaces can be obtained almost exclusively in ultra-high vacuum conditions. Isted, et al. [81] prepared ultra-flat Au(110) and Cu(110) surfaces from mechanically polished crystals by repeated cycles of Ar ion bombardment and subsequent annealing to 580 K for Au (110), and 860 K for Cu (110). Such procedures yielded surfaces clean and ordered on the atomic scale, but their morphology observed by STM varied over micrometer scale areas. To prepare surfaces with well-defined terraces terminated by monoatomic height steps, these authors used "thermal roughening annealing" at 900 K and 1,100 K, respectively. Subsequent exposure of Cu to ambient air resulted in granular surface structure with the steps no longer observed; the rms roughness was 0.4 nm.

Yet some metallic surfaces with very small roughness could be reportedly produced outside of vacuum chambers [82]. A layer of metal was consecutively polished with the colloidal slurries of silica and alumina. This procedure was followed by a multi-step cleaning from the slurry and residual metal particles by solutions of some chemicals. The roughness of a freshly deposited Pt surface characterized by grain sizes of 3 nm could be reduced to a rms of 0.1 nm.

Wadsak, et al. [83] studied various preparation methods for copper sheets, like chemical etching and subsequent polishing. Mechanical polishing with a monocrystalline diamond paste, followed by electrochemical polishing of the sheet, proved most suitable: rms-roughness of a scanned area of $5 \times 5\,\mu m^2$ and $1 \times 1\,\mu m^2$ was 2.6 nm and 1.7 nm, respectively; scratches that were 30 nm in width and 5 nm in depth were evident afterwards. After electrochemical polishing the figures for rms were 2.4 and 1.0 nm; the surface was smooth but with visible grain boundaries.

An adsorbed molecule would experience different adsorption potential when attached to a terrace, step or kink. On the other hand, on some metals (platinum as an example) the heats of adsorption of simple gases (ethylene, CO, H_2) vary little with surface structure. Lee, et al. [84] examined the effect of surface reconstruction and relaxation on the electronic coordination numbers of the surface atoms. They found that highly stepped surfaces relax into a configuration where the surface atoms have about the same electronic coordination number as Pt(111) surface atoms. It means that they lack the coordinative unsaturation of the atoms, which would take place on unaltered stepped surfaces.

5.3.5 Modification of Metal Surfaces

There are few works on the modification of the surface of metals by haloginating reagents, which are as detailed as the studies of silica. In the meantime, a sole work compares thermochromatographic behavior of molecular bromides in the columns made of nickel and of silica; the observed deposition temperatures happened to be equal [32]. This finding is very difficult to rationalize if the microscopic picture of adsorption is such as shown in Figs. 5.4 and 5.5.

Fishlock, et al. [85] studied the interaction of bromine with Ni(110) by STM. To obtain a defined surface, they polished the surface with diamond paste (down to 0.25 μm size) and then cleaned it in ultra-high vacuum by repeated cycles of Ar ion sputtering and subsequent annealing at 700 °C. Images of sizes of $30 \times 32 \text{Å}^2$ in their work show an ideal terrace on the surface of Ni(110). The initial adsorption of molecular bromine happened to be dissociative. The coverage with bromine was investigated up to saturation, which took place at 6×10^{14} Br atoms per cm^2 and was accompanied by roughening of the surface with formation of etch pits. Annealing to 200 °C at the densest coverage did not cause any pronounced desorption, but progressively (with time) changed the topography (reconstructed the surface) without more desorption. Earlier [86] the same laboratory published similar, less extensive data on the interaction of chlorine with the Ni(110) surface.

5.4 Lateral Migration of Adsorbate

5.4.1 Surface Diffusion

The model of localized adsorption essentially implies that each adsorption center is isolated; that is, completely surrounded by a potential barrier with the height equal to the desorption energy. Then the adsorbed entity can visit some other sites only through repeated desorption-adsorption steps. In Sect. 4.3 it was shown that even the shortest displacements between the two events are of the mean molecular free flight. At STP, it means mostly hundreds of distances between adjacent adsorption sites, not just a few as it is highly schematically shown in Fig. 5.21.

Actually as early as in the 1930s, direct evidence was obtained showing that the molecules migrate across the surface without being released into gas; cf. [87] and a brief history in [88]. It was concluded that owing to the thermal motion and because the energetic barriers between the sites are relatively low, an adsorbed molecule is hopping between neighboring sites, while staying in the monolayer. These concepts laid the foundation for the later extensive studies of surface mobility. The multi-author book [89] is devoted to various aspects of the process, which can be mathematically described as two-dimensional diffusion, though dissimilar from the diffusion of the two-dimensional gas.

Detailed studies of the microscopic picture of surface diffusion have mostly dealt with the adsorption on metals. Barth [88] extensively reviewed the diffusion of nonmetallic adsorbates at metal surfaces. He outlines a more sophisticated microscopic picture of the processes involved. Gaseous molecules (atoms) strike the surface of the metal and thermalize with the phonon heat bath of the crystal. The sites are separated by energy barriers ε_b, which are significantly smaller than the energy of desorption. The magnitude of the thermal energies, with respect to the migration energy barrier, is decisive for the possibility of real diffusive motion. At relatively low temperatures, when $k_B T \ll \varepsilon_b$, the atoms are confined to the adsorption sites; in this particular sense, they practically rest immobile. However, they vibrate (10^{12} to 10^{13} s^{-1}) about the potential energy minima, the vibrational amplitude of the surface atoms at room temperature is ≈ 0.1 Å. (Notice that it is small compared

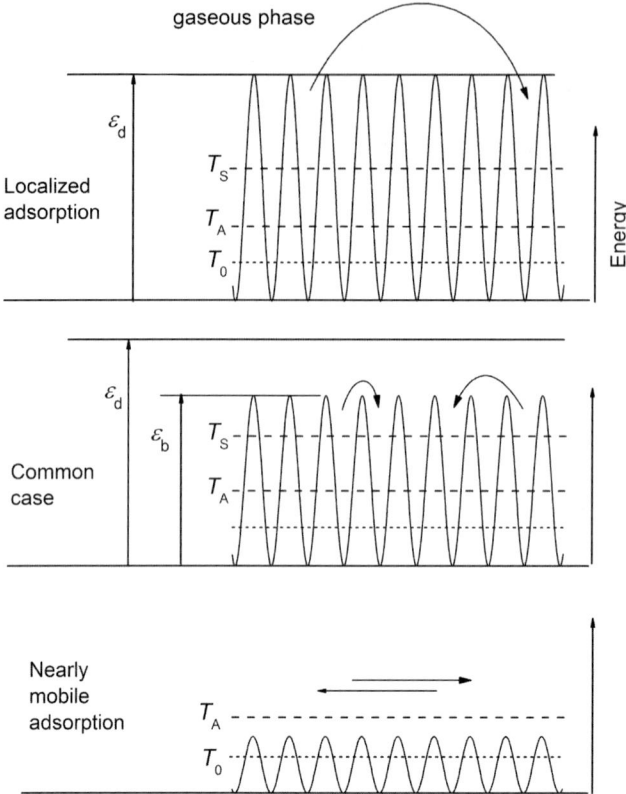

Fig. 5.21 A schematic of various conditions for the surface diffusion on homogeneous surfaces with periodic adsorption potential. T_0 is the ambient temperature, (dotted line), T_A and T_S relate to the TC experiments, T_c in IC experiments can be identified with T_A and T_S in the figure.

with the interatomic distances, and so with the size of the well and the molecule). With higher temperature of the surface, the probability increases that the adsorbed molecule will receive sufficient energy to pass over the migration barrier. Thus, the energy fluctuations result in stochastic jumps from one energy minimum to another, and the adsorbate moves laterally across the surface. Due to the Boltzmann factor such jumps are rare. Hence, subsequent displacements are uncorrelated and proceed with a definite average frequency. If the mean jump lengths are also known, the process can be described in terms of diffusion (see below).

On the other hand, if the thermal energies are close to or exceed the desorption energy, like in the bottom of Fig. 5.21, its undulation has smaller influence on the lateral motion of the adsorbate. The migration becomes less restricted and the adatoms move rather freely, being unconfined to specific sites. The situation can be better described as two-dimensional Brownian motion [88], in which a coefficient of viscous friction simulates dumping by the substrate [90]. The surface diffusion

5.4 Lateral Migration of Adsorbate

in the transition regime has not yet been well studied. The regime does not seem to occur in gas-solid chromatography of the heavy element compounds.

The potential energy schematic shown in the middle of Fig. 5.21 seems to outline the most common relations between the thermal energies and the characteristics of desorption and lateral diffusion which take place in IC and TC experiments. The schematic does not illustrate the geometry of the lateral movement. Indeed, even if an adsorbed entity approaches the level of the desorption energy, it continues to stay within the monolayer thickness. This follows from the fact that the attractive Lennard-Jones potential is inversely proportional to a high power of the distance. The one-dimensional graphs in Fig. 5.21 are also oversimplified in the sense that the adsorption potential is a function of both lateral dimensions. Nevertheless, such sketches allow useful qualitative conclusions.

Modern scanning techniques enable direct watching of the migration of atoms or molecules. For physisorbed molecules, data of this kind are scarce. The highest resolution studies were reported for chemisorbed light molecules of non-metals, like nitrogen or oxygen, and especially for the adsorption of metallic atoms on metals. In the latter case, usually, $k_B T$ was much smaller than ε_b. The conclusions from theoretical studies of the hopping mechanism are reportedly valid up to $k_B T < 0.2\varepsilon_b$. As to the diffusion, the frequency of attempts is obviously the lattice vibration frequency times the Boltzmann factor so that the effective diffusion coefficient D_a is

$$D_a = \frac{1}{2}\nu_0 e^{-\frac{\varepsilon_b}{RT}} \lambda_b^2 \tag{5.62}$$

where λ_b is the mean jump length. It seems that the hops to the nearest neighbor sites prevail so that λ_b is close to the surface lattice constant, typically ≈ 3 Å. Longer jumps, two to three lattice spacing in length, are not negligible. They must contribute to diffusion of adatoms with masses smaller than the mass of adsorbent atoms, and only if the Debye frequency of the adsorbent is either very small or very high [90, 91]. In such situations, the damping interaction is ineffective and allows longer hops. For Pd atoms on the surface of tungsten, minor occurrence of the jumps over two and more intersite separations was experimentally observed by FIM technique at room temperature in [92].

Recent experimental works [93] showed that the values of the Arrhenius prefactor in the surface diffusion coefficient are invariably of the order of 0.001 or so. It is in agreement with Eq. 5.62, if we substitute into it the typical values of frequency and jump length. An empirical rule for the diffusion of different adatoms on the surface of Re, Mo, W, Ir, and Rh is that the barriers amount to about one-tenth of the sublimation energy of the adsorbate and are higher on rougher crystal planes [94]. The barriers for self-diffusion on the fcc(100) surfaces of Rh, Ir, Ni, Pd, Pt, Cu, Ag and Au are one-sixth of the corresponding bulk cohesive energy. The factor is considered as a standard to judge whether a measured diffusion barrier is uncommon. Thus, the lateral diffusion of foreign metals seems to be reasonably well predictable.

The specific regularities in adsorption of metallic adatoms on metal surfaces, like the small diffusion barriers compared with the sublimation energies, are due to the peculiar nature of metallic bonds. They cannot be extended to the case of physisorption of molecular compounds on both metallic and non-metallic surfaces. The available data for such situations are scarce. It seems that in physisorption, the height of diffusion barriers makes a considerable part of the desorption energy.

5.4.2 Surface Diffusion and Entropy of Adsorbate

The periodic adsorption potential of surfaces and the potential barriers between adjacent sites lower than the desorption energy must result such that the state of adsorbates is intermediate between the ideal mobile and localized models. The lateral migration across the surface must make a positive contribution to the entropy of the adsorbate.

An illustrative schematic of ingredients of the adsorbate entropy in different states of the system is shown in Fig. 5.22. We wish to accept a reference state that would unite the two extreme models of adsorption. Let us start, on one side, with the closely packed monolayer of non-interacting entities at $0\,\mathrm{K}$ on a smooth surface as a possible *reference state for the mobile adsorption*. On the other side, there is *the* layer of completely filled adsorption sites at $0\,\mathrm{K}$ as *the reference state for the localized adsorption*. Both states can be realized only by one way, so that their entropies are zero. Now, for the purpose, let us abandon the possible difference between the capacities of the two filled layers (probably, they do not differ more than 10 times). Upon heating, the molecules released from the smooth surface would make two-dimensional gas with the appropriate translational entropy, plus entropy of the induced vibrations; they also acquire internal entropy of rotation and vibration. In the case of the distinct adsorption potential wells, heating activates the three-dimensional vibrations of the adsorbate in the potential wells and internal degrees of freedom, while incomplete coverage gives origin to positive configurational entropy of the adsorbed layer.

The internal degrees of freedom may be "less free" in the molecule captured in a potential well than in the gaseous molecule, so that the internal entropy is eventually different in the two phases. This possibility is indicated by the "<" notions in Fig. 5.22. In practice, the difference cannot be estimated; commonly, it is assumed that the internal entropy remains unchanged upon adsorption. The left side of the diagram shows how the entropy change $\Delta_{des}S^{mb}$, which has been assumed in the Third Law evaluation of the desorption enthalpy, relates to S_a^{lc}, S_a^{mb}, and S_g — the true entropies of the localized, mobile-adsorbed and gaseous states of the molecules, respectively. Shown is also $\Delta_{des}S^{lc}$, an analog of $\Delta_{des}S^{mb}$, which has not yet been employed in such evaluations.

As it can be seen from Fig. 5.22, when moving from the localized adsorption towards the mobile model, we can expect smooth decrease in the entropy of desorption. The entropy of the adsorbate which experiences lateral diffusion was discussed, in particular, by Patrikiejew, et al. [95]. They approached the problem by assuming that a fraction of the molecules is in completely mobile state, while the others are completely localized. Then they suggested that the canonical partition function Q^{ml}

5.4 Lateral Migration of Adsorbate

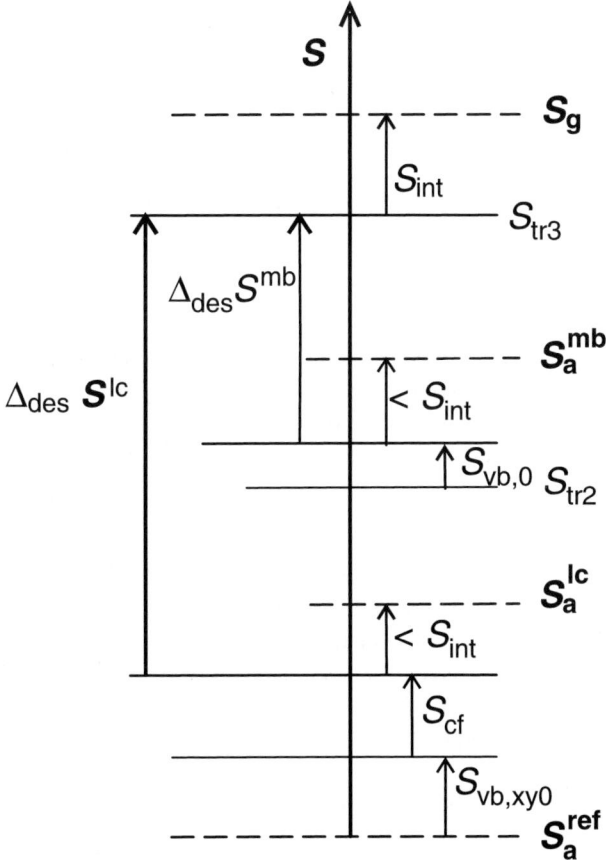

Fig. 5.22 Diagram of the entropies related to the evaluation of the desorption enthalpy from the experimental desorption constant. The dashed levels have not yet been taken into account in evaluation of the experimental data.

for the given number of molecules N^{ml} adsorbed on certain surface area may be written as the product of the corresponding functions for the mobile and localized molecules:

$$Q^{ml}(N^{ml}, T) = Q^{mb}\left[(N^{ml} - N^{lc}), T\right] \times Q^{lc}\left(N^{lc}, T\right) \quad (5.63)$$

It means that the resulting entropy is the sum of two contributions. They derived this and other thermodynamic functions for the whole range of fractional coverage value, as well as the Langmuir isotherm, by using the procedures of statistical thermodynamics.

Let us look for a simple way to approximate the entropy of the adsorbed state as a function of ε_b. We will assume that ε_d is larger enough than ε_b to neglect the frequency of desorptions.

The formula 5.32 for the standard entropy of the adsorbate in the mobile adsorption model can be rewritten as:

$$\frac{S_{\text{tr2}}^{\text{mb}} + S_{\text{vib}}^{\text{mb}}}{R} = \ln\left(e^3 \frac{A}{N_A} \frac{1}{\Lambda_m^2} \frac{k_B T}{h\nu_0}\right) = \ln\left[16e^3 \frac{A}{N_A} \frac{\nu_{xy}^2}{u_m^2} \left(\frac{k_B T}{h\nu}\right)^3\right] \quad (5.64)$$

It changes with a larger activated diffusion barrier to finally become:

$$\frac{S_{\text{vib}}^{\text{lc}} + S_{\text{cf}}^{\text{lc}}}{R} = \ln\left[\frac{A C_a}{N_A}\left(\frac{k_B T}{h\nu}\right)^3\right] \quad (5.65)$$

The last factor under the logarithm to the right (in parentheses) in the two above equations is the same, making it necessary to follow only the changes of the other factors. We know that the molecules diffuse across the surface by making hops, mostly between the adjacent adsorption sites, with the frequency ν_b, which is:

$$\nu_b = \nu_{xy} e^{-\frac{\varepsilon_b}{k_B T}} \quad (5.66)$$

About the same is the fraction of molecules that are "in flight" – in the state of two-dimensional gas – in any given moment of time. It is because, by chance, the time to pass the distance between adjacent sites, which is about $\sqrt{1/C_a} \approx 0.5$ nm, with the typical thermal speed 10^4 cm s^{-1} is some 10^{-13} s $\approx 1/\nu$. Therefore, we can expect that the entropy of the adsorbate in the state intermediate between the mobile and localized adsorption is:

$$\frac{S_a^{\text{ml}}}{R} = e^{-\frac{\varepsilon_b}{k_B T}} \frac{S_{\text{tr2}}^{\text{mb}} + S_{\text{vib}}^{\text{mb}}}{R} + \left(1 - e^{-\frac{\varepsilon_b}{k_B T}}\right) \frac{S_{\text{vib}}^{\text{lc}} + S_{\text{cf}}^{\text{lc}}}{R} \quad (5.67)$$

Explicitly,

$$\frac{S_a^{\text{ml}}}{R} = e^{-\frac{\varepsilon_b}{k_B T}}\left[\ln\left(16e^3 \frac{A}{N_A}\right) + \ln\frac{\nu_{xy}^2}{u_m^2}\right] + \left(1 - e^{-\frac{\varepsilon_b}{k_B T}}\right)\ln\frac{1}{\theta°} + \ln\left(\frac{k_B T}{h\nu}\right)^3 \quad (5.68)$$

which can be rearranged to:

$$\frac{S_a^{\text{ml}}}{R} = e^{-\frac{\varepsilon_b}{k_B T}} \ln\left(16e^3 \frac{A}{N_A} \frac{\nu_{xy}^2}{u_m^2} \theta°\right) + \ln\left[\frac{1}{\theta°}\left(\frac{k_B T}{h\nu}\right)^3\right] \quad (5.69)$$

For the accepted standard states given by 5.39 we obtain:

$$\frac{S_a^{\text{o,ml}}}{R} = \frac{S_a^{\text{o,lc}}}{R} + e^{-\frac{\varepsilon_b}{k_B T}} \ln\left(16e^3 \frac{\nu_{xy}^2}{C_a u_m^2}\right) \quad (5.70)$$

Hence, in the first approximation, the extra entropy of the mobile adsorbate compared with that of the localized adsorbate exponentially decreases with the higher barrier of diffusion. The formula is probably not very accurate for the barriers smaller than some $3RT$, when the true mobility contributes significantly. It should also be so because, at lower barriers, the mean displacement must increasingly exceed the above postulated $\sqrt{1/C_a}$. On the other hand, in the TC experiments yet discussed, typical desorption energies were about 20 times larger than the thermal energies at the temperature of the TC zone. Therefore, what happens if the diffusion barrier is lower than $3RT$ is not important for the present purpose. It seems that the lateral diffusion only slightly contributes to the entropy calculated for the localized adsorption model.

There is an obvious benefit from the phenomenon of lateral diffusion — a just adsorbed molecule usually visits a number of other sites before it gets desorbed. As will be discussed in Sect. 6.1, in general, the distribution between various phases can be evaluated from the behavior of single atoms if the concentrations are replaced by the corresponding probabilities. The laterally diffusing molecule samples much more adsorption sites than it would through mere adsorption–desorption events. This is extremely important in the case of adsorption on strongly heterogeneous surfaces, which will be discussed soon. In a practical sense, the diffusion accelerates reaching the adsorption equilibrium.

5.5 Evaluation of Adsorption Enthalpies on Real Surfaces

We have seen above that the microscopic picture of real column surfaces is so complex that it has little in common with the earlier assumed ideal homogeneous surface. The present knowledge, especially as the silica surface is concerned, can be summarized as follows:

- Most real surfaces are geometrically rough, from the micrometer down to the nanometer scale. The true surface area of a macroscopic object is usually larger than its apparent geometric area.
- The adsorption potential is never flat, which would be the rationale for the model of ideal mobile adsorption; there invariably occur distinct potential wells.
- The above wells are never completely separated, which would be the rationale for the localized adsorption model. Actually, the potential barriers between adjacent minima are smaller than the local desorption energies, and this results in lateral diffusion of the adsorbate.
- Most real surfaces are heterogeneous in adsorption properties. Even the regular structure of a particular crystal face never extends over a macroscopic area. There are numerous factors, primarily the roughness, which give rise to imperfections. For the surface of amorphous silica, heterogeneity is an inherent fundamental property.
- The bonds between the surface atoms of adsorbents (in particular, of silica) are often strained and some multivalent atoms are unsaturated. Because of this, a fresh surface may react with the atmosphere and with the chemically active components of the carrier gas.

These findings necessarily pose the question of how to account for the real heterogeneity when deducing the values of the adsorption parameters from the experimental data. The formulae of widespread use were originally derived exclusively for the ideal models of adsorption. Too straightforward application of these relations might have resulted in errors.

We will attempt an analysis of the possible misconceptions. To begin with, let us consider only the heterogeneity, independent of whether the surface is "bare" or chemically modified. The topography of the adsorption centers is usually described as "patch-wise," "random" and intermediate. It seems that the surfaces employed in the radiochemical studies have been almost solely of the random type. In the strict sense of the term, such a surface consists of sites with uncorrelated values of the adsorption energy. However, it seems that the potential barrier between two adjacent adsorption sites cannot be of completely random height; it must be somewhat correlated with the depths of the partner wells. As to the adsorbate, we suppose that each molecular entity takes only one site and that the particles occupying adjacent sites do not interact with each other.

Thus, because we lack knowledge of many details of each particular structure, there again necessarily emerges a model, though more sophisticated than the earlier ones. Lateral diffusion on a heterogeneous surface clearly results in that a molecule, which got adsorbed on the particular site, usually desorbs from a site with different coordinates and adsorption energy. For this reason it seems logical to characterize the adsorption sites and the adsorption–desorption equilibria by the parameters of desorption. Next, when discussing a more realistic but more complex picture of adsorption, we will again have to make some assumptions and approximations. In the first place, it seems reasonable to abandon the difference between the enthalpy H and the internal energy U and accept the notations $E_d \equiv \Delta_{des} U \approx \Delta_{des} H$ and $E_a \equiv \Delta_{ads} U \approx \Delta_{ads} H$ for the changes in enthalpy and internal energy. For consistency, we will also prefer the entropy change in desorption $\Delta_{des} S$. Unlike the parameters of adsorption, the values E_d and $\Delta_{des} S$ are positive. This will also make the equations more readable and help the reader to differentiate between the more rigorous but unrealistic formulae suitable only for the ideal models and the less rigorous but more helpful relationships.

The heterogeneity in the model of one-center localized adsorption is characterized by the normalized distribution of the occurrence of the sites as a function of the desorption energy. Generally, the spectrum of the desorption energies E_d^* may look like a set of discrete lines, a histogram, a continuous function or a combination of these varieties. Measurement of the spectra is a specific and difficult problem. It is discussed in a number of books and reviews; unfortunately, most of them deal with porous adsorbents and seldom contain information matching our present interest. An exception is the recent application of the inverse gas chromatography for characterization of heterogeneous surfaces [96], which might help in solving our specific problems. The method consists of careful measurements of the profiles of the elution curves under a variety of conditions. The data are decoded in terms of equilibrium and kinetics of adsorption. Bakaev, et al. [97] managed to also use the approach for nonporous adsorbents.

A principally different method is based on "thermal desorption." In it the temperature of an adsorbent–adsorbate sample is steadily raised and the desorption rate of the adsorbate is measured. The resulting graph is interpreted in the parameters of the adsorption heterogeneity of the surface. The method seems promising for problems of our interest, though to the best of our knowledge, it has not yet been used to study systems like ours.

5.5.1 Thermodynamic Parameters of Adsorption on Heterogeneous Surface

A spectrum of desorption energies from a heterogeneous surface may consist of a few narrow lines. In reality, it is seldom so, except maybe for an almost ideal metallic crystal with patchwise heterogeneity, which is illustrated in Fig. 5.20. In the Henry's region, the isotherm on a heterogeneous surface with the line spectrum of E_d^* values is the weighted sum of the corresponding partial isotherms. Most real spectra of E_d^* seem to be much more complicated, approaching continuous (as a rule, single-humped) probability density distribution. We denote them by $\rho(E_d^*)$, implying that this function is determined in an interval $[E_d^{\min}, E_d^{\max}]$. The only natural limitation is that real distributions do not extend to infinity (like e.g., the Gaussians). Generalization of the approach to the line spectra allows us to calculate an effective desorption energy E_d^{het} for the heterogeneous surface with a continuous spectrum of E_d^*. The appropriate equation is:

$$e^{\frac{E_d^{\text{het}}}{RT}} = \int_{E_d^{\min}}^{E_d^{\max}} e^{\frac{E_d^*}{RT}} \rho(E_d^*) dE_d^* \tag{5.71}$$

Figure 5.23 shows the results of numerical evaluation of the function under the above integral at certain temperature for several different $\rho(E_d^*)$. As could be expected, the difference between $E_d^{\max}$ and the E_d^* value at the peak of the curve critically depends on the behavior of $\rho(E_d^*)$ in the vicinity of $E_d^{\max}$. The characteristics of the curves are listed in Table 5.4. Distributions are peaking at E_d^* values smaller than the appropriate E_d^{het} because of the asymmetry of the curves. Notice that FWHM of Gaussian distribution is 2.355σ; thus, in the case of the two normal pdfs, the dispersion of the central part of $\rho(E_d) \times \exp(E_d/RT)$ equals that of $\rho(E_d^*)$.

An important result is that the effective desorption energy for the heterogeneous surfaces depends on temperature! Nevertheless, let us for a while abandon it and suppose that E_d^{het} does not depend on temperature. Another assumption will be that the entropy change is the same for all partial isotherms, independent of E_d^*. Then we take into account Eq. 5.3 for the experimental constant of adsorption and move to the characteristics of desorption from heterogeneous surfaces. It follows that the measurements yield an equilibrium constant, which is to be interpreted as the entropy factor multiplied by the expectation value of the desorption energy factor:

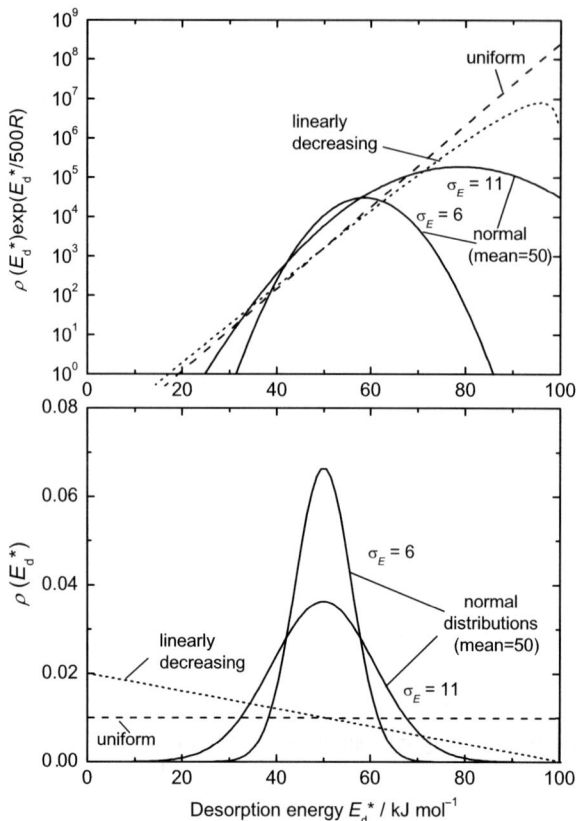

Fig. 5.23 Bottom: various $\rho(E_d^*)$, which were substituted into the integral in Eq. 5.71. Top: the functions $\exp(E_d^*/RT) \times \rho(E_d^*)$ under the integral at $T = 500$ K.

Table 5.4 Characteristics of the Distribution Functions of $\exp(E_d^*/RT) \times \rho(E_d^*)$ and the Values of E_d^{het} Calculated from Eq. 5.71 (all values are in kJ mol^{-1}).

Distribution (cf. Fig. 5.23)	400 K			500 K			600 K			2nd law[a]
	E_d^{het}	FWHM	E_d at peak	E_d^{het}	FWHM	E_d at peak	E_d^{het}	FWHM	E_d at peak	E_d^{het}
uniform	88.7	-	-	86.7	-	-	85.0	-	-	96.5
linearly decreasing	79.6	8.1	96.5	76.4	10.2	96.0	73.5	12.2	95.0	92.1
normal, $\sigma = 11$[b]	67.8	25.9	86.4	64.4	25.9	78.9	62.0	25.9	74.4	79.6
normal, $\sigma = 6$	55.4	14.13	60.8	54.3	14.13	58.8	53.6	14.13	57.3	59.2

[a] In the range 400 to 600 K
[b] The distribution was truncated at $E_d = 100$ kJ mol^{-1}

5.5 Evaluation of Adsorption Enthalpies on Real Surfaces

$$K_d^{het}(T) = e^{-\frac{E_d^{het}}{RT}} e^{\frac{\Delta_{des} S}{R}} \quad \text{if} \quad \frac{\partial E_d^{het}}{\partial T} = 0; \quad \frac{\partial \Delta_{des} S}{\partial T} = 0; \quad \frac{\partial \Delta_{des} S}{\partial E_d} = 0 \tag{5.72}$$

An illustrative example is the uniform density distribution of the desorption energy $\rho(E_d^*) = E_d^*/E_d^{max}$ for $0 \le E_d^* \le E_d^{max}$; cf. Fig. 5.23. For it, then the expectation value of the desorption energy factor is:

$$e^{E_d^{het}/RT} = \int_0^{E_d^{max}} \frac{e^{E_d^*/RT}}{E_d^{max}} dE_d^* = \frac{RT}{E_d^{max}} \left(e^{E_d^{max}/RT} - 1\right) \approx \frac{RT}{E_d^{max}} e^{E_d^{max}/RT} \tag{5.73}$$

It leads to

$$E_d^{het} = E_d^{max} - RT \ln(E_d^{max}/RT) \tag{5.74}$$

which, of course, agrees with the very first line of the data in Table 5.4.

Let us apply the simplest form of van't Hoff's equation to the data from various distributions of E_d^* given in Table 5.4. Notice that if Eq. 5.72 is valid, then the ratio of the equilibrium constants equals that of the desorption energy factors calculated from 5.71. Then the Second Law value of the effective E_d^{het} would come from:

$$\ln \frac{E_d^{het}(400\,K)}{E_d^{het}(600\,K)} = \frac{E_d^{het,2nd}}{R} \left(\frac{1}{400} - \frac{1}{600}\right) \tag{5.75}$$

The results are in the last column of Table 5.4. They show that employing the Second Law procedure gives higher values of E_d^{het} than the calculations based on the knowledge of $\rho(E_d^*)$. It is an interesting observation which deserves more attention.

Another possible approach to the evaluation of the required effective quantities is to start with the effective entropy change. Hill [98, 99] considered the statistical mechanics of the localized unimolecular adsorption on heterogeneous surfaces to propose formulae for the configurational entropy.

5.5.2 Adsorption Entropy on Heterogeneous Surfaces with Surface Diffusion

The next question is how to account for the entropy of surface diffusion on heterogeneous surfaces. While, in principle, $\rho(E_d^*)$ can be found from certain kind of experimental data, no measurements have yet been reported for the pdfs of diffusion barriers $\rho(E_b^*)$. Stallons and Inglesia [44] performed remarkable calculations of the potential energy surface for N_2 adsorbed on the bare (in particular, relaxed) surface of silica, which were illustrated by Fig. 5.12, and is also an illuminating picture of connectivity between the adsorption sites. Their calculations also described the diffusion barriers, and made it possible to follow surface migration of nitrogen using molecular-dynamic methods. This way the authors [44] found the effective activation energy to be 7.8 kJ mol^{-1}, that is about two-thirds of the average desorption energy of nitrogen. They concluded that the surface diffusivities are more sensitive to the chemical and geometric details of the surface than are the binding energies of the adsorbate. Their work well demonstrated that each adsorption site is connected to several adjacent potential wells over barriers of dissimilar height. In accordance with expectation, the migrating molecules mostly chose the paths over the lowest available barriers.

Let us try to estimate how the surface diffusion might affect the effective desorption constant and then desorption energy. The complicated situation with heterogeneous surfaces again necessitates making some assumptions and simplifications which do not completely agree with the conclusions of the previous paragraph. It seems reasonable to assume that the barrier height to be overcome is a function of the local (at the site) desorption energy. A suitable relationship may be a constant difference between the two quantities. For simplicity, not loosing much of generality, let $\rho(E_d^*)$ be truncated right at the value of the deduction; it means that we assume $E_b^* = E_d^* - E_d^{min}$. Then $\rho(E_b^*)$ extends from 0 to $E_d^{max} - E_d^{min}$ and $\rho(E_b^* + E_d^{min}) = \rho(E_d^*)$.

Pursuing more realistic formulae for calculation of the adsorption enthalpy, we will incorporate the mixed adsorption entropy S_a^{ml}/R. From Eqs. 5.42 and 5.70 it follows that the entropy change for the each partial adsorption isotherm is:

$$\frac{\Delta_{des} S^{ml}}{R} = \frac{\Delta_{des} S^{lc}}{R} + e^{-\frac{E_d^* - E_d^{min}}{RT}} \ln\left(16e^3 \frac{v_{xy}^2}{C_a u_m^2}\right) = \\ = \ln\left(\frac{C_a u_m^3}{64 e^{5/2} v^3}\right) + e^{-\frac{E_d^* - E_d^{min}}{RT}} \ln\left(16e^3 \frac{v_{xy}^2}{C_a u_m^2}\right) \quad (5.76)$$

Let the combined quantities under the logarithms in the above formula be B_{lc} and B_{mb}, respectively. Then the equilibrium constant, cf. Eq. 5.72, can be evaluated from:

$$K_a^{het,ml} = \frac{B_{lc}}{(E_d^{max} - E_d^{min})} \int_{E_d^{min}}^{E_d^{max}} e^{\frac{E_d^*}{RT}} \exp\left[\ln B_{mb} \cdot e^{-\frac{E_d^* - E_d^{min}}{RT}}\right] dE_d^* \quad (5.77)$$

It requires numerical integration. Another possibility is to suppose that the barrier is a constant fraction b_b of the desorption energy, $E_b^* = b_b E_d^*$; then no condition is imposed on E_d^{min}. The resulting formula, analogous to 5.77, is equally cumbersome.

A simpler, though non-rigorous way to account for the combined effect of heterogeneity and surface diffusion is to ignore the functional relationship between $E_d{}^*$ and $E_b{}^*$ and replace the integral in Eq. 5.77 with the product of the expectation values of E_d and E_b. The formulae derived this way will be valid for a homogeneous surface consisting of sites with equal desorption energy and equal barriers between them. The only step forward is that the characteristics of the actual heterogeneous surface take part in the evaluation. It would require more analysis to reveal whether doing so would create a larger error in the final results than the uncertainties resulting from other assumptions and simplifications.

In the case of the constant deduction and the uniform $\rho(E_d^*)$ we come this way to:

$$e^{\frac{E_b^{het}}{RT}} = e^{-\frac{E_d^{min}}{RT}} \int_{E_d^{min}}^{E_d^{max}} \rho(E_d^*) e^{\frac{E_d^*}{RT}} dE_d^* = e^{\frac{E_d^{het}}{RT}} e^{-\frac{E_d^{min}}{RT}} \qquad (5.78)$$

When the barrier height is proportional to the desorption energy and $\rho(E_d^*)$ is again uniform, then:

$$e^{\frac{E_b^{het}}{RT}} = \int_{E_d^{min}}^{E_d^{max}} \rho(E_d^*) e^{\frac{b_b E_d^*}{RT}} dE_d^*$$

$$= \frac{RT}{(E_d^{max} - E_d^{min}) b_b} \left(e^{\frac{b_b E_d^{max}}{RT}} - e^{\frac{b_b E_d^{min}}{RT}} \right) \approx \frac{RT}{(E_d^{max} - E_d^{min}) b_b} e^{\frac{b_b E_d^{max}}{RT}} \qquad (5.79)$$

Now, the effective equilibrium constants for the two variants of the dependence of E_b^* on E_d^* are:

$$K^{het,ml} = e^{\frac{E_d^{het}}{RT}} \exp\left[\frac{-\Delta_{des} S^{lc}}{R} + \ln B_{mb} \cdot e^{-\frac{E_b^{het}}{RT}} \right]$$

$$= e^{\frac{E_d^{max}}{RT}} \frac{B_{lc} RT}{E_d^{max} - E_d^{min}} \exp\left[\ln B_{mb} \frac{E_d^{max} - E_d^{min}}{RT} e^{-\frac{E_d^{max} - E_d^{min}}{RT}} \right] \qquad (5.80)$$

and

$$K^{het,lm} = e^{\frac{E_d^{max}}{RT}} \frac{B_{lc} RT}{E_d^{max} - E_d^{min}} \exp\left[\ln B_{mb} \frac{(E_d^{max} - E_d^{min}) b_b}{RT} e^{-\frac{b_b E_d^{max}}{RT}} \right] \qquad (5.81)$$

5.6 Revised Approach to Interpretation of the Data on Transactinoid Halides

5.6.1 Microscopic Picture of the Modified Silica Surface

The actual picture of the surface of fused silica described in the preceding sections differs much from that explicitly or implicitly assumed earlier. We will try to apply this knowledge for analysis of the experiments on gas-solid chromatography of halides and oxyhalides of the first transactinoid elements in fused silica columns. By way of example, we shall frequently use $ZrCl_4$ or $ZrBr_4$ as the adsorbate and $TiCl_4$ or BBr_3, respectively, as the halogenating and modifying agent. For typical values of the involved quantities we shall take: $C_a = 3 \times 10^{14}$ cm^{-2}; $\nu_0 = 10^{13}$ s^{-1}; $\nu_{xy} = 1.5 \times 10^{12}$ s^{-1}; and $\nu = 2.5 \times 10^{12}$ s^{-1}. Let the energy of desorption be $E_d \approx 80$ kJ mol^{-1}, so that $E_d/RT \approx 20$ at 500 K.

A most important conclusion is that the vapors of all halides and halogenating agents, which are commonly used for stabilizing certain chemical forms of the tracers, seem to modify the surface of the column to a great extent. They react with the silanol and siloxane groups to produce a variety of results: a halogen atom becomes directly attached to a silicon atom on the surface; a grafted molecule of the polyatomic agent (like BBr_3 or $TiCl_4$) replaces hydrogen in a silanol, and hydrogen atoms of two close silanols are replaced by a grafted molecule. All this is schematically shown in the right side of Fig. 5.19. Notice that the $-OTiCl_3$ group has a much larger volume and cross section than the original hydroxyl. It means that, on average, only 2.8 such groups per 100 $Å^2$ suffice to cover the surface and prevent further chemisorption merely for geometrical reasons [72], even if not all the available silanol groups have reacted. The molecules of heavy element chlorides are still larger than $-OTiCl_3$; hence, they would be all the more prevented from chemisorption. The cross section of a chlorine atom attached to the surface silicon is probably also significantly larger than that of $-OH$.

Therefore, when interpreting the gas-solid chromatography experiments with the transactinoid halides, we should allow for the modification of the column surface. Let us imagine a truly smooth, flat homogeneous silica surface fully covered by the directly attached halogen atoms and fragments of a halogenating reagent. The energy required to desorb a dissimilar molecular halide from such a surface might be expected to equal a fraction, half or so, of the vaporization energy. It is so because only a few halogens bonded to the surface atoms are involved in the interaction with the adsorbate molecule. However, as was suggested in [65] and illustrated by Fig. 5.16, the rough real surfaces should also contain, besides other structures, geometric wells or nanopores, a few molecular diameters in width and depth. Now, if the surface is deeply modified, a molecule adsorbed in such a well will be in a situation which closely resembles that in the bulk liquid or solid. Hence, a geometric well appears to also become a potential pocket. At some sites the analogy may be so close that the energy of desorption can be similar to the energy of vaporization or even sublimation. However, it seems that it cannot be higher! The geometry is visualized in Fig. 5.24, which shows the real relative sizes of the ions involved in the discussed picture.

5.6.2 Rationale for the Correlation of Adsorption and Sublimation Energies

The above considerations concerning the special potential wells might appear to be the required rationale for the similar values of energies, which characterize macro and microvolatility. In reality, the deepest adsorption pockets should not be very abundant. Let us estimate whether there is enough time to establish equilibrium for a nuclide living a second or so. In the experiments, the characteristic temperature (at the TC peak or the 50 percent yield in IC) is such that $E_d/RT \approx 20$; then the mean adsorption sojourn is $10^{-12} \times \exp(\varepsilon_d/k_B T) \approx 10^{-4}$ s. Assume that the surface

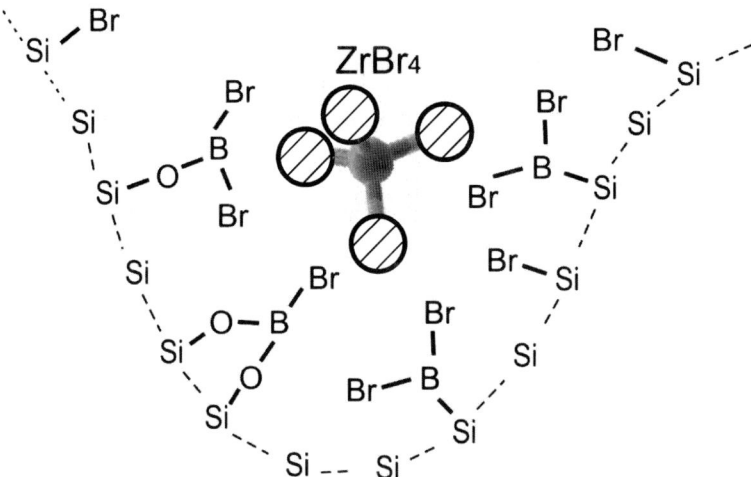

Fig. 5.24 Realization of the maximum possible adsorption energy of ZrBr$_4$ on silica surface, fully modified by BBr$_3$.

oscillations perpendicular and parallel to the surface have the same frequency and that $\varepsilon_b = 0.5\varepsilon_d$. Then each molecule of the adsorbate visits some 10^4 adsorption sites before it desorbs from the surface. As we saw earlier, the effective desorption energy on a heterogeneous surface is close to the maximum in the spectrum. We have expected the latter to equal the sublimation energy. Thus, we may believe that we found the sought-for rationale. However, thus far, we have taken into account only superficially outlined selected features of the novel picture of the column surfaces. The problem seems to call for more careful examination.

Remember, in the first place, that the adsorption enthalpies in the discussed correlations were evaluated upon postulating strict and exclusive validity of the mobile adsorption model. Now we must analyze for each particular system whether one of the two simplest models of adsorption prevails or the situation is intermediate. On a truly smooth, flat and fully modified homogeneous silica surface, which was mentioned above, the surface diffusion barrier might be low. At least it seems to be so when imagining a large, nearly spherical molecule resting on a flat array of almost closely packed smaller spherical halogen atoms. However, the inherent heterogeneity of vitreous silica and its surface makes such an ideal picture improbable. If ever, it probably takes place only on small patches and does not prevail. Notice that the surface diffusion of the completely mobile tracer (two-dimensional gas) in long lasting IC or TC experiments would considerably broaden the profile of the peak because $D_{1,2}$ must be about the same as in three-dimensional gas, some 0.3 cm^2 s^{-1}. The diffusion coefficient is much less affected by the surface temperature than the adsorption sojourn time. Hence, in one-hour experiments, the mean diffusion displacement could reach several centimeters. There have not been any reported experimental data that would suggest such rapid diffusion. Our conclusion is that, to evaluate the effective value of the entropy change, the localized model is preferable to the mobile one, though a contribution from the latter cannot be excluded. Then the actual entropy change in desorption is larger than that calculated for the mobile adsorption model, and the desorption energies reported to date are underestimated. According to Eq. 5.45, we have:

$$\frac{E_d^{lc}}{RT} - \frac{E_d^{mb}}{RT} = \ln\left(\frac{16e^3 \, v_{xy}^2}{C_a \, u_m^2}\right) \quad (5.82)$$

This holds for the isothermal chromatography. Turning to thermochromatography, we can make use of Eq. 5.58. By analogy with it, while recalling the equations for heterogeneous surfaces, we obtain

$$\frac{E_d^{eff}}{RT_A} = \frac{E_d^{max}}{RT_A} - b_{het} \approx 1.047 \left[\ln \frac{7.85 t_R^{TC} Q_0 g}{a_c T_0} + \frac{\Delta_{des} S^\circ}{R} \right] \quad (5.83)$$

where the positive b_{het} depends on the concrete $\rho(E_d^*)$. Its value, as was exemplified by Eq. 5.74, probably does not exceed a few units. Equation 5.83 shows that the experimental effective desorption energy similarly depends on the accepted adsorption model, as in the case of IC data.

It seems that $\approx 10RT$ is the upper limit for the difference $E_d^{lc} - E_d^{mb}$, which follows from Eq. 5.82. In typical conditions, for $E_d = 80 \, \text{kJ mol}^{-1}$, the difference makes $40 \, \text{kJ mol}^{-1}$. This and related figures indicate, at least at the first sight, that the actual adsorption energies of the molecular halides may be considerably larger than those plotted in Fig. 5.3. If so, our superficial basis for the sought-for rationale can be highly questioned.

Another factor which might make the desorption entropy larger than its simple $\Delta_{des} S^{mb}$ value, is our ignoring of the inner degrees of freedom when evaluating the entropy change in adsorption. It is generally recognized that rotation especially may be hindered in the localized adsorption state. There are large reserves of this sort of entropy in the gaseous state; for example, the molar internal entropies (they are absolute) S_{int}/R of the gaseous $ZrCl_4$, and $ZrBr_4$ are 22.7 and 28.1, respectively, at 298.15 K. Each unaccounted-for unity subtracted from S_{int}/R by the adsorption makes the calculated energy of desorption underestimated by additional $1.047RT$; cf. Eq. 5.83. This possibility is commonly ignored because it is not clear how to estimate the losses of the internal entropy in the adsorption processes of our concern.

Finally, we possibly overestimate C_a. Then according to Eq. 5.82, we again obtain E_d smaller than its accurate value. Though we deal with a continuous $\rho(E_d^*)$, we must see it as a sum of partial isotherms. The latter have been supposed to differ from each other only by the energy of desorption and by the abundance; the C_a values have been taken equally. However, it may not be true in the case of the potential pockets shown in Fig. 5.24. Such sites must have larger size than others and, if tightly arranged, their number per unit area would be much less than a typical value of $3 \times 10^{14} \, \text{cm}^{-2}$ accepted earlier. However small the percentage of the pockets in the E_d^* spectrum might be, the contribution of the particular isotherm to the overall configurational entropy is similar to its weight in determining E_d^{eff}. A factor of 10 in C_a changes E_d^{eff} by $2.3RT$.

In the meantime, leaving the typical vibration frequencies of the localized adsorbate in Eq. 5.82 unchanged when considering the special pocket sites might result in over-estimating E_d^{eff}. The calculated values of ν_0 cannot be directly used for ν_{xy} because the latter is generally somewhat less than ν_0. Moreover, now the adsorbate contacts with a layer of grafted modifier rather than with silica. The values straightforwardly calculated from Eq. 2.22 are $\nu_0 = 1.05 \times 10^{13}$ s^{-1} for silica, 2.0×10^{12} s^{-1} for solid TiCl$_4$ and 1.4×10^{12} s^{-1} for BBr$_3$. The silica frequency is markedly higher than that of the halides –modifiers, and if it does not dominate in desorption, the calculated E_d will be correspondingly overvalued.

An uncertain value in the experimental equilibrium constant is a_z. It is usually underestimated by taking it equal to the apparent (geometrical) surface of the column of unit length. Actually a_z can only be larger and so it makes the true E_d^{eff} smaller.

5.6.2.1 Why can E_d^{eff} be Larger than $\Delta_{\text{sub}} H$?

The above-discussed factors are clearly dominated by those which add to the probability that E_d^{max} does considerably exceed E_{sub} of the compound. Thus, we must discuss the fundamental phenomena which might be responsible for it.

For a fully modified surface, we should consider reactions like:

$$\equiv\text{Si-O-TiCl}_3 + \text{ZrCl}_4 = \text{TiCl}_4 + \equiv\text{Si-O-ZrCl}_3 \text{ or}$$
$$\equiv\text{Si-O-BBr}_2 + \text{ZrBr}_4 = \text{BBr}_3 + \equiv\text{Si-O-ZrBr}_3.$$

This way, the tracer would locally take the place of the modifier through a sort of exchange reaction. The energetic effects of such reactions can be estimated from the regularities in energies of the appropriate chemical bonds. The required thermochemical data for zirconium tetrahalides can be found in Ref. [100], for other halides in Refs. [101, 102]. The following relations between the bond energies can be noted:

- The M-Cl bond dissociation energy is by some 50 to 70 kJ mol^{-1} larger than that of M-Br.
- The energy of a single M-O bond is close to that of M-Cl and higher than the energy of M-Br.
- The enthalpy change of the gaseous reactions MX$_4$ = MX$_3$ + X at 500 K is 364, 447 and 479 kJ mol^{-1} for Ti, Si and Zr tetrachlorides, respectively, as well as 299, 370 and 412 kJ mol^{-1} for the corresponding tetrabromides; for BBr$_3$ = BBr$_2$ + Br, it is 404 kJ mol^{-1}.

From this side, the processes seem possible because the absolute value of the energy of replacing the modifier must be rather small. However, the bonds to be broken and then restored are several times stronger than the desorption energies of the involved molecules. Because of this, such reactions may experience severe steric hindrance and require large activation energy; combination of these factors makes the process hardly probable.

Reactions like

$$\equiv\text{SiCl} + \text{ZrCl}_4 \rightarrow \equiv\text{Si-ZrCl}_3 + \text{Cl}_2$$

seem impossible because, to date, any molecules with Si–Zr bonds have not been reported.

5.6.2.2 Incomplete Modification of the Surface

The studies performed so far did not explicitly care about the modification as such; the papers did not describe preconditioning of the columns. Meanwhile, the most probable origin of the huge desorption energies seems to be incomplete modification of the surface. We rule out the possibility of chemical interaction of the tracer with the surface OHs — one cannot expect that the silanols, if sterically available, would be left intact by the modifier. In the meantime, we recall the work of Stallons and Inglesia [44] who calculated the spectra of the energy of desorption of CH_4, N_2 and Ar from the bare surface of fused silica. Their results are shown in Fig. 5.25. Both the span of the spectra and the most abundant energies, when compared with E_{sub} of the adsorptives, are remarkable. Notice that the $\rho\left(E_d^*\right)$ for CH_4, N_2 and Ar are peaking at about 9, 11 and 12 kJ mol^{-1}, while the corresponding values of E_{sub} are 9.6, 6.9 and 7.7 kJ mol^{-1}, respectively. Thus, even the most probable values of E_d^* are equal or larger than E_{sub}, and the spectra extend to some twice-as-large energies.

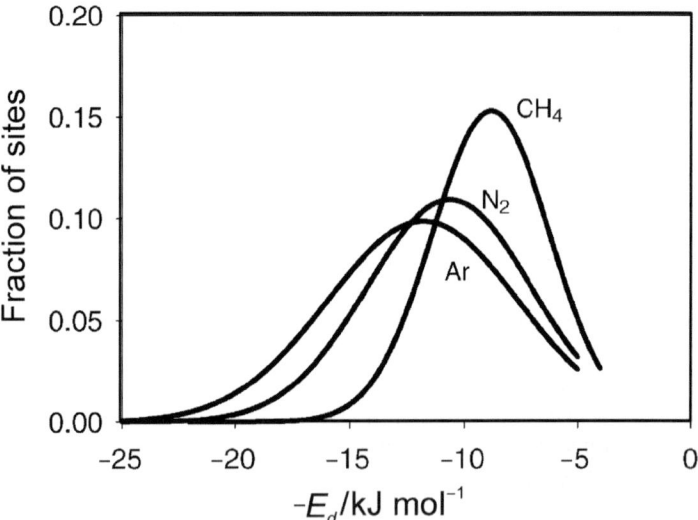

Fig. 5.25 Adsorption energy distributions for CH_4, N_2, and Ar on a relaxed simulated silica surface [44].

Reprinted from Chemical Engineering Science, 56(14), Stallons JM, Inglesia E, Simulations of the structure and properties of amorphous silica surface, 4205–4216, © 2001, with permission from Elsevier.

The simulations were done using the Lennard–Jones potential of molecular interactions. Interaction with the silicon atoms could be neglected; most binding energy came from the oxygen atoms of silanols and siloxanes. The data for the average desorption energies in Henry's law region reasonably agreed with the available experimental values for CH_4, N_2 and Ar.

We cannot just multiply the above data by the ratio of sublimation energies to obtain estimates for the adsorbates like $ZrCl_4$ on bare surface. However, we can expect that, on an incompletely modified silica surface, a closer contact of the tracer molecules with oxygen atoms will result in stronger attractive forces than those of physisorption on the fully modified surface. It might be the most fundamental rationale for the excess of E_d^{max} over the sublimation energy, provided that the modification of the surface is incomplete.

5.6.3 Required New Experimental Data

The above considerations may seem too speculative. However, it is necessary to emphasize the uniqueness of the goal. Adsorption on heterogeneous surfaces, as such, remains an extremely complex, still poorly investigated problem. Meanwhile, we attempt to obtain the adsorption energy on a real surface having only single experimental value of the effective equilibrium constant. Such a task has never been undertaken before. We essentially try to solve it by calculating the entropy change from the first principles and subtracting it from the Gibbs free energy.

In the meantime, the whole field of transactinoid studies, with its peculiar and nontrivial features, is still young — if not by years since birth, then by the number of experiments performed to date. The major motivation — the quest for still new elements and demands for their chemical identification — greatly stimulated the development of the experimental instruments and techniques. In the first place, these were high efficiency spectroscopic low-level measurements of the particle radioactivity of short-lived nuclides in the specific conditions of chemical experiments. Necessarily, the fundamental chemical and physicochemical problems behind the employed methods, as well as evaluation of uncertainties of the results, have not been paid adequate attention. Much more could be learned (but was not) in off-line studies of long-lived radioisotopes of common elements with good statistics. As a result, some conclusions in the literature are not well founded, and important details of the experimental conditions are not given.

The present situation with the observed trends and with the precision of the data is better illustrated by Fig. 5.26 than by the simple linear correlations. It can be seen that the desorption energies (obtained by conventional approach) of very volatile molecular liquids tend to approach the sublimation energies (calculated for mobile adsorption), while those of more ionic liquids make only about two-thirds of $\Delta_{sub}H$. The scatter of the experimental $\Delta_{des}H$ values is very large. Evidently, if the data for less volatile and ionic liquids were not known, one would not insist on any linear correlation. The scatter of values obtained for a particular compound by different

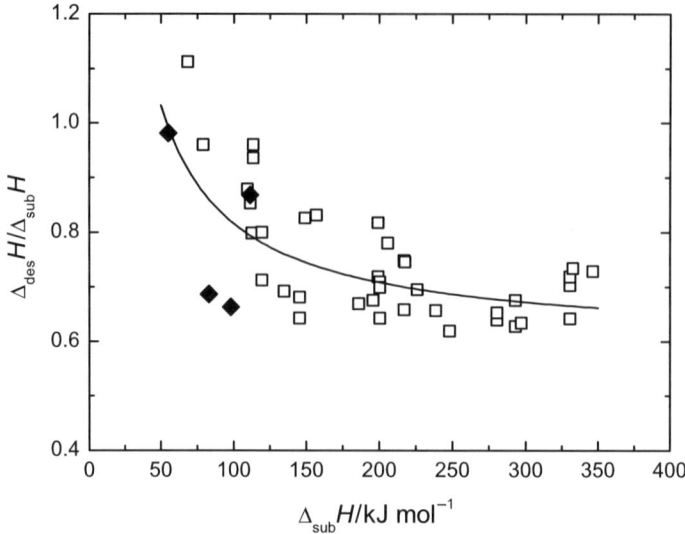

Fig. 5.26 Ratio of the desorption to sublimation enthalpy of various chlorides, versus the sublimation enthalpy. Solid curve is the linear correlation given by Eq. 5.61 in the novel coordinates. Black squares are the data used to evaluate the correlation. Rhombs are some data by Bächmann, et al. from Table 5.1.

researchers (some are included in Fig. 5.26) is similar to the scatter in parameters for different compounds of similar volatility. Particular cases cannot be analyzed here in detail.

A correlation between the parameters of adsorption and sublimation (vaporization) must not necessarily be a straight line. However, what it must show is nearly zero adsorption energy of the compounds with very low sublimation energy. It is not the case with our present day linear correlations; it is their fundamental flaw because there is no good reason for a constant contribution to the adsorption energy. These observations issue a strong warning that it is a dangerous misconception to take the correlations for functional relationships and make some far-going conclusions about the quantitative measures of the properties of new elements. What can be properly done is, for example, a comparative TC study of behavior of the new element and its actual homolog, if the elements can be produced simultaneously and have similar half-lives.

Pershina [103] believes that a better way to compare the volatility of analogous compounds is to take the ratio of the experimental adsorption equilibrium constants, rather than to evaluate the adsorption enthalpies. The ratio should be corrected for the involved half-lives, and the contribution from the difference in internal entropy of the compared molecules (owing to unequal masses of the metal atoms and different bond lengths) can estimated by calculations from first principles. The approach was used for analysis of the experimental data on hassium and osmium tetroxides, which were reported in [104].

5.6.3.1 How to Improve our Understanding of the Methods and Results?

On the experimental side, we have learned that the original surface of silica has variable morphology and inherent heterogeneity. Meanwhile, it is important to know the true surface area of our columns and avoid too specific a morphology. The tubes produced by different manufacturers show a broad variety of surface patterns. Quantitative characterization of the surface properties to compare different batches of capillaries is, at present, unrealistic. Even the measurement of the true surface area per unit lengths of the column is not easy because of the relatively small values. Moreover, there is not yet any agreed-upon technique or procedure to quantitatively characterize the roughness and heterogeneity of the surfaces.

Next, the qualitative and quantitative compositions of the carrier gas, especially as to the most chemically active components, have not been the same in different laboratories. As a result it is not known whether the surfaces were modified to a comparable degree in different studies. An even more important question is, how complete was the coverage of the surface by the grafted molecular fragments? These factors critically affect the resulting surface heterogeneity and the spectrum of the desorption energies.

While it is difficult to avoid all possible uncertainties, an obvious step forward would be a purposeful, careful conditioning of the columns before use. The data reviewed in Sects. 5.3.2 and 5.3.3 allow us to propose a tentative scheme of processing the tubes; it might include the following steps:

- Preliminary thorough cleaning by solvents, acids, etc.
- Chlorination of the surface by $SOCl_2$ – the most efficient reagent serving to cover the surface with chlorine atoms
- Hydrolysis by water to obtain the highest possible concentration of silanols
- Heating to remove physically adsorbed water
- Modification by reagents whose radicals have strong enough bonds with oxygen atoms, and
- Chlorination by $SOCl_2$ to remove OH groups remaining under possible "umbrellas" formed by the radicals attached as above

In the experiments with bromides, conditioning would require using $SOBr_2$ or $CO + Br_2$ to fulfil the last two steps. The conditioning of the metallic surfaces (nickel, copper?) will, after common cleaning, probably require reduction of the surface oxides by hydrogen gas, followed by treatment with gaseous halogen. Such surfaces will be covered only by halogen atoms, thus being in every respect simpler than the modified surface of silica; though being also heterogeneous, they promise better reproducibility and less variety in the appearance of the adsorption sites.

The pre-treatment of the columns might be followed by some standard IC or TC (off-line) experiments with easily available radionuclides in well-established chemical state to test whether the columns perform in a reproducible way. Implementation of such procedures in the various laboratories in the field would help to make the measurements more precise and to find the origin of eventual disagreements in the data.

With the columns "improved" as recommended above, it is desirable to repeat many off-line experiments performed earlier. To avoid necessity of extra corrections, it is preferable to simultaneously chromatograph as many elements as possible, and to run experiments with this mixture under various experimental conditions. In particular, because our major concern is the column surface, of great interest would be experiments in the columns made of metals or temperature-resistant fluorinated polymers. The polymeric columns remain unemployed in gas-phase radiochemistry. Their surface will probably not be modified, which can lead to unexpected phenomena or regularities.

Additional on-line experiments with simultaneously produced isotopes of an element with different half-lives are needed, to investigate the actual relationship between the (mean) temperature of the TC zones and the retention time.

To complete the topic, we should mention that correlations between the parameters of adsorption and sublimation were also observed for elements [105], oxides and hydroxides [106] and sulfides [107].

5.6.4 Real Picture of Adsorption and Monte Carlo Simulations

In the Monte Carlo simulations of the chromatographic and related processes we should also account for the new knowledge about the deep heterogeneity of surfaces, about the role of localized adsorption and about the occurrence of surface diffusion. Evidently, now the molecular desorption energy accepted for a concrete simulation is to be understood as the effective value over the spectrum of possible energies in the sense of Eq. 5.71. As such, it is related to a concrete form of the $\rho(E_d^*)$. The mean adsorption sojourn time of a molecule is obviously $\tau_0 \exp\left(E_d^{\text{eff}}/RT\right)$. The probability distribution of the elementary sojourn time and that in a random series of the adsorption–desorption events may become a mathematical problem. The surface diffusion, even if otherwise negligible, helps faster equilibration of molecules between phases and is a positive factor. It does not seem to affect the simulations directly.

Provided that the localized adsorption clearly prevails, it can be accounted for in a simple way — by introducing the effective vibration frequency $\nu_{\text{MC}}^{\text{lc}}$, which would make the mobile entropy equal to the localized one. From Eqs. 5.64 and 5.65 it follows that:

$$\frac{16e^3}{u_m^2} \frac{1}{\nu_{\text{MC}}^{\text{lc}}} = \frac{C_a}{\nu^3} \quad \text{and} \quad \nu_{\text{MC}}^{\text{lc}} = \frac{16e^3}{u_m^2} \frac{\nu^3}{C_a} \tag{5.84}$$

Thus $\nu_{\text{MC}}^{\text{lc}}$ is of the order of 10^{16}–10^{17} s^{-1}.

5.7 Non-trivial Mechanisms in Gas-Solid Chromatography

The problems met in evaluating the gas-solid chromatography data get more complicated when the underlying mechanisms are other than the reversible physisorption. It may be necessary to admit participation of novel compounds of known elements

5.7 Non-trivial Mechanisms in Gas-Solid Chromatography

which do not exist as condensed matter. It also concerns, of course, the expected and unexpected compounds of new elements. To rationalize the experimental findings by some mechanisms and chemical formulae, we need estimates of the thermodynamic state functions of the anticipated compounds in different states. Numerous recipes are available in the literature; choosing from them is a matter of taste. Some interesting ideas concerning the "nonexistent" (as solids or liquids) compounds can be found in an older book [108]. Most of the concrete estimates are based on empirical correlations, which are then interpolated or extrapolated. Choices of the appropriate "variable" in correlation are numerous, starting with the atomic number or the number of the period in the Periodic Table. Eichler and co-workers estimated a number of thermodynamic values for oxides and hydroxides [109] as well as for (oxy)halides [30, 110] on the basis of correlations with the sublimation energy of the metal. They have also published estimates of the adsorption energy of the atoms of known metals, as well as of superheavy nuclides (elements) [111] on a variety of metallic surfaces.

In the case of the simplest mechanism of repeated adsorption–desorption events of the unaltered molecules, the retention time and the peak shape are insensitive to the composition of the carrier gas — now we would add: "provided that it constantly modifies the column surface." Some of the more complex mechanisms of migration are indistinguishable "from the outside" because they are analogously insensitive. It is, for instance, simple chemisorption, when the initial electronic structure of the adsorptive substantially changes upon adsorption, but is restored at the desorption stage. Such is also the microscopic history of metallic adatoms in metal columns. Another case is the chromatography of molecular halides in columns loaded with alkali halides; the adsorbed state is a surface complex between the two halides; see Sect. 1.5.1. In both examples the structure of the original adsorption sites is not necessarily restored. It is, of course, unimportant in the experiments with tracers.

There are more possibilities. Some experimental results obtained in IC and TC studies of (oxy)halides of common elements looked irregular. Eichler [112] interpreted these observations as the result of reversible chemical reactions superimposed on the adsorption – desorption mechanism of the compounds. "Reaction gas chromatography" [113] seems to be an appropriate general term, close in definition to the corresponding IUPAC recommendation on terminology [114]. According to Ref. [112], there are different varieties of such processes, which are characterized below. Needless to say, the equations below are only for homogeneous surfaces; the accepted standard state for the adsorbed species is again $V/A = 1\,\text{cm}$, standard concentration for bulk gaseous reagents is that at $p = 1$ bar.

5.7.1 Dissociative Adsorption – Associative Desorption

This mechanism exists when, for example, a gaseous halide (or oxide) upon adsorption releases a halogen (oxygen) atom and desorbs when regenerated. It successfully describes the TC behavior of the chlorides of Pu, Cr, Ru and Ce, and recently also Bk [115] in gases of various compositions. The first observation of volatilization of plutonium chloride tracer in the atmosphere of chlorine in tracer studies [116] was quite unexpected. Generally, in the absence of strong chlorinating agents, the TC deposition temperature of Pu was rather high — in the range typical for known trichlorides of related elements. Meanwhile, with stronger chlorinating agents (e.g., $CCl_4 + Cl_2$) in the gas, the adsorption peaks unexpectedly shifted to lower temperatures, which are more characteristic for tetrachlorides. The trend was most pronounced for Pu and smallest for Ce. The proposed transportation process was:

$$PuCl_4(g) \leftrightarrow PuCl_3(a) + 1/2\, Cl_2(g)$$

The resulting formula is presented here in the style of Eq. 5.54 to show the change due to the chemical mechanism. So,

$$Ei^* \left(\frac{-\Delta_r H}{RT_A} \right) = \frac{t_R Q_0 g}{a_z T_0} \exp\left(\frac{-\Delta_r S}{R} \right) \left[\frac{c(Cl_2)}{c^\circ(Cl_2)} \right]^{1/2} \qquad (5.85)$$

where $c(Cl_2)$ is the concentration of gaseous chlorine and $c^\circ(Cl_2)$ is its standard value.

Mechanisms of this type seem to also occur under certain conditions in thermochromatography of platinum metal oxides. In particular, the reaction

$$IrO_3(g) \leftrightarrow IrO_2(a) + 1/2\, O_2(g)$$

was discussed [109].

Rather extensive studies have been devoted to the reaction chromatography of volatile hydroxides. They are characteristic for elements of groups six to eight and were experimentally detected for a TAE – seaborgium. Vahle, et al. [113, 117] developed the corresponding equations; for Mo and W they considered the process:

$$MoO_2(OH)_2(g) \leftrightarrow MoO_3(a) + H_2O(g)$$

They attempted its Monte Carlo simulation by the appropriately modified approach described in Sect. 4.3.2. The change replaced τ_0 by an effective value τ_0^{ds-ad} (much longer than τ_0). The latter depends on the frequency of collisions of the adsorbed MoO_3 with H_2O molecules, and so on the concentration of the water vapor (see Eq. 2.7 for n_a) and entropy change in the reaction. Then the "reaction sojourn" time is:

$$\tau_a = \tau_0^{ds-ad} e^{-\frac{\Delta_r H}{RT}} = \frac{4}{u_m(H_2O)} \frac{c_a^\circ(MoO_3)}{c_g(H_2O)} e^{-\frac{\Delta_r S}{R}} e^{-\frac{\Delta_r H}{RT}} \qquad (5.86)$$

The experimental and simulated profiles of TC zones reportedly agreed well.

The dissociative adsorption – associative desorption type of reaction also takes place when Ln or An trichlorides are chemically volatilized into the carrier gas containing a rather large concentration of Al_2Cl_6 vapors. Chromatography evidently proceeds due to the equilibria like:

$$La(Al_2Cl_7)_3(g) \leftrightarrow LaCl_3(a) + 3\, Al_2Cl_6(g)$$

The stoichiometry of the gaseous complex can vary for any particular element, depending on the partial pressure of Al_2Cl_6 and on the temperature. The complexes have been studied quite exhaustibly, though the only application in radiochemistry has yet been the spectacular IC interseparation of lanthanoids [118] and actinoids [119]; see also Sect. 1.5.2.

5.7.2 Associative Adsorption – Dissociative Desorption

This mechanism seems to be behind the unexpectedly high deposition temperature of Ag in a chlorinating gas [112]. This data, and the fact that AgCl is highly dissociated in the gaseous state, allows us to conclude that the reaction thermochromatography proceeds through

$$Ag(g) + \tfrac{1}{2} Cl_2(g) \leftrightarrow AgCl(a)$$

that is, by attachment of a chlorine upon adsorption. Now, the formula is:

$$Ei^* \left(\frac{-\Delta_r H}{RT_A} \right) = \frac{t_R Q_0 g}{a_z T_0} \exp\left(\frac{-\Delta_r S}{R} \right) \left(\frac{c^\circ(Cl_2)}{c(Cl_2)} \right)^{1/2} \qquad (5.87)$$

5.7.3 Substitutive Adsorption – Substitutive Desorption

Under certain experimental conditions this mechanism affects the thermochromatographic behavior of $ZrCl_4$– the compound with its place in the history of transactinoid chemistry. It was anticipated, and the very first data indicated, that thermochromatography of $ZrOCl_2$, would not be possible, at least at moderate temperatures. The basis for such judgment was the expected low volatility, characteristic of the halides of metals MX_3. Moreover, the bulk solid oxychloride $ZrOCl_2$ does not sublime because it disproportionates into dioxide and tetrachloride. In an IC column, with $SOCl_2$ as the chlorinating agent in the carrier gas, radionuclides of zirconium deposited at a low temperature, which conclusively evidenced formation of $ZrCl_4$. However, the progressive addition of free oxygen to the gas resulted in a gradual rise of the deposition temperature, rather than an abrupt change [120]. Later [121] such regularity was also observed for Hf and Rf. In Ref. [112], these findings were explained by suggesting that $ZrOCl_2$ molecules exist only in the adsorbed state and are involved in the reversible reaction:

$$ZrCl_4(g) + \tfrac{1}{2} O_2(g) \leftrightarrow ZrOCl_2(a) + Cl_2(g)$$

The basic relationship is now:

$$Ei^* \left(\frac{-\Delta_r H}{RT_A} \right) = \frac{t_R Q_0 g}{a_z T_0} \exp\left(\frac{-\Delta_r S}{R} \right) \frac{c(Cl_2)}{[c^\circ(Cl_2) \cdot c(O_2)]^{1/2}} \qquad (5.88)$$

Evidently, because the gaseous reagents are involved in the adsorption mechanism, the behavior of the elements is smoothly affected by the quantitative composition of the gas. It would not be the case if the reagent merely irreversibly chlorinated the tracer.

5.7.4 Physical Adsorption – Substitutive Desorption

An interesting phenomenon was observed by the Dubna group in the course of work on thermochromatographic chemical identification of the 0.9-s ^{263}Sg with SOCl$_2$ as the chlorinating agent. The transactinoid yielded a distinct peak around 250 °C, while the 2.5-h ^{176}W, the only tungsten activity which could be watched simultaneously, was deposited as low as about 100 °C. Such a large difference in the deposition temperatures could not be ascribed to the difference in mean lifetimes. It seemed to indicate dissimilar chemical states of the congeners Better understanding of the chemical processes was expected from separate experiments on comparative TC behavior of short-lived $^{164-169,174}$W. These were simultaneously obtained by the reaction ^{24}Mg + ^{144,147,nat}Sm [122]. Their thermochromatograms were measured after a long on-line experiment. For the shortest-lived activities it was possible through detecting radiation of their Ln descendants. The latter did not migrate because of the low volatility of trichlorides. Except for ^{176}W, two zones of adsorption were observed — at about 250 °C and around 160 °C. The distribution of different isotopes of W between the zones correlated with their half-lives; namely, the percentage in the low temperature zone increased with increasing $t_{1/2}$. The isotopes with half-lives of a few seconds decayed to >50 percent in the first zone, while the activities with $t_{1/2}$ > 1 min were adsorbed mostly as the second zone; see Fig. 5.27. This seemed to provide evidence of the formation of two compounds, the less volatile of which reacts relatively slowly in the adsorbed state to form the more volatile one. The characteristic time of the transformation evaluated from the data was 45 ± 15 seconds (1σ). To assign the chemical states, the volatility of known (oxy)halides of tungsten were taken into account, the thermodynamic equilibria were calculated for the system and some fundamental regularities in the kinetics of the supposed reactions.

The otherwise inert carrier gas contained a few volume percents of both SOCl$_2$ and O$_2$. The thermodynamic equilibrium calculations predict complete oxidation of SOCl$_2$ to SO$_3$ + Cl$_2$, and WO$_2$Cl$_2$ as the chemical state of W. However, the gas exiting the column showed a very small degree of oxidation or pyrolysis of SOCl$_2$, evidently because the latter passed the column in mere 0.3 seconds. Thus, thermodynamic considerations proved to be of limited use in this particular case. The "equilibrium preference" for WO$_2$Cl$_2$ over the almost equally stable WOCl$_4$ follows from the reaction

$$WOCl_4(g) + 0.5O_2(g) = WO_2Cl_2(g). + Cl_2(g),$$

which is characterized by $\Delta_r G / \text{J mol}^{-1} = -40,000 - 80T$. Because SOCl$_2$ was preserved in the real carrier gas, the reaction

5.7 Non-trivial Mechanisms in Gas-Solid Chromatography

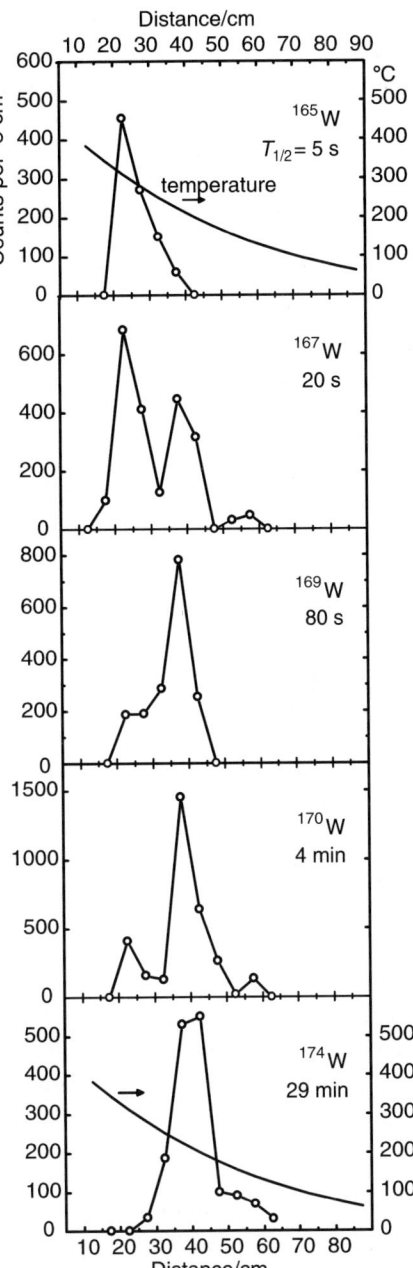

Fig. 5.27 Deposition of short-lived tungsten isotopes along a TC column as measured through their lanthanoid descendants [122].

Reproduced from Czechoslovak Journal of Physics, 49 (Suppl 1/II), Lebedev VYa, Yakushev AB, Timokhin SN, Vedeneev MB, Zvara I, 589–595, © 1999, with permission of the Institute of Physics, Academy of Sciences of the Czech Republik.

$$WO_2Cl_2(g) + SOCl_2(g) = WOCl_4(g) + SO_2(g)$$

with $\Delta_r G / \text{J mol}^{-1} = -45{,}000 + 22T$ is to be considered as well.

Fig. 3.4 displayed the enthalpy data for the bimolecular reactions of Mo and W species with the actual active components of the carrier gas ($SOCl_2$ and O_2), which might determine the chemical state of the atoms. Remember that, in most cases, the negative enthalpy means that the reaction is fast enough for the present purpose; see Sect. 3.2.2. Obviously, WCl_6 cannot emerge when there are many exothermic ways to convert the intermediate W(IV) compounds into WO_2Cl_2 or $WOCl_4$ as the end products. However, the thermodynamic preference for WO_2Cl_2 is due to the formally trimolecular reaction $2WOCl_4 + O_2 = 2WO_2Cl_2 + 2Cl_2$. As such, it cannot play any role from the point of view of kinetics, because two tracer molecules never meet. A bimolecular mechanism in principle exists; it is $WOCl_4 + O = WO_2Cl_2 + Cl_2$. However, in the particular experimental conditions, the maximum degree of dissociation of O_2 could be only about 10^{-15}. It is by some six to seven orders of magnitude less than the level which would ensure some encounters of the tracer molecule with a free oxygen atom. Thus, the conclusions from the thermodynamic analysis must be drawn with caution.

As to the bulk volatility of the oxychlorides, the equilibrium vapor pressure of $WOCl_4$ at 500 K is at least 100 times larger than that of WO_2Cl_2. The above considerations corroborate the suggestion that the initial stage of the TC processing is the formation of WO_2Cl_2. The compound has relatively high deposition temperature and, when adsorbed, slowly reacts to yield the more volatile oxychloride. That is,

$$WO_2Cl_2(g) \rightarrow WOCl_2(a)$$

is followed by:

$$WO_2Cl_2(a) + SOCl_2(g) = WOCl_4(g) + SO_2(g)$$

Thus, the process is seen as physical adsorption with subsequent substitutive desorption. The characteristic time of the reaction could be roughly evaluated from the ratio of areas of the two peaks as 45 ± 15 s (1σ). It indicates an activation energy of the reaction of 80 kJ mol^{-1} or so. The enthalpy of the gaseous chlorination of WO_2Cl_2 is -40 kJ mol^{-1} and an estimate for the adsorption enthalpy of WO_2Cl_2 is about -100 kJ mol^{-1}; these three values are consistent with the proposed mechanism. The reason why the chlorination does not proceed in gas, though, is its quite exothermic may be the short time, not more than 0.1 seconds.

The data for short-lived tungsten nuclides were subject to Monte Carlo simulations [29]. The original model was appropriately changed to describe the slow transmutation of the "primary" compound into a more volatile one in terms of the activation energy. Both homogeneous and heterogeneous surfaces were assumed. The simulations described the overall experimental picture reasonably well; a better agreement was achieved for the heterogeneous surfaces by fitting their characteristics to the experimental data.

5.7.5 Existence of Yet Unknown Compounds

The most difficult problem in the radiochemistry of TAE is possible formation of a compound which has no analogs in chemistry of the known elements. The only example of a satisfactory solution to a similar problem in modern radiochemistry was the assignment of a cationic form of astatine [123]. It had required many years of experimentation despite the fact that the utilized At isotope had convenient half-life and was readily available in high enough activity to allow thorough experimentation.

The researchers of a TAE would certainly enjoy finding that it forms a unique compound, not observed in chemistry of congeners. However, in practice, it would produce a stalemate situation with insuperable barriers, merely because it would be impossible to realize the required great number of different chemical experiments. Thus, with the TAEs, we are — and, in the near future, will be — concerned mostly with quantitative measurements of not too large differences in common properties of the congeners. Rare exceptions might be the experiments that simply ask, "yes or no?" Still, we will have to be cautious about statistics.

A little easier situation emerges when a compound of the element seems to obey the simplest gas-solid chromatography mechanism, but behaves differently from its expected homolog(s), approaching properties of a certain compound of a non-congener. Unfortunately, it might also result from a nontrivial mechanism of chromatography, like those discussed in the previous section. Reported conclusions about violation of the expected homology have been scarce and even scarcer have been the attempts to confirm or reject such claims in independent studies; they were mentioned in Chapter 1.

References

1. Merinis J, Boussieres G (1961) Anal Chim Acta 25:498
2. Chuburkov YuT, Zvara I, Shilov BV (1969). Radiokhimiya 11:174
3. Zvara I, Chuburkov YuT, Belov VZ, Buklanov GV, Zakhvataev BB, Zvarova TS, Maslov OD, Caletka R, Shalayevsky MR (1970) J Inorg Nucl Chem 32:1885; Soviet Radiochem 12:530; Radiokhimiya 12:565
4. Eichler B, Zvara I (1982) Radiochim Acta 30:233
5. Gäggeler HW, Jost DT, Baltensperger U, Weber A, Kovacs A, Vermeulen D, Türler A (1991) Nucl Instr Meth Phys Res, A 309:201
6. Rudolph J, Bächmann K (1979) J Chromatogr 178:459
7. Zvara I (1973) In: XXIVth Internat congress pure applied chem, vol 6. Butterworth, London, p. 73
8. Zvara I, Belov VZ, Domanov VP, Shalaevski MR (1976) Radiokhimiya 18:371; Soviet Radiochem 18:328
9. Rudolph J, Bächmann K (1976) J Radioanal Nucl Chem 32:245
10. Rudolph J, Bächmann K (1980) J Chromatogr, A 187: 319
11. Rudolph J, Bächmann K (1979) Microchim Acta 71:477
12. de Boer JH (1953) The dynamical character of adsorption. Clarendon, Oxford, Sect. 79
13. Zvara I, Tarasov LK, (1962) Zh Neorg Khim 7:2665

14. Zvara I (1966) Dissertation theses. Rep JINR 2591, Dubna. Reproduced in Zvara I (1990) Isotopenpraxis 26:251
15. Exner O (1964) Coll Czech Chem Commun 26:1094
16. Thorn RJ (1971) High Temp Sci 3:197
17. Krug RR, Hunter WG, Grieger RA (1976) J Phys Chem 80:2335
18. Krug RR, Hunter WG, Grieger RA (1976) J Phys Chem 80:2341
19. Cornish-Bowden A (2002) J Biosci 27:121
20. Steffen A, Bächmann K (1978) Talanta 25:677
21. Steffen A, Bächmann K (1978) Talanta 25:551
22. Hill TL (1960) An Introduction to statistical thermodynamics. Addison-Wesley, Reading
23. Irikura KK (1998) In: Irikura KK, Frurip DJ (eds) Computational thermochemistry: Prediction and estimation of molecular thermodynamics (ACS Symposium Series 677). ACS, Washington DC; *http://www.nist.gov/compchem/irikura/docs/app_b2_rev.pdf*. Cited 15 May 2007
24. Christmann K (1991) Introduction to surface physical chemistry. Steinkopff, Darmstadt; Springer, Berlin Heidelberg New York, p 24
25. Zhuikov BL (1982) Report R12-82-63. JINR, Dubna
26. Fremont-Lamouranne R, Legoux Y, Merinis J, Zhuikov BL, Eichler B (1985) Thermochromatography of the actinides and their compounds. In: Freeman AJ, Keller C eds. Handbook on the physics and chemistry of the actinides. Elsevier, Amsterdam, p 331
27. Jin KU, Zvara I (1986) Report R6-86-228. JINR, Dubna
28. Eichler R, Eichler B, Gäggeler HW, Jost DT, Piguet D, Türler A (2000) Radiochim Acta 88:87
29. Eichler B, Türler A, Gäggeler HW (1999) J Phys Chem, A 103:9296
30. Shannon RD (1976) Acta Crystalogr, A 32:751
31. Zvara I Unpublished results
32. Zvara I, Keller OL, Silva RJ, Tarrant JR (1975) J Chromatog 103:77
33. Iler RK (1979) (ed) The Chemistry of Silica. Wiley, New York
34. Papirer E (2000) (ed) Adsorption on Silica Surfaces. Dekker, New York
35. Legrand AP (1998) (ed) The Surface Properties of Silicas. Wiley, New York
36. Kiselev AV, Yashin YaI (1969) Gas-adsorption chromatography. Plenum, New York
37. Timokhin SN, Orelovich OL, Zvara I (2001) Unpublished results
38. Bonvent JJ, Barberi R, Bartolino R, Capelli L, Righetti PG (1996) J Chromatogr, A 756:233
39. Knaupp S, Steffen R, Watzig H (1996) J Chromatogr, A 744:93
40. Gupta PK, Inniss D, Kurkjian CR, Zhong Qian (2000) J Non-Cryst Solids 262:200
41. Henke L, Nagy N, Krull UJ (2002) Biosens Bioelectron 17:547
42. Greenwood NN, Earnshaw A (1989) Chemistry of the elements 1st ed. Pergamon, Oxford, p 396
43. Poggemann JF, Gos A, Heide G, Radlein E, Frischat GH (2001) J Non-Cryst Solids 281:221
44. Stallons JM, Inglesia E (2001) Chem Eng Sci 56:4205
45. Kiselev AV (1936) Kolloidnyi Zh 2:17
46. Bakaev VA, Steele WA (1999) J Chem Phys 111:9803
47. Yates JT Jr (2004) Surf Sci 565:103
48. Wendt S, Frerichs M, Wei T, Chen MS, Kempter V, Goodman DV (2004) Surf Sci 565:107
49. Armistead CG, Tyler AJ, Hambleton FH, Mitchell SA, Hocley JA (1969) J Phys Chem 73:3947
50. Zhuravlev LT (2000) Colloids Surf, A 173:1
51. Jal PK, Patel S, Mishra BK (2004) Talanta 62:1005
52. Leonardelli S, Facchini L, Fretigni C, Tougne P, Legrand AP (1992) J Am Chem Soc 114:6412
53. Ceresoli D, Bernasconi M, Iarlori S, Parrinello M, Tosatti E (2000) Phys Rev Lett 84:3887
54. Souza SD, Panatano CG (2002) J Am Ceram Soc 85:1499
55. Masini P, Bernasconi M (2002) J Phys: Condens Matter 14:4133
56. Yang JJ, Meng S, Xu LF, Wang EG (2005) Phys Rev B 71:035413
57. Chuiko A (1987) Teoret Eksper Khim 23:597; J Theor Exp Chem 23:551

References

58. Zhuravlev LT (1987) Langmuir 3:316
59. Zhuravlev LT (1989) Pure Appl Chem 61:1969
60. Sneh O, George SM (1995) J Phys Chem 99:4639
61. Peri JB, Hensley AL (1968) J Phys Chem 72:2926
62. Hamann DR (1997) Phys Rev B 55:14784
63. Lopez N, Vitiello M, Illas F, Pacchioni G (2000) J Non-Cryst Solids 271:56
64. Rarivomanantsoa M, Jund P, Jullien R (2001) J Phys: Condens Matter 13:6707
65. Hambleton FH, Hockey JA (1966) Trans Farad Soc 62:7
66. Lygin VI (2002) Zh Ross Khim O-va im DI Mendeleeva 46(3):12
67. Lygin VI (1997) Zh Fiz Khim 71:1735
68. Griffin S (2002) LCGC North America 20:928
69. Vansant EF, van der Voort P, Vrancken KC (1995) Characterization and chemical modification of the silica surface. Elsevier, Amsterdam
70. Laskorin BI, Strelko VV, Strazhesko DN, Denisov VI (1977) Sorbenty na osnove silikagelya v radiokhimii (Silica gel-based sorbents in radiochemistry). Atomizdat, Moskva. Sect. 2.2
71. Tertykh VA, Belyakova (1991) Khimicheskie reaktsii s uchastiem poverkhnosti kremnezyoma (Chemical reactions mediated by silica). Kiev, Naukova Dumka
72. Schrijnemakers K, Van Der Voort P, Vansant EF (1999) Phys Chem Chem Phys 1:2569
73. Ellestad OH, Blinheim U (J Mol Catal (1985) 33:275
74. Haukka S, Lakomaa EL, Root A (1993) J Phys Chem 97:5085
75. Gorlov YuI, Melnichenko GN, Nazarenko VA (1985). Teor Eksp Khim 20:754; J Theor Exp Chem 20:712
76. Lang SJ, Morrow BA (1994) J Phys Chem 98:13314
77. Hair ML, Hertl W (1973) J Phys Chem 77:2070
78. McDaniel MP (1981) J Phys Chem 85:532
79. Satoh S, Susa K, Matsuyama I, Suganuma T, Matsumure H (1997) J Non-Cryst Solids 217:22
80. McDaniel MP (1981) J Phys Chem 85:537
81. Isted GE, Martin DS, Marnell L, Weightman P (2004) Surf Sci 566–568:35
82. Islam MS, Jung GI, Ha T, Stewart DR, Chen Y, Wang SY, Williams RS (2005) Appl Phys A 80:1385
83. Wadsak M, Schreiner M, Aastrup T, Leygraf C (2000) Appl Surf Sci 157:39
84. Lee WT, Ford L, Blowers P, Nigg HL, Masel RI (1998) Surf Sci 416:141
85. Fishlock TW, Pethica JB, Oral O, Egdell RG, Jones FH (1999) Surf Sci 426:212
86. Fishlock TW, Pethica JB, Jones FH, Egdell RG, Foord JS (1997) Surf Sci 377–379:629
87. Lennard-Jones JE (1937) Proc Phys Soc 49:140
88. Barth JV (2000) Surf Sci Rep 40:75
89. Tringides MC (ed) (1997) Surface diffusion. Springer, Berlin Heidelberg New York
90. Braun OM, Ferrando R (2002) Phys Rev E 65:061107
91. Chen LY, Ying SC (1993) Phys Rev Lett 71:4361
92. Senft DC, Ehrlich G (1995) Phys Rev Lett 74:294
93. Wang SC, Ehrlich G (1988) Surf Sci 206:451
94. Kellog GL (1994) Surf Sci Rep 21:1
95. Patrykiejev A, Jaroniec M, Rudzinski W (1978) Tin Solid Films 52:295
96. Charmas B, Leboda R (2000) J Chromatog, A 886:133
97. Bakaev VA, Bakaeva TI, Pantano CG (2002) J Chromatog, A, 969:153
98. Hill TL (1952) Theory of physical adsorption In: Frankenburg WG, Komarewski, Rideal EK (eds) Advances in catalysis, Vol IV. Academic, New York, p 211
99. Hill Tl (1949) J Chem Phys 17:762
100. van der Vis MGM, Cordfunke EHP, Konings RJM (1997) Thermochim Acta 302:93
101. Gurvich LV, Veitz IV, Alcock ChB (1989) (eds) Thermodynamic properties of individual substances, 4th ed. Hemisphere, New York; original (1982) Termodinamicheskiye svoistva individualnykh veshchestv Nauka, Moskva
102. McBride BJ, Zehe MJ, Gordon S (2002) NASA Glenn coefficients for calculating thermodynamic properties of individual species. NASA/TP—2002-211556. Glenn Research Center, Cleveland. http://gltrs.grc.nasa.gov/reports/2002/tp-2002-211556.pdf. Cited May 15 2007

103. Pershina V (2005) Radiochim Acta 93:125
104. Düllmann ChE, Brüchle W, Dressler R, Eberhardt K, Eichler B, Eichler R, Gäggeler HW, Ginter TN, Glaus F, Gregorich KE, Hoffman DC, Jager E, Jost DT, Kirbach UW, Lee DM, Nitsche H, Patin JB, Pershina V, Piguet D, Qin Z, Schädel M, Schausten B, Schimpf E, Schott H-J, Soverna S, Sudowe R, Thorle P, Timokhin SN, Trautmann N, Türler A, Vahle A, Wirth G, Yakushev AB, Zielinski PM (2002) Nature 418:859
105. Eichler B, Kim SC (1985) Isotopenpraxis 21:180
106. Eichler R, Eichler B, Gäggeler HW, Jost DT, Dressler R, Türler A (1999) Radiochim Acta 87:151
107. Korotkin YuS, Jin KU, Timokhin SN, Orelovich OL, Altynov VA (1988) Report P6-88-595. JINR, Dubna
108. Dasent WE, (1965) Nonexistent compounds. Arnold, London; Dekker, New York
109. Eichler B, Zude F, Fan W, Trautmann N, Herrmann G (1993) Radiochim Acta 81:90
110. Gärtner M, Boettger M, Eichler B, Gäggeler HW, Grantz M, Hübener S, Jost DT, Piguet D, Dressler R, Türler A, Yakushev AB (1998) Radiochim Acta 78:59
111. Eichler B, Eichler R (2003) Gas-phase adsorption chromatographic determination of thermochemical data and empirical methods for their estimation. In: Schädel M (ed) The chemistry of superheavy elements. Kluwer, Dordrecht, p 205
112. Eichler B (1996) Radiochim Acta 72, 19
113. Vahle A, Hübener S, Dressler R, Eichler B, Türler A (1997) Radiochim Acta 78:53
114. IUPAC Compendium of chemical terminology - the Gold book (1997) 2nd edn. http://goldbook.iupac.org/. Cited 15 May 2007
115. Yakushev AB, Eichler B, Türler A, Gäggeler HW, Peterson J (2003) Radiochim Acta 91:123
116. Merinis J, Legoux Y, Boussieres G (1970) Radiochim Radioanal Lett 3:255
117. Vahle A, Hübener S, Funke H, Eichler B, Jost DT, Türler A, Brüchle W, Jager E (1999) Radiochim Acta 84:43
118. Zvarova TS, Zvara I (1969) J Chromatogr 44:604
119. Zvarova TS, Zvara I (1970) J Chromatogr 49:290
120. Domanov VP, Zin KU (1989) Radiokhimiya 31:12
121. Türler A, Buklanov GV, Eichler B, Gäggeler HW, Granz M, Hübener S, Jost DT, Lebedev VYa, Piguet D, Timokhin SN, Yakushev AB, Zvara I (1998) J Alloys Compounds 271–273:287
122. Lebedev VYa, Yakushev AB, Timokhin SN, Vedeneev MB, Zvara I (1999) Czech J Phys 49 (Suppl 1/II):589
123. NorseevYuV, Khalkin VA (1999) Radiokhimiya 41:318

Chapter 6
Validity and Accuracy of Single Atom Studies

Abstract Validity of the results obtained in experiments with single atoms has been discussed in the literature from different points of view. The few dedicated experimental works did not reveal deviations in behavior of tracers with decreasing amount down to hundreds of atoms. Kinetic considerations suggest that the data are essentially valid unless the relative statistical uncertainty is comparable with the number of detected atoms. Thermodynamic quantities can be reasonably represented by probabilistic terms. The statistical limitations to quantitative conclusions were visualized by repetitive Monte Carlo simulations of gas-solid thermochromatographic experiments. With poor statistics of measurements, the Bayesian approach seems conceptually preferable to obtain information about the sought parameters. Some Bayesian confidential intervals for the difference and ratio of two Poisson-distributed quantities from the literature are tabulated. Several difficult situations which were met in real experiments are analyzed using Monte Carlo simulations to obtain the likelihood function and then the Bayesian posterior function. The presented examples concern the adsorption enthalpy from chromatographic experiments, and the half-life from a background free decay curve, if the measurement was terminated before reaching the zero counting rate.

6.1 Validity of Single Atom Chemistry

From the early studies of gas-phase chemistry of transactinoids, the experimenters have faced the question of whether the ultimately low number of short-lived atoms would bring new fundamental problems in obtaining reliable data and their interpretation. Classical radiochemistry dealt mostly with aqueous solutions and widely used batch realization of the partition methods, like coprecipitation or extraction. The goal was to find known element(s) with chemical behavior like the new radionuclide, at least in some particular chemical systems; the homology was then used for isolation, concentration and assignment of the chemical state of the activity. Researchers in the field occasionally noted some peculiar, even erratic behavior of the tracer elements. At first, these observations might hardly be rationalized; later, such findings could often be attributed to adsorption of the tracers on the surface of containers, which was of foreign nature to the solution systems. A well-known study of the problem of few atoms in solution experiments was performed by Reichsmann, et al. [1, 2]. They investigated coprecipitation of ^{210,218}Po with Te and As sulfides, while taking gradually smaller numbers of the Po atoms in the system. They did not observe dependence of the Po behavior on the concentration of the element.

6.1.1 Monte Carlo Simulation of Single Atom Experiments

Gas-solid chromatography is essentially a dynamic partition experiment in which the column surface plays the role of one of the phases. It was why the Dubna researchers were concerned with making the surface properties reproducible and free of too-active adsorption sites. To that end, they introduced chemical reagents into the carrier gas. With the lighter homologs of the transactinoids they obtained similar results independent of the number of atoms present in the particular chemical system; the investigated range was down to some 10^3 atoms. Compared with the liquid chromatography, the microscopic picture of gas-solid chromatography seems simpler. It enables Monte Carlo simulation of tiny details of the individual trajectories of molecules in space and time; the procedure was discussed in Sect. 4.3.1. In the late 1970s, the Dubna group realized the first thermochromatographic studies of the element 105 bromide [3, 4]. Twelve spontaneous fission decay events of dubnium were detected within an extended adsorption zone; their mean deposition temperature was about 450 K, and the temperature gradient over the zone range was nearly linear. To visualize the possible effects of small numbers, these experimental results were simulated by a Monte Carlo technique like that described in Sect. 4.3.2. Ten separate "runs" were Monte Carlo-simulated, taking a half-life of two seconds and assuming an appropriate value of the desorption energy. The data of separate experiments are shown in the top of Fig. 6.1. In each of them, the number of molecules was taken at random from the Poisson distribution with a parameter of 12. These numbers are shown at the lines; the circles indicate the temperature at the particulate coordinate of the decay event, and the short vertical bars show an average of the temperatures. All the mean values are within 30 K range. Such a scatter is acceptable for characterization of the properties of the studied compound with reasonable accuracy. It is also evidenced by the cumulative distributions of the total of 115 events (solid line) from the 10 simulated runs, as well as by the distribution of 2,500 additional individual histories (broken line). Their mean deposition temperature values are again the 30 K range. It also happened for the medians, though by chance — the particular combination of the experimental parameters yielded symmetric zone profiles.

The Dubna group also discussed the kinetics of reactions in gases to look for the fundamental reasons which might cause deviations in behavior of a tracer with the decreasing number or, better, concentration of its species. Only one inherent reason could be seen: a deviation may happen when two entities practically cannot

6.1 Validity of Single Atom Chemistry

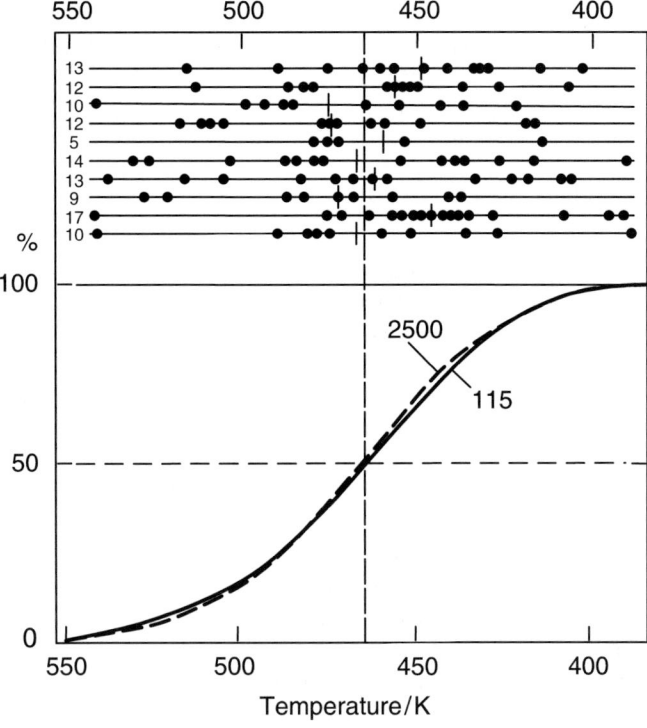

Fig. 6.1 Monte Carlo simulation of the thermochromatographic experiment [3,4] with the bromide of 2-s ^{260,261}Db.

The simulated runs were supposed to produce on average 12 detectable nuclei. Top: Distribution of decay events along the column in the runs. Bottom: Cumulative distributions for good statistics; see text.

Reproduced (adapted) from Radiokhimiya, 18(3), Zvara I, Belov VZ, Domanov VP, Shalaevski MR, Chemical separation of nielsbohrium, 371–373, © 1976, with permission from Academizdatcenter Nauka Publishers.

collide during the time of the experiment or during the lifetime of the nuclide, if the latter is shorter. Thus, the reactions like dimerization cannot proceed any more. The dividing concentration level is not low. Indeed, a gaseous molecule collides with an identical one only once per second when the concentration is 10^{10} cm^{-3} and once a day at 10^5 cm^{-3}. In liquids, the respective values are 10^7 and 10^2 cm^{-3}. Hence, mutual collisions are completely irrelevant to the chemistry of single atoms and molecules. In the meantime, the above figures are also the minimal necessary concentrations of the reagents, which we purposely introduce into the system to achieve or support a required chemical state of short-lived tracers. The reactions of a tracer with macroscopic reagents are formally of the first order and the mean time of conversion is inversely proportional to the reagent concentration. Hence, the agents do not work if present in lesser concentrations. Consistently, these figures indicate the highest allowable ("safe") concentration of unwanted or unaccounted for contaminations in the system. Above these limits, the impurities might harmfully change the chemical state of the tracer. Some of these points were already discussed in Sect. 3.2.

6.1.2 Theoretical Kinetic Limits

More formal and general approaches to the problem deal not only with what happens with a handful of the atoms, but also how to evaluate and interpret the experimental data. They are motivated by the wish to adhere as closely as possible to statistical thermodynamics and to the conventional kinetics. The first dedicated attempt was done by Borg and Dienes [5], who put the question, "Are ordinary chemical reactions adequate to describe and distinguish one element from another when the mean nuclear lifetime is comparable to, say, the mean lifetime in a single chemical state?" They noted that, when too few atoms are present, most of the available states are unoccupied and dynamic equilibrium is not maintained at each instant. A long-lived atom can pass through all states within an ensemble and thus conform to the ergodic hypothesis. However, the latter also demands that the fraction of time spent by the atom in each state be equal to the fraction of systems of an equilibrium ensemble in that state. Thus, a short-lived atom can hardly conform to the ergodic hypothesis. The authors analyzed the applicability of chemical kinetics in the combination with the kinetics of radioactive decay. They illustrated it by a binary reaction with an impurity (exchange of ligands), which would yield an unforeseen chemical state of the tracer element. In a general form, they considered the free energy of activation (in the sense of the transition state theory of reaction rates), rather than just the activation energy. Further on, a standard probability treatment was used to show that "a surprisingly small number of atoms" are required to establish chemical identity under the conditions of typical solution or gas-phase experiments. They did not specify the number, but came to the conclusion that with 50 atoms "an erroneous conclusion as to the chemical nature is most unlikely." In addition, "an ensemble containing as few as 10 atoms gives nearly the same result as a truly statistical ensemble." The latter figure is very close to the conclusions "empirically" drawn from the Monte Carlo simulations of a thermochromatographic experiment.

6.1.3 Equivalent to Law of Mass Action

Later, Guillamont, et al. [6,7] considered some aspects of the environmental solution chemistry of tracers (complexation, redox reactions) and of solving the speciation problems using partition methods. They paid most attention to the concentrations at which the mutual encounters of the tracer entities are somewhat probable. These conditions affect stoichiometry of the reactions and the kinetic laws of the interactions. Below this limit, if two different chemical forms of the tracer were introduced into the system, they would coexist as they are prevented from getting involved in dimerization (polymerization), isotopic exchange or disproportionation. For the problem of single atoms, they introduced a specific thermodynamic function, "single-particle free enthalpy," and suggested an equivalent to the law of mass action in terms of certain probabilities, rather than concentrations. Accordingly, for a tracer element M, "the distribution coefficient of M between two phases is correctly defined in terms of the total probabilities of finding M (whatever its species) in one phase or the other." Hence, multiple static partition experiments are needed to measure this coefficient. Meanwhile, dynamic partition methods can be considered as a series of repetitive static partitions isolated in space. Thus, "the displacement of M, in itself, is a statistical result and only one experiment is necessary." This is a general conclusion, obviously valid also for gas-solid chromatography. As to "the fundamental question whether the law of mass action really holds" at single atom level, the authors of Ref. [6] came to the conclusion that "So far, no satisfactory answer to the question is available and this point remains open to discussion."

6.1.4 More Considerations

A recent review of the problem is due to Trubert and Le Naour [8] who also focus on solution chemistry. When considering the kinetic aspects of the problems, they present a criticism of the approach of Borg and Dienes [5]. The argument is that the concept of the energy of the activated complex is inherently statistical, and so inapplicable. They pay attention to Ref. [9], which reports evaluation of the fluctuations in energies of individual entities around the mean value to estimate the relative mean deviation of a property of the system as a function of the number of particles. The deflection happens to be equal to the inverse square root of the number. The result is of general validity and importance. Thus, the characteristic fluctuation of a property of the system (like total energy) of 10 entities is some 30 percent, which is consistent with what was found or estimated in Refs. [4, 5]. However, for a single atom or species, the fluctuation is equal to the value of the property.

Evidently, there is a common approach in the theoretical works cited above. They formulate the basic question like "are the concepts of thermodynamics and kinetics used in classical chemistry relevant at tracer scale...and for a single atom?" [8]. Actually, from the very beginning, statistical mechanics and thermodynamics clearly recognized their applicability only to large numbers of particles. They were not concerned with what would happen at ultimately low numbers — the issue became urgent relatively recently. Attention to kinetics and to the probabilistic ideas appears quite naturally. At present, it would seem logical to *start* with single short-lived entities and with their accompanying *randomness*, and then *proceed* to larger numbers and longer times. The result would probably be about the same as that resulting from the history of the question. However, such analysis of the peculiar range of the ultimately low numbers might result in novel insights.

In the literature, there sometimes occur inaccurate statements about the "validity" of Stirling's approximation in connection with the validity of single atom chemistry.

The two points do not seem to be directly interrelated. Notice that there is the Stirling's *series* — an asymptotic expansion of the logarithm of a factorial — written as:

$$\ln n! = n \ln n - n + \ln \sqrt{2\pi n} + \frac{1}{12n} - \frac{1}{360n^3} + \ldots \ldots \quad (6.1)$$

The Stirling's *approximation* consists in the first two terms: $\ln n! \approx n \ln n - n$. Its absolute error increases with larger n. Hence, the approximation never becomes perfectly valid, though its relative error becomes negligible for numbers like N_A. In the meantime, a few more terms of Eq. 6.1 yield enough precision to meet any practical demand. Already four terms give an error less than $1/360n^3$: for $\ln 1! = 0$ it is less than 0.0023, for ln3! – less than 0.0001, etc.; the relative error obviously decreases with n even faster.

The present author believes that many practical problems can be illuminated and even solved with the use Monte Carlo simulations like those presented above in this section. The simulations are inherently probabilistic and can incorporate both thermodynamic and kinetic factors. The simulations do not provide formulae, but their great advantage is the possibility to take into account real experimental conditions. Simulations do not take any significant time, even on standard PC workstations. On the other hand, deriving analytical formulae as a rule requires assuming idealized experimental conditions, which are often far from reality.

6.2 Analysis of Poor-Statistics Data

When studying chemical properties of TAEs on the one-atom-at-a-time (per shift, day, week or month) level – which requires not only large intellectual and material efforts but also a lot of patience – the experimenters are inevitably tempted to draw more definite and numerous conclusions than the poor statistics actually permit. It is important to be aware of this danger and pay due attention to the statistical uncertainties of the obtained data, especially when counting is not free of background.

The traditional statistics are not a proper way to evaluate the accuracy or precision of the results obtained on the basis of detecting only a few decay events. This is because the Gaussian approximation for the data distribution is not valid and the "errors" come from the statistical uncertainty, rather than from the imperfect measurements. To allow more rigorous treatment of "low-level" counting data, some authors updated the traditional approach by taking into account the inherent Poisson distribution [10, 11].

Another approach to the data based on low-level counting uses the method of maximum likelihood. The likelihood of a set of data is the probability of obtaining the particular set, given the chosen probability distribution model. The idea is to determine the parameters that maximize the likelihood of the sample data. The methodology is simple, but the implementation may need intense mathematics [12]. The method has been used, for instance, to treat data on production rates [12] and half-lives [13–17]. Bukin [18] compares the methods of obtaining confidence intervals for the half-lives by maximum likelihood, Bayesian (see below) and traditional approaches.

The poor statistics seem to call for a different philosophy of interpretation of the evaluated uncertainties. There is increasing attention to the Bayesian statistics [19] and standing discussions take place on the relative merits of various approaches and the philosophy behind them. "Modern statistical practice is dominated by two principal schools of thought....the 'traditional' school and the 'Bayesian' school. The division between these two (is)...the meaning given to the term probability frequency...(and)....belief (respectively)" [20]. With the standard approach one essentially asks what will the probability density function of the data (in terms of the parameters of normal distribution) be if the experiment is repeated many times? With mere two or three detected counts, the question is obviously difficult to answer and does not sound correct. This is especially true if one has little real hope or opportunity to repeat the experiment in the future. However, even in the case of better statistics there is a reason for preferring the Bayesian approach. According to Porter [21]: "A Confidence Interval (CI), in frequentist statistics, tells us about our data; for example, a small interval indicates a more precise measurement than a wide interval...(but it) tells us nothing about the true value of sought-for parameter Θ. The utility of the CI is in summarizing, objectively, the quality of the measurement; the true value of Θ is irrelevant."

With the Bayesian approach, the question is about the interval *inferred by the observation* of the data k, which contains the true value of Θ at the confidence level α. The Bayesian interval (BI) for Θ is the interval, which contains the fraction α of the area under the appropriate Bayes' distribution. Again from Porter [21], "The motivation for BI is that we really are interested in making some statement about the value of Θ. The interpretation of the Bayes' distribution is that it expresses a 'rational degree of belief' in where the parameter lies. The Bayes' distribution is mathematically a probability distribution, but it does not have a frequency interpretation. We cannot sample from it. The utility of the Bayes' distribution/interval is that it gives us formalism with which to make decisions, based on available knowledge." Therefore, in the Bayesian spirit, the meaning of the common 68 percent, or "1σ," confidence interval is that one can bet only 2:1 that the true value is within this range. It is illuminating evidence that one cannot rely upon such a narrow interval when trying to experimentally answer fundamental questions.

6.2.1 Bayesian Approach to Statistical Treatment

The Bayes' distribution or the *posterior* probability density function $F_{\text{pos}}(\Theta|k)$, which serves the evaluation of BIs, is:

$$F_{\text{pos}}(\Theta|k) = \frac{L(k|\Theta) \cdot F_{\text{pri}}(\Theta)}{\int L(k|\Theta) \cdot F_{\text{pri}}(\Theta) d\Theta} \quad (6.2)$$

Here $L(k|\Theta)$ is the likelihood of observing the concrete data k given the parameter value. The pdf-like $F_{\text{pri}}(\Theta)$ is a non-negative function called prior distribution, the choice of which is discussed below. The integration is performed over the range of the prior distribution.

When the parameter is a discrete quantity, Eq. 6.2 should be consistently rewritten for summation:

$$F_{\text{pos}}(\Theta|k) = \frac{L(k|\Theta) \cdot F_{\text{pri}}(\Theta)}{\sum L(k|\Theta) \cdot F_{\text{pri}}(\Theta)} \quad (6.3)$$

For example, if k decays or a rare short-lived nuclide were detected in the course of a long bombardment and we need to know the Bayesian confidence interval for the yield rate (per the duration of the bombardment) Θ, then $L(k|\Theta)$ is the Poisson distribution:

$$L(k|\Theta) = \frac{\Theta^k}{k!} e^{-\Theta} \quad (6.4)$$

In another common problem, Θ is the number of radioactive nuclei which were originally present in a sample and gave k counts when measured with the detection efficiency p_λ until they decayed. Now the likelihood is given by the binomial distribution:

$$L(k|\Theta) = \frac{\Theta!}{k!(\Theta - k)!} p_\lambda^k (1 - p_\lambda^k)^{\Theta - k} \quad (6.5)$$

Because both the Bayesian statistics and the above-mentioned method of maximum likelihood operate with the likelihood function, concrete applications of these approaches have some common features.

The next Bayesian step is the choice of $F_{\text{pri}}(\Theta)$. It is evidently possible to outline a finite, physically acceptable range of the parameter values, outside of which, in the "unphysical region," $F_{\text{pri}}(\Theta)$ vanishes — either necessarily or practically. For example, in the problem of the production rate, Θ cannot be zero; it can be both less and more than k, though hardly many times. This suggests an interval like $[0.1k, 10k]$ to more than guarantee inclusion of all possible values, but with $k = 0$ the range is to be like $[0, 10]$.

In the problem of the original number of nuclei in the sample, Θ is a positive integer. It cannot be less than k (if background free) and hardly can exceed k many times; so reasonable values of Θ are integers in $[k, 10k]$.

A more difficult problem is the shape of $F_{\text{pri}}(\Theta)$. Generally, it may be any function satisfying certain simple mathematical conditions. However, to justify an a priori preference to some narrower range of Θ is mostly difficult. Commonly, one suggests that nothing is known about the shape of the prior distribution and accepts it as being uniform (flat). The approach is called "complete (or total) ignorance." Practice shows that the results depend little on the shape of $F_{\text{pri}}(\Theta)$. Still, one must pay attention to the following warning: "Complete ignorance in Θ is not complete ignorance in all functions of Θ, for example, Θ^2. Ignorance is parameterization dependent!...My (philosophical!) perspective: Go ahead and use a flat, or other smooth, prior for whatever parameter you care about. But Be Warned: if it makes much difference in any decisions you're going to make!" [21].

6.2 Analysis of Poor-Statistics Data

The flat prior need not be normalized so that:

$$F_{\text{pos}}(\Theta|k) = \frac{L(k|\Theta)}{\int L(k|\Theta)d\Theta} \tag{6.6}$$

It allows obtaining analytical expressions for the BIs in simple cases like the difference between two counting results or their ratio.

Solutions to relatively simple problems at various confidence levels by the Poisson classic, maximum likelihood and Bayes' inference approaches are available in literature as formulae and tables. Currie [10] re-examined the terms detection and determination limits, which are widely used in analytical chemistry, paying special attention to Poisson-distributed data of radioactivity measurement; the corresponding formulae still do not help in ultra low-level counting. Strom and MacLellan [11] took a contemporary look of the problem within the same approach, as well as a criticism of the previous works. James and Roos [22] presented Tables of 68 percent and 90 percent CIs for the ratios of 0 to 10 Poisson-distributed counts when the numerator is a subset of the denominator (combined Poisson and binomial distributions). De Angelis and Iori [23] considered the same situation for up to 16 events and, in addition, the ratio of the Poisson-distributed numerator and denominator. Helene [19] presented tables of the BIs (68 percent and 95 percent) for the small number of counts (up to 20), for the difference of two small numbers (the background problem; mean background exactly known) and for the ratio of two numbers. Later on, Prosper [24] evaluated the Bayesian formulae for the difference and ratio problems when the background mean is known with finite precision. His discussion of the Bayesian approach is very instructive. Recently the problem of the difference and ratio of two Poisson-distributed quantities was re-examined by Brüchle [25]. He tabulated the confidence intervals up to 10 counts in the numerator and denominator, each for the confidence levels 65 percent, 90 percent and 95 percent, and analyzed the origin of the differences with the Helene's [19] evaluations. He also discusses the choice between the central and shortest (highest probability density) intervals. Dressler, et al. [26] presented formulae of some (complete ignorance) Bayes distributions $F_{\text{pos}}(\Theta|k)$ to help solve real tasks faced in TAE studies.

Tables 6.1 and 6.2 display sample results from [19,25]. Generally, the confidence limits necessarily depend on the accepted approach and its details; it is not surprising that they slightly differ.

There are more complicated cases, like measurements with non-zero background, measurement times comparable with the activity lifetimes, variation of counting efficiency in time and so forth. Then the likelihood $L(k|\Theta)$ cannot be evaluated "analytically," but only using Monte Carlo simulations of a model of the data; this will be demonstrated below. It is possible to describe data dependent on more parameters.

An experimenter often tries to answer the very serious questions, like whether there is a significant difference in properties of a very heavy element and its lighter congener(s). Then the 68 percent confidence level for a measured quantity is hardly convincing. Indeed, as will be seen below, the Bayes' distribution may be strongly

Table 6.1 The Difference Between the Poisson-Distributed Quantity K and the Background Bkg. The mean value of Bkg is known. Presented are the shortest intervals for K − Bkg at the 68 percent and 95 percent confidence levels obtained by Bayesian approach [19].

Excerpts from tables in Nuclear instruments and methods in physics research, 228(1), Helene O, Errors in experiments with small numbers, 120–128, © 1984, with permission from Elsevier.

K	68.3 percent confidence level						95 percent confidence level						K
	Bkg = 0[a]		Bkg = 1		Bkg = 2		Bkg = 0[a]		Bkg = 1		Bkg = 2		
0[a]	0.0	1.15	0.0	1.15	0.0	1.15	0.0	3.00	0.0	3.00	0.0	3.00	0[a]
1	0.27	2.50	0.0	1.79	0.0	1.57	0.04	4.77	0.0	4.12	0.0	3.82	1
2	0.86	3.86	0.0	2.66	0.0	2.17	0.30	6.40	0.0	5.41	0.0	4.83	2
3	1.55	5.15	0.58	4.08	0.0	2.95	0.71	7.95	0.0	6.78	0.0	5.99	3
4	2.29	6.40	1.30	5.39	0.39	4.21	1.21	9.43	0.23	8.33	0.0	7.24	4
5	3.06	7.63	2.06	6.63	1.09	5.56	1.76	10.87	0.76	9.85	0.0	8.54	5
6	3.85	8.84	2.85	7.84	1.86	6.82	2.35	12.26	1.35	11.26	0.40	10.12	6
7	4.65	10.0	3.65	9.03	2.65	8.03	2.97	13.63	1.97	12.63	0.99	11.60	7
8	5.47	11.2	4.47	10.2	3.47	9.21	3.62	14.98	2.62	13.98	1.62	12.97	8
9	6.30	12.3	5.30	11.3	4.30	10.38	4.29	16.30	3.29	15.30	2.29	14.30	9
10	7.14	13.54	6.14	12.54	5.14	11.54	4.98	17.61	3.98	16.61	2.98	15.61	10

[a] data for K = 0 and Bkg = 0 are from Ref. [25].

Table 6.2 The ratio of the Poisson-distributed Numerator *Num* and Denominator *Den*. Presented are most probable values of *Num/Den* and the shortest intervals for *Num/Den* at the 68 percent and 95 percent confidence levels. Data of Helene [19] and Bruchle [25].

Excerpts from tables in Nuclear instruments and methods in physics research, 228(1), Helene O, Errors in experiments with small numbers, 120–128, © 1984, with permission from Elsevier.

Excerpts from tables in Radiochimica Acta, 91 (2), Brüchle W, Confidence intervals for experiments with background and small number of event, 71–80, © 2003, with permission from Oldenbourg Wissenschaftsverlag.

Num	Den	Most prob- able [25]	95 percent [25]		68.3 percent [25]		68.3 percent [19]		95 percent [19]		Num	Den
0	1	0	0	3.366	0	0.767	0	0.776	0	3.47	0	1
0	2	0	0	1.650	0	0.461	0	0.467	0	1.71	0	2
0	3	0	0	1.072	0	0.329	0	0.333	0	1.11	0	3
0	4	0	0	0.787	0	0.255	0	0.258	0	0.821	0	4
0	5	0	0	0.621	0	0.209	0	0.211	0	0.648	0	5
0	6	0	0	0.513	0	0.177	0	0.178	0	0.534	0	6
0	7	0	0	0.436	0	0.153	0	0.154	0	0.454	0	7
0	8	0	0	0.380	0	0.135	0	0.136	0	0.395	0	8
0	9	0	0	0.336	0	0.121	0	0.122	0	0.349	0	9
0	10	0	0	0.301	0	0.109	0	0.110	0	0.313	0	10
1	1	1.0	0.104	9.620	0.337	2.962	0.038	1.68	0.002	6.39	1	1
1	2	0.307	0.032	3.206	0.092	1.097	0.037	1.01			1	2
1	3	0.225	0.015	1.955	0.057	0.748	0.034	0.718			1	3
1	4	0.180	0.009	1.404	0.043	0.572	0.031	0.557			1	4
1	5	0.151	0.006	1.092	0.035	0.463	0.028	0.456	0.003	1.09	1	5
1	6	0.130	0.005	0.892	0.030	0.391	0.026	0.385			1	6
1	7	0.114	0.004	0.753	0.026	0.337	0.024	0.334			1	7

(Continued)

6.2 Analysis of Poor-Statistics Data

Table 6.2 Continued.

Num	Den	Most probable [25]	95 percent [25]		68.3 percent [25]		68.3 percent [19]		95 percent [19]		Num	Den
1	8	0.102	0.003	0.651	0.024	0.297	0.022	0.294			1	8
1	9	0.093	0.003	0.574	0.022	0.266	0.020	0.263			1	9
1	10	0.085	0.003	0.513	0.020	0.240	0.019	0.238	0.002	0.513	1	10
2	2	1.0	0.171	5.840	0.422	2.370	0.137	1.56	0.032	4.30	2	2
2	3	0.540	0.103	3.034	0.236	1.305	0.122	1.11			2	3
2	4	0.397	0.070	2.058	0.166	0.942	0.109	0.860			2	4
2	5	0.320	0.053	1.562	0.131	0.746	0.097	0.703			2	5
2	6	0.272	0.042	1.259	0.109	0.620	0.088	0.594	0.024	1.23	2	6
2	7	0.237	0.035	1.052	0.094	0.532	0.080	0.515			2	7
2	8	0.211	0.030	0.905	0.083	0.466	0.074	0.454			2	8
2	9	0.189	0.027	0.795	0.075	0.414	0.068	0.406			2	9
2	10	0.173	0.024	0.706	0.068	0.374	0.063	0.368	0.02	0.702	2	10
3	3	1.0	0.225	4.439	0.478	2.090	0.224	1.49	0.08	3.46	3	3
3	4	0.674	0.164	2.837	0.334	1.394	0.198	1.16			3	4
3	5	0.521	0.126	2.081	0.256	1.061	0.177	0.945			3	5
3	6	0.432	0.102	1.648	0.211	0.866	0.160	0.798			3	6
3	7	0.371	0.086	1.363	0.180	0.734	0.145	0.691	0.057	1.31	3	7
3	8	0.326	0.074	1.165	0.157	0.638	0.133	0.610			3	8
3	9	0.292	0.065	1.015	0.141	0.566	0.123	0.545			3	9
3	10	0.264	0.059	0.901	0.128	0.509	0.114	0.493			3	10
4	4	1.0	0.268	3.724	0.520	1.923	0.294	1.45	0.126	3.01	4	4
4	5	0.749	0.213	2.658	0.401	1.421	0.262	1.18			4	5
4	6	0.609	0.174	2.066	0.326	1.134	0.236	0.999			4	6
4	7	0.516	0.147	1.690	0.277	0.949	0.215	0.865	0.098	1.58	4	7
4	8	0.449	0.128	1.433	0.240	0.819	0.197	0.762			4	8
4	9	0.399	0.113	1.242	0.214	0.721	0.181	0.681			4	9
4	10	0.359	0.101	1.097	0.192	0.645	0.168	0.616	0.08	1.06	4	10
5	5	1.0	0.304	3.284	0.552	1.811	0.350	1.42	0.169	2.72	5	5
5	6	0.802	0.254	2.516	0.451	1.425	0.315	1.20			5	6
5	7	0.671	0.216	2.037	0.381	1.179	0.286	1.04			5	7
5	8	0.577	0.188	1.707	0.330	1.007	0.262	0.913			5	8
5	9	0.512	0.166	1.475	0.292	0.882	0.242	0.816			5	9
5	10	0.459	0.150	1.298	0.263	0.786	0.224	0.737			5	10
6	6	1.0	0.335	2.982	0.578	1.728	0.395	1.40	0.207	2.52	6	6
6	7	0.834	0.288	2.396	0.490	1.421	0.359	1.21			6	7
6	8	0.715	0.252	2.001	0.425	1.206	0.329	1.06			6	8
6	9	0.628	0.223	1.713	0.375	1.049	0.304	0.950			6	9
6	10	0.562	0.201	1.502	0.336	0.930	0.282	0.858	0.155	1.41	6	10
7	7	1.0	0.362	2.762	0.600	1.666	0.433	1.38	0.24	2.38	7	7
7	8	0.858	0.318	2.295	0.521	1.409	0.465	1.21			7	8
7	9	0.751	0.283	1.962	0.461	1.223	0.366	1.08			7	9
7	10	0.670	0.255	1.713	0.413	1.080	0.340	0.978			7	10
8	8	1.0	0.384	2.599	0.619	1.615	0.534	1.36	0.271	2.26	8	8
8	9	0.875	0.344	2.213	0.548	1.398	0.429	1.21			8	9
8	10	0.778	0.310	1.926	0.491	1.233	0.398	1.10			8	10
9	9	1.0	0.405	2.465	0.635	1.575	0.493	1.35	0.298	2.17	9	9
9	10	0.890	0.366	2.141	0.570	1.388	0.458	1.22			9	10
10	10	1.0	0.423	2.361	0.648	1.544	0.517	1.33	0.323	2.10	10	10

skewed and, at a higher confidence level, one of the BI limits may even become uncertain!

The Bayesian approach is almost universally applicable when combined with the Monte Carlo simulations on computers. In practice, because of many external circumstances, almost every long-lasting chemical experiment has specific features which can be traced only by the simulations. Some examples are given and discussed below. The outlined simulation procedures were not optimized because, on conventional workstations, even very good statistics require little CPU time.

6.2.2 Half-life from Fraction of Decay Curve

Schmidt [16] thoroughly analyzed the problem of uncertainties in measuring the mean lifetime with a small number of detected nuclei, including the presence of a stochastic background. The crucial point was that the measurement was supposed to last until complete decay of the nuclide. The treatment was based on the maximum likelihood approach; the 90 percent confidence intervals were tabulated.

In 1964, in the first experiment of the Dubna group on chemical identification of element 104 as ekahafnium, the SF events of the isolated element were detected in a long, flat, flow-through chamber, the walls of which were covered with track detectors of fission fragments; cf. Fig. 1.3. By that time another Dubna group had reported synthesis and identification of an isotope of element 104 with a half-life of about 0.3 seconds by nuclear physical methods. Such a short half-life dictated high flow rate of the carrier gas in the radiochemical experiments to secure fast transportation from the accelerator target to the experimental setup. The gap between the detectors had to be possibly narrow to enhance registration of both fragments from an SF event occurring in gas. On the other hand, the area of the detectors could not be very large. As a compromise, the chosen volume of the chamber provided the carrier gas hold-up time of 0.7 seconds. The time distribution of 14 detected events [27], which is shown in Fig. 6.2, did not seem to contradict the assumed half-life. However, later, the reference half-life proved to be erroneous; actually, the concrete bombardment produces 3-s ^{259}Rf with a considerable s.f. branch. Thus, the chamber could detect only a small part of the decay curve of this nuclide. With 14 counts, when more decays could happen beyond the chamber, it was impossible to determine the half-life. However, it was important to analyze whether the measured distribution is *compatible* with the three-second half-life at a reasonable confidence level. Such analysis [28] is not a trivial exercise; it has little in common with the problem mentioned in the beginning of this section.

To simulate the likelihood function for employing the Bayesian approach it was necessary to choose a reasonable range for the uniform prior distribution of the mean lifetime t_λ. A minimum of 0.1 seconds was safe, taking into account the time distribution of the 14 decay events. The maximal t_λ leaned upon the total effective production cross section of s.f. nuclei, which was measured in physical experiments. Obviously, 14 decays in 0.7 seconds with the upper t_λ should not correspond to many more events than was the observed total.

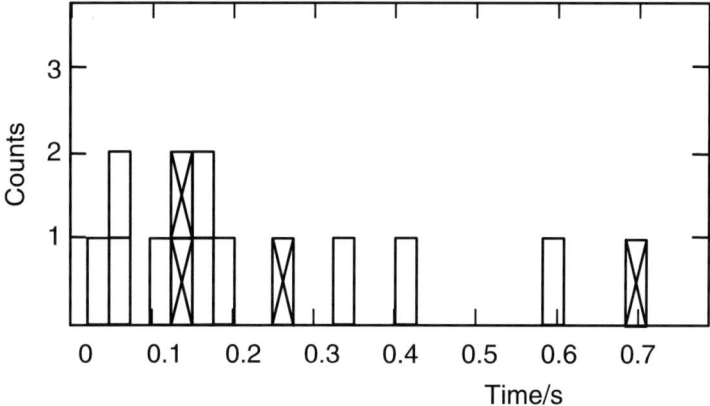

Fig. 6.2 Incomplete decay curve of an s.f. isotope of element 104 [27].

The crosses mark the fission events in which both fragments were registered.

Reproduced (adapted) from Radiokhimiya, 11(2), Zvara I, Chuburkov YuT, Caletka R, Shalaevskii MR, Experiments on chemistry of element 104 II. Chemical study of the spontaneous fission isotope, 163–174, © 1969, with permission of Academizdatcenter "Nauka" Publishers.

The likelihood function was obtained by repeated Monte Carlo simulations of the real experiment at different values of the mean lifetime. These served to find the frequency of "successful" outcomes at each preset value. Because the mean time mark of the 14 counts in Fig. 6.2 was about 0.24 seconds, a simulation was the "success" when the average of the random lifetimes of the first 14 nuclei *which decayed within the initial 0.7 seconds* was 0.24 ± 0.01 seconds.

In greater detail, the values of prior t_λ were incremented from minimum to maximum by 0.1 seconds; then, for each value, the following was repeated 5,000 times:

- Pick up a random lifetime, exponentially distributed with parameter t_λ;
- If it is shorter than 0.7 seconds, store its time mark — otherwise, return to start;
- After collecting 14 such events, calculate their average time mark;
- If the average is the "success," increment the counter at the running t_λ — otherwise, return to start

The normalized histogram of the fractional successes was fitted by a continuous function. It was the Bayesian distribution for t_λ required to find the Bayesian confidence intervals for the true value. The analysis showed that, indeed, the three-second half-life falls within the 95 percent confidence interval. In a similar later work [29], the measurement time was 1.5 seconds and as many as 50 counts were registered. The results, as such, clearly indicated that more decays had taken place beyond 1.5 seconds, so that the half-life is of the order of several seconds.

6.2.3 Adsorption Enthalpy from IC Experiment

In the isothermal chromatography the column is kept at constant temperature. At its exit one detects the atoms which have not decayed during the retention time. Measurements should be done at several decreasing column temperatures to find the interval in which the survival yield is considerably less than the initial 100 percent and so the retention time is comparable with t_λ. Let Y_S be the normalized (per fluence unit) yield at a "high" temperature T_S and Y_A – such yield at a "low" temperature T_A. Their ratio r_λ^{IC} is a direct measure of the retention time, t_R because:

$$r_\lambda^{IC} \equiv Y_A / Y_S = e^{-\frac{t_R^{IC}}{t_\lambda}} \qquad (6.7)$$

In its turn, cf. Sect. 5.1.1.1:

$$t_R^{IC} = \frac{a_z l_c}{Q} e^{\frac{\Delta_{ads} S^\circ}{R}} e^{\frac{-\Delta_{ads} H^\circ}{RT_c}} \qquad (6.8)$$

Let us put:

$$A^{IC} \equiv \frac{a_z l_c}{t_\lambda Q} e^{\frac{\Delta_{ads} S^\circ}{R}} \qquad (6.9)$$

Then

$$r_\lambda^{IC} = \exp\left[-\left(A^{IC}/T_A\right) \times \exp\left(-\Delta_{ads} H^\circ / RT_A\right)\right] \qquad (6.10)$$

and:

$$-\Delta_{ads} H^\circ = RT_A \ln(-T_A \ln r_\lambda^{IC}/A^{IC}) \qquad (6.11)$$

We will restrict our discussion to the case of homogeneous surfaces. If the number of theoretical plates in the IC column is not small, Eq. 6.7 must be quite accurate.

With poor statistics the problem is the accuracy of the experimental r_λ^{IC}. It is demonstrated by the analysis [30] of the studies of seaborgium oxochloride reported in [31–33]. Figure 6.3 shows some relevant data from these papers. The experimental point at 350 °C for Sg is actually the sum of the measurements at 300, 350 and 400 °C placed at the average 350 °C (in Fig. 6.3, this temperature range is indicated by the horizontal bar). The smooth curves (they correspond to Eq. 6.7) were obtained by Monte Carlo simulations based on principles presented in Sect. 4.2. They are supposed to be the best fits to the data.

For a thorough analysis it is necessary to start with unfolded raw data: the numbers of counts detected at the particular temperatures and the expected background. They are given in Fig. 6.4 with the background in parentheses. The beam doses were different at each of the temperatures; Figure 6.4 shows the yield rates (per unit beam dose), which were evaluated in [30] using the counting efficiencies reported in [31–33]. The rate at 300 °C, obtained with a very low beam dose, required the logarithmic scale to be presented. It is questionable whether the data can be described by the shapes drawn in Fig. 6.3.

6.2 Analysis of Poor-Statistics Data

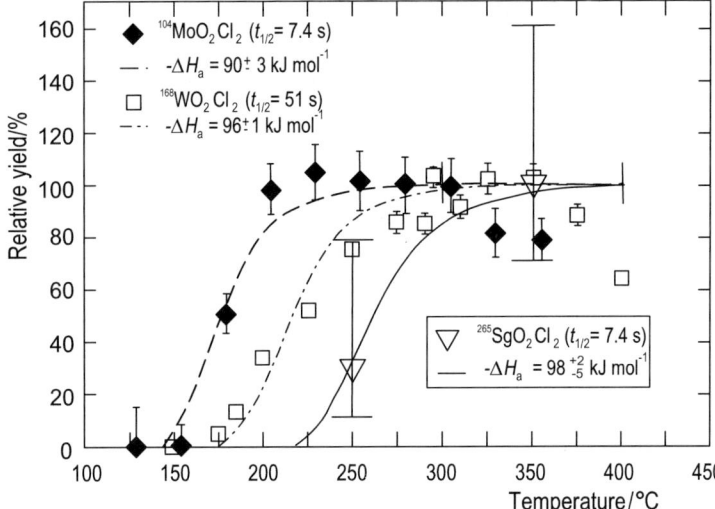

Fig. 6.3 Isothermal chromatography experiments with Mo, W and Sg isotopes [32].
Sg decays were identified by observing correlations of mother-daughter decay events (α or s.f.).
Reproduced from Angewandte Chemie International Edition, 38(15), Türler A, Brüchle W, Dressler R, Eichler B, Eichler R, Gäggeler HW, Gartner M, Gregorich KE, Hübener S, Jost DT, Lebedev VY, Pershina VG, Schädel M, Taut S, Timokhin SN, Trautmann N, Vahle A, Yakushev AB, 2212–2213, © 1999, with permission from John Wiley & Sons Ltd.

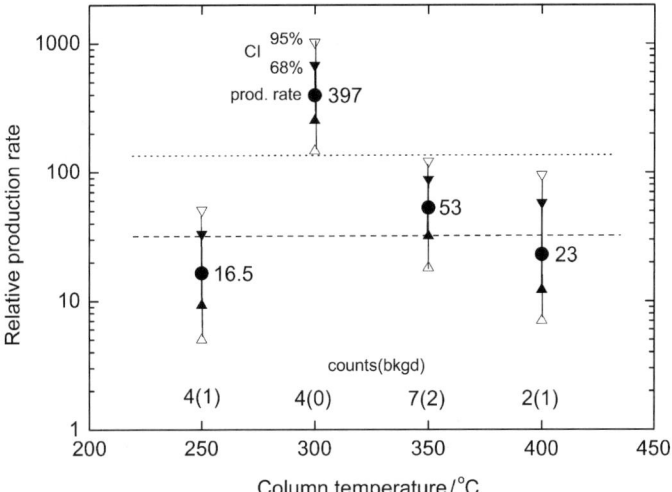

Fig. 6.4 Numbers (per some unit beam dose) of the Sg atoms surviving at the column exit as a function of temperature [30].
The indicated confidence intervals are weighted 68 percent and 95 percent BI.
Reproduced from Physics Atomic Nuclei (Yadernaya fizika), 66(6), Zvara I, Accuracy of the chemical data evaluated from one-atom-at-a-time experiments, 1161–1166, © 2003, with permission from Pleiades Publishing.

Because r_λ^{IC} can be evaluated from only two measurements, one may sum up the data obtained at high temperatures to calculate N_S. However, with such a low statistics, it is difficult to tell the "high" and "low" temperature points apart, and the assignment in the intermediate region becomes arbitrary. Careful analysis is required in considering all possible variants:

- When folding the three points to the right into one at 350 °C, the authors of Ref. [32] claimed that the production rate was "about the same." It is not correct — even the huge 95 percent error bars of the point at 300 °C in Fig. 6.4 do not overlap with those at the other two temperatures. Meanwhile, even the 68 percent BIs at 250, 350 and 400 °C do overlap; thus there is no firm ground to believe that the 250 °C value is significantly smaller. Despite that folding the data like in [32] is not justified, the variant was examined.
- The strange result at 300 °C was omitted and the two measurements at the highest temperatures were combined.
- To avoid any prejudice, the total of the two left points in Fig. 6.3, as well as that of the two right points, were taken in consideration.

The bottom part of Fig. 6.5 graphically illustrates the three above variants of folding and gives the pairs of numbers, which were examined by the simulations. The strategy of simulations can be exemplified for the first variant [28]. In this case, the beam dose in the 250 °C experiment was nearly the same as the total dose in the three higher temperature tests. Hence, this particular simulation did not allow for different fluences. The high temperature point in Fig. 6.3 stems from 13 events with an expected background of three; at 250 °C the numbers are four and one, respectively. The background rates were estimated by the authors of [32, 33] from the α-particle spectra. They were contaminated by radiation of unwanted nuclides with energies similar to Sg and its daughters, which could result in false mother–daughter correlation events due to random coincidences.

A reasonable extent of the uniform prior distribution of the production rate comes from the considerations which follow.

First, it is necessary to admit that all the events at high temperature originated from background, however small the probability of that might be. As an opposite extreme, one must account for the possibility that the yield rate of detectable atoms was much larger than the concrete random 13. Therefore, a reasonably safe interval for N_S is like [0, 30]. The interval for r_λ^{IC} also starts from zero — it cannot be excluded that all low temperature counts are due to background. Its should be extended above unity, again because both measurements can contain background counts; then the reasonable interval is [0, 5]. Thus,

$$F_{pri}(r_\lambda^{IC}) = 0.2 \quad \text{for} \quad 0 \le r \le 5 \quad F_{pri}(r_\lambda^{IC}) = 0 \quad \text{for} \quad r > 5$$
$$F_{pri}(N_S) = 0.02 \quad \text{for} \quad 0 \le N_S \le 50 \quad F_{pri}(N_S) = 0 \quad \text{for} \quad N_S > 0.02$$
(6.12)

The goal of Monte Carlo simulations is to obtain a histogram of the likelihood of observing 13 and 4 counts (including the background) for different values of r_λ^{IC}. Because experimenters were only interested in BIs for r_λ^{IC}, the simulation procedure

6.2 Analysis of Poor-Statistics Data

could be simplified. Its core consisted in that for each value of r_λ^{IC} (incremented by 0.05) from the interval [0, 5]; the procedure below was repeated 5,000 times to guarantee good statistics:

For N_S taken at random from the above interval the outcome was simulated in terms of the number of counts detected at 250 and 350 °C. When they were four and 13, it was a "success." The flowchart of simulations involved the following instructions:

- Pick up a random value of N_S from the interval [0, 50];
- Pick up a random k_S, Poisson-distributed with parameter N_S;
- Pick up a random b_S, Poisson-distributed with parameter three;
- Pick up a random $k_{S,r}$, Poisson-distributed with parameter $N_S r_\lambda^{IC}$;
- Pick up a random $b_{S,r}$, Poisson-distributed with parameter one;
- If $k_S^* + b_S^* = 13$ and $k_{S,r}^* + b_{S,r}^* = 4$, increment the success counter by one — otherwise, return to start.

An efficient and fast subroutine for obtaining Poisson-distributed values like k_S is [34]:

$$k_S = \min\left\{n_P : \prod_{i=0}^{n_P} \xi_i < e^{-N_S}\right\} \qquad (6.13)$$

It means calculating the product of consecutive random numbers ξ_i, uniformly distributed in (0, 1], until the result gets smaller than e^{-N_S}. Then the random k_S is the number of necessary multiplications; if the first ξ_i is already smaller, then $k_S = 0$.

The normalized histogram of the frequency of successes was again fitted by a smooth curve to evaluate the confidential intervals. The other two possible combinations of the data points, which are shown in the bottom of Fig. 6.5, were simulated in a similar way.

The concrete weighted [26] confidence intervals of r_λ^{IC} for the three different ways of folding the data are shown by the solid circles with error bars in the middle of Fig. 6.5. The numerical confidence intervals for the enthalpy of adsorption are listed in the top of the figure. They were obtained using Eq. 6.11. One can see that the 95 percent BIs extend above $r_\lambda^{IC} = 1$ even in the "most favorable" case of four and 13 counts, so that the difference between the yield rates at higher and lower temperatures is not statistically significant! The other two choices, which are more reasonable, gave even less significant difference. Now, using Eq. 6.11, it was possible to evaluate the BIs for the values of $-\Delta_{ads}H$; they are presented in the top lines of Fig. 6.5. It shows that, at a confidence level of 95 percent, one can set only the upper limit for the $-\Delta_{ads}H$, but not the lower one. Molybdenum and tungsten showed decreasing yield at the highest temperatures, cf. Fig. 6.3. It probably resulted from lesser efficiency of re-clustering (see Sect. 3.4.1). An empirical correction for it might consist in multiplying the number of counts observed at 400 °C by a factor of 1.2 or so. However, it would negligibly affect the quantitative conclusions of the above analysis.

There might also be "systematic errors" like, for example, erroneous assignment of mother–daughter decay correlations or random coincidences. Such problems are

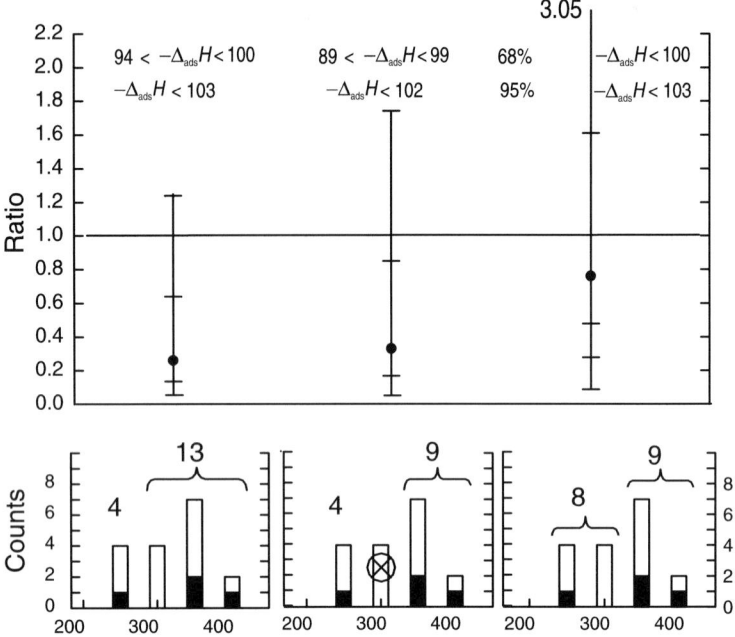

Fig. 6.5 Evaluation [30] of the data [5] in Fig. 6.3.

Bottom: Schematic of different combinations of the data points. The expected contribution of background is indicated by the black portion of the histogram bars. Middle: Confidence intervals, 68 percent and 95 percent, for the N_A/N_S values (bars) for different combinations of the experimental points. Top lines: see text.

Reproduced from Physics Atomic Nuclei (Yadernaya fizika), 66(6), Zvara I, Accuracy of the chemical data evaluated from one-atom-at-a-time experiments, 1161–1166, © 2003, with permission from Pleiades Publishing.

demonstrated by the fact that the first publication [31], after preliminary evaluation of the experimental data, reported firm observation of zero counts at 250 °C with a 68 percent CI of $0^{+0.3}_{-0}$. The above statistical analysis concerns the later papers [32, 33], which gave 3^{+7}_{-2} at the same temperature.

6.2.4 Adsorption Enthalpy from TC Experiment

The theory of "ideal" thermochromatography suggests that, at any point, there exists thermodynamical (adsorption) equilibrium between the gas-phase and surface concentrations of the adsorbable molecules. The ideal *elution* TC, like the ideal IC of nonradioactive species, would yield a narrow band at some T_A^{id} – a function of $\Delta_{ads}H$ and $\Delta_{ads}S$, as well as of the processing time and temperature profile. Figure 4.1 illustrates different situations.

In the ideal *frontal* thermochromatography, the high temperature branch of the TC peak must obviously approach the adsorption isobar

$$\rho^{TC}(z) \sim e^{\frac{-\Delta_{ads}H}{RT_z}} \qquad (6.14)$$

and the zone must abruptly terminate at the same T_A^{id} as above. The integral of the isobar from T_S to T_A^{id} equals the total number of detectable atoms.

Let us put (by analogy with Eq. 6.9 for IC)

$$A^{TC} \equiv \frac{a_z T_A^{id}}{t_R^{TC} Qg} e^{\frac{\Delta_{ads}S^\circ}{R}} \qquad (6.15)$$

where $g > 0$ is the temperature gradient. Remember that we assume $\Delta_{ads}S$ is known and is independent of T. Then,

$$\text{Ei}^*(-\Delta_{ads}H/RT_A^{id}) = 1/A^{TC} \qquad (6.16)$$

After approximations:

$$-\Delta_{ads}H = RT_A^{id} \cdot \ln\left[\frac{-\Delta_{ads}H/RT_A^{id}}{A^{TC}(1 - RT_A^{id}/\Delta_{ads}H)}\right] \qquad (6.17)$$

Here we mostly discuss long continuous experiments with short-lived nuclides, in which the "adsorption zone" is seen as the longitudinal distribution of the detected decays. As an approximation, we can consider it as a result of frontal chromatography lasting the mean lifetime. The randomness of t_λ in itself must broaden the zone. Moreover, the short half-lives require a high linear velocity of the carrier gas, which prevents reaching the equilibrium as the temperature drops. All this makes it difficult to calculate the shapes of the TC adsorption zones. In the meantime, any concrete experiment can be simulated by Monte Carlo techniques, like those described in Sect. 4.3, which provide satisfactory fits to real TC zones. In the case of sufficient statistics, like a total of 40 seaborgium decays in Fig. 6.6 from Ref. [35], one can perform Monte Carlo simulations of many molecular histories under given experimental conditions to obtain smooth profiles for various values of $\Delta_{ads}H$. The profiles, parameterized by a (semi)empirical formula, can serve to obtain the best $\Delta_{ads}H$ value to fit the experimental data.

6.2.5 Adsorption Enthalpy from Corrupted Thermochromatogram

If the statistics are poor, the measured distribution of the rare counts over sections of the column can be directly simulated a great many times to obtain the likelihood function suitable for the Bayesian approach to the problem.

A more complicated case is the corrupted thermochromatogram obtained in the recent chemical identification of hassium [36]. The tetroxide (presumably) of the element was detected in a TC column of narrow rectangular cross section. The working surface of the column was that of the flat spectrometric detectors of charged particles, which registered decays in real time. The raw experimental data were shown in Fig. 1.9. Seven decay events were detected, but only three of them, – those due to ^{269}Hs – were used for evaluation of $\Delta_{ads}H$ because the half-lives of other isotopes were then not known. The first pair of the detectors ("first section") did not operate properly, so it was not known whether some decays of Hs occurred. This ambiguity makes interpretation of the data especially difficult. All three "useful" counts of ^{269}Hs occurred in the third section of the column. By using the Monte Carlo simulations of thermochromatographic peaks, the authors of Ref. [36] arrived at $\Delta_{ads}H = -46 \pm 3$ kJ mol^{-1} (95 percent CI) for HsO$_4$; no additional information was given.

A Bayesian analysis (with uniform prior for $-\Delta_{ads}H$) was performed by the present author in [30]. The likelihood was obtained for 15 values of the desorption enthalpy in the interval 40 to 70 kJ mol^{-1}. The "success" was when a series of consecutive individual histories resulted in any number of decays within the first section, zero in the second, three events within the third section and zero decays beyond. The frequencies are plotted in Fig. 6.6. A peculiar feature is the constant "tail" toward high $-\Delta_{ads}H$. Therefore, the BIs cannot be determined as they depend on the choice of the upper $-\Delta_{ads}H$ in its prior distribution.

Some considerations suggest the necessity of additional experiments. For example, a considerable fraction of the successes in simulation was due to the molecules that reached the third detector pair through the sole first jump from the column inlet. This fraction becomes constant with higher desorption energy because adsorption is getting irreversible. Remember that the molecules were transported to the column at ambient temperature, through a long capillary of polymeric material. Hence, highly adsorbable molecules may have already been deposited in the duct and could not reach the column. It would be possible to account for it by appropriately lowering the upper $-\Delta_{ads}H$ in the uniform prior. However, it would require knowledge of relative adsorbability of the tetroxides on all the dissimilar contacted surfaces, which is not available.

Another factor is the strong perturbation of the flow at the column inlet. For two first column sections the effective Monte Carlo jump lengths must be shorter than those in the developed flow (cf. Fig. 3.7). The simulations in [30] did not take it into account; as a result, the likelihood of reaching the third pair of detectors by the first jump was overestimated.

From the above we can alternatively suppose that the deposition of tetroxides during transportation is negligible, and that in the beginning of the channel the TC regime is more equilibrated. Then we can subtract the first jump contribution to the frequency of successes to obtain another extreme for this function. It is shown by the dotted curve in Fig. 6.6.

The above qualitative considerations could be reasonably quantified if the deposition of a nonvolatile long-lived gamma-active nuclide were also measured with a resolution of some 1 cm or better. Such data would allow more realistic simulations, accounting for the failure of the first detector and zero counts observed

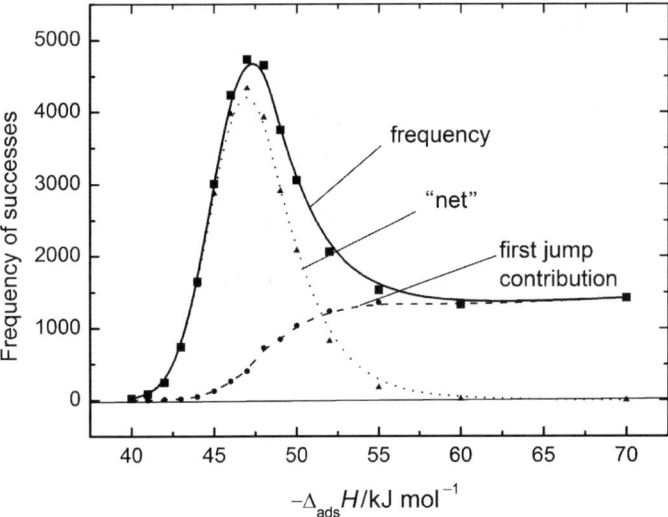

Fig. 6.6 The Bayesian posterior for the adsorption enthalpy of HsO_4 [30]. The "net" curve is the difference between two others.

Reproduced from Physics Atomic Nuclei (Yadernaya fizika), 66(6), Zvara I, Accuracy of the chemical data evaluated from one-atom-at-a-time experiments, 1161–1166, © 2003, with permission from Pleiades Publishing.

on the second one. Actually, they are highly desirable each time the adsorption zones are close to the column inlet.

The author's feeling [30] is that $-\Delta_{ads}H$ on the detector surface is hardly less than $42 \, kJ \, mol^{-1}$, but that a well-founded quantitative estimate of the upper limit from the original data is impossible. The limit given in [36] cannot be justified as such because of the information about the assumptions behind the reported numbers is incomplete.

6.2.6 Conclusions

We conclude that an experiment with poor statistics can yield conclusive results of crucial importance only when there is a *qualitative* difference in behavior of the compared elements. For example, if the chemical system can so strongly distinguish an expected congener and the new element that they are resolved into individual fractions, and the few available atoms of TAE are found only in one of them. Zero counts in the other fraction provide the result of highest statistical significance possible in such experiment, though they do not evidence firm zero distribution coefficient. Such systems seem to seldom occur.

Obtaining a good *quantitative* estimate of the adsorption enthalpy in an experiment like those discussed above is even more difficult. First, the inaccuracy of the result stems not only from the poor statistics, but also from possible systematic errors (they were discussed in detail in Sect. 5.6). It is highly desirable that the new element be produced simultaneously with an equally short-lived isotope of its chemical homolog. It allows reducing the necessary corrections of the raw data for the differences in half-lives, zone temperatures and similar parameters to minimum. Otherwise it is questionable to draw any serious conclusions, like whether the "experimental" and "theoretical" (quantum chemistry) values of a characteristic agree or disagree. Finding a statistically significant difference in the characteristics of a TAE and its homolog is a fundamental result. It requires accurate statistical treatment of each particular experiment.

References

1. Reischmann FJ, Trautmann N, Herrmann G (1984) Radiochim Acta 36:139
2. Reischmann FJ, Rumler B, Trautmann N, Herrmann G (1986) Radiochim Acta 39:185
3. Zvara I, Belov VZ, Domanov VP, Shalaevski MR (1976) Radiokhimiya 18:371; Soviet Radiochem 18: 328
4. Zvara I (1976) One-atom-at-a-time chemical studies of transactinide elements. In: Muller W, Lindner R (eds) Transplutonium elements – Proc 4th Internat transplutonium elements symposium, Baden Baden. North Holland, Amsterdam, p. 11
5. Borg RJ, Dienes GJ (1981) J Inorg Nucl Chem 43:1129
6. Guillaumont R, Adloff JP, Peneloux A (1989) Radiochim Acta 46:169
7. Guillaumont R, Peneloux A (1990) J Radioanal Nucl Chem 143: 275
8. Trubert D, Le Naour C (2003) Fundamental aspects of single atom chemistry. In: Schädel M (ed) The chemistry of superheavy elements. Kluwer, Dordrecht, p.95
9. Koudriavtsev AB, Linert W, Jameson RF (2001) The law of mass action. Springer, Berlin Heidelberg New York, p. 67
10. Currie LA (1968) Anal Chem 40:586
11. Strom DJ, MacLellan JA (2001) Health Phys 81:27
12. Cleveland BC (1983) Nucl Instrum Methods 214:451
13. Zlokazov VB (1978) Nucl Instrum Methods 151:303
14. Hall P, Selinger B (1981) J Phys Chem 85:2941
15. Schmidt KH, Sahm CC, Pielenz K, Clerc HG (1984) Z Phys A 316:19
16. Schmidt KH (2000) Eur Phys J A8:141
17. Gregorich KE (1991) Nucl Instrum Methods Phys Res, Sect A 302:135
18. Bukin A (2003) A comparison of methods for confidence intervals. In: Lyons L, Mount R, Reitmeyer R (eds) Proceedings of Phystat 2003 conference. SLAC, Menlo Park, p. 148
19. Helene O (1984) Nucl Instrum Methods Phys Res 228:120
20. Barlow R, Cahn R, Cowan G, Di Lodovico F, Ford W, de Monchenault GH, Hitlin D, Kirkby D, Le Diberder F, Lynch G, Porter F, Prell S, Snyder A, Sokoloff M, Waldi R (2002) Recommended statistical procedures for BABAR. BABAR analysis document #318, Version1 *http://replicator.phenix.bnl.gov/phenix/WWW/publish/lixh/BaBar-stat-report.ps* Cited 15 May 2007
21. Porter F (2000) SLUO statistics lecture 3. *http://www-group.slac.stanford.edu/sluo/Lectures/stat_lecture_files/sluolec3.pdf*. Cited 15 May 2007
22. James F, Roos M (1980) Nucl Phys B 72: 475
23. De Angelis A, Iori M (1987) Nucl Instrum Methods Phys Res, Sect A 260:451
24. Prosper HB (1985) Nucl Instrum Methods Phys Res, Sect A 241:236
25. Brüchle W (2003) Radiochim Acta 91:71

References

26. Dressler R, Türler A, Schumann D (1999) The method of total ignorance. In: 1st International conference on the chemistry and physics of the transactinide elements, Sept 1999, Seeheim. Extended abstracts, P-M-9
27. Zvara I, Chuburkov YuT, Caletka R, Shalaevskii MR (1969) Radiokhimiya 11:163
28. Zvara I Unpublished results
29. Zvara I, Chuburkov YuT, Belov VZ, Buklanov GV, Zakhvataev BB, Zvarova TS, Maslov OD, Caletka R, Shalayevsky MR (1970) Radiokhimiya 12:565; Soviet Radiochem 12: 530, J Inorg Nucl Chem 32:1885
30. Zvara I (2003) Physics Atomic Nuclei (Yadernaya fizika) 66:1161
31. Gäggeler HW (1998) J Alloys Compd 271–273:277
32. Türler A, Brüchle W, Dressler R, Eichler B, Eichler R, Gäggeler HW, Gartner M, Gregorich KE, Hübener S, Jost DT, Lebedev VY, Pershina VG, Schädel M, Taut S, Timokhin SN, Trautmann N, Vahle A, Yakushev AB (1999) Angew Chem Int Ed 38:2212
33. Türler A, Dressler R, Eichler B, Gaggeler HW, Jost DT (1998) Phys Rev, C 57:1648
34. Yermakov SM, Mikhailov GA (1982) Statisticheskoye modelirovaniye (Statistical modeling). Nauka, Moskva
35. Zvara I, Yakushev AB, Timokhin SN, Xu HG, Perelygin VP, Chuburkov YuT (1998) Radiochim Acta 81:179
36. Düllmann ChE, Brüchle W, Dressler R, Eberhardt K, Eichler B, Eichler R, Gäggeler HW, Ginter TN, Glaus F, Gregorich KE, Hoffman DC, Jager E, Jost DT, Kirbach UW, Lee DM, Nitsche H, Patin JB, Pershina V, Piguet D, Qin Z, Schädel M, Schausten B, Schimpf E, Schott H-J, Soverna S, Sudowe R, Thorle P, Timokhin SN, Trautmann N, Türler A, Vahle A, Wirth G, Yakushev AB, Zielinski PM (2002) Nature 418:859

Author Index

A
Acosta, J.J.C., 23
Altynov, V.A., 20
Aris, R., 95

B
Bächmann, K., 10, 19, 28, 30, 126, 128, 138, 178
Bakaev, V.A., 148, 166
Baltensperger, U., 11
Barberi, R., 144
Barth, J.V., 159
Bartolino, R., 144
Belov, V.Z., 8, 26, 64, 126, 193
Berg, E.W., 23
Bernasconi, M., 150, 154
Blachot, L.C., 28
Bombi, G.G., 94
Bonvent, J.J., 142, 144
Borg, R.J., 194, 195
Boussières, G., 3
Brüchle, W., 15, 16, 199, 200, 205
Bukin, A., 197
Buklanov, G.V., 8, 64

C
Caletka, R., 6, 63, 78, 203
Capelli, L., 144
Chelnokov, L.P., 8, 64
Chepigin, V.I., 29
Chuburkov, Yu.T., 4, 6, 63, 78, 203
Chun, K.S., 29
Currie, L.A., 199
Czerwinski, K.R., 13

D
De Angelis, A., 199
De Boer, J.H., 127, 130, 141

Debye, P., 42, 161
Dienes, G.J., 194, 195
Di Marco, D.B., 94
Domanov, V.P., 8, 18, 29, 64, 126, 193
Dressler, R., 15, 16, 199, 205
Düllmann, Ch.E., 14–16, 40, 59

E
Eberhardt, K., 15, 16
Eichler, B., 15, 16, 26, 28, 29, 113, 138, 181, 205
Eichler, R., 15, 16, 117, 205
Evans, M.G., 67

F
Fedoseev, E.V., 23
Fehnse, H.F., 10
Fishlock, T.W., 158
Folden, C.M., 59
Frischat, G.H., 146, 147

G
Gäggeler, H.W., 11, 13, 15–16, 28–29, 113, 124, 205
Gäggeler-Koch, H., 29
Gartner, M., 205
George, S.M., 151
Giddings, J.C., 41, 93
Gilliland, E.R., 40, 41
Ginter, T.N., 15, 16
Glaus, F., 15, 16
Gnielinski, W., 50
Gorlov, Yu.I., 156
Gormley, P.G., 47
Goss, A., 146, 147
Gregorich, K.E., 13, 15, 16, 205

Greulich, N., 28, 29
Guillamont, R., 194
Gupta, P.K., 142, 144

H

Hamann, D.R., 151
Hambleton, F.H., 151, 152
Haukka, S., 157
Heide, G., 146, 147
Helene, O., 199, 200
Henderson, R.A., 13
Henke, L., 145
Herrmann, G., 29
Hickmann, U., 10, 28, 29
Hill, T.I., 169
Hockey, J.A., 151, 152
Hoffman, D.C., 15, 16
Hohn, A., 117
Hübener, S., 19, 23, 24, 205
Hussonnois, M., 8, 26, 64

I

Illas, F., 152
Inglesia, E., 147, 148, 169, 176
Inniss, D., 144
Iori, M., 199
Isted, G.E., 158

J

Jäger, E., 15, 16
James, F., 199
Jin, K.U., 137
Jonsson, J.A., 93
Jorgensen, J.W., 95
Jost, D.T., 11, 13, 15, 16, 205

K

Kadkhodayan, B., 12
Kennedy, M., 47
Kim, U.J., 20, 21
Kirbach, U.W., 15, 16
Kiselev, A.V., 148
Knaupp, S., 142
Knudsen, M., 87, 112, 114, 115
Kolatchkowski, 97
Korotkin, Yu.S., 8, 20, 26, 64
Kosanke, K.L., 79
Kovacs, A., 11
Kovacs, J., 13
Krivanek, M., 4
Krull, U.J., 145
Kurkijan, C.R., 144

L

Lan, K., 95
Le Naour, C., 195
Lebedev, V.Y., 205
Lebedev, V. Ya., 185, 205
Lee, D.M., 15, 16
Lee, W.T., 158
Leonardelli, S., 149
Lindemann, F.A., 42
Lopez, N., 151, 152
Lygin, V.I., 153

M

MacLellan, J.A., 199
Masini, P., 150, 154
McDaniel, M.P., 157
Merinis, J., 3

N

Nagy, N., 145
Nitsche, H., 15, 16
Novgorodov, A.F., 97

O

Orelowich, O.L., 20

P

Pacchioni, G., 152
Pantano, C.G., 150
Patin, J.B., 15, 16
Patrikiejew, A., 162
Pershina, V., 15, 16, 178, 205
Piguet, D., 15, 16
Poggemann, J.F., 146, 147, 150
Polanyi, M., 67
Pollard, W.G., 114
Porstendorfer, J., 81
Porter, F., 197
Prosper, H.B., 199

Q

Qin, Z., 15, 16

R

Radlein, E., 146, 147
Rarivomanantsoa, M., 151
Reichsmann, 192
Rengan, K., 28
Righetti, P.G., 144
Roos, M., 199
Rudolph, J., 28, 30

S

Samhoun, K., 22
Schädel, M., 15, 16, 113, 205

Schausten, B., 15, 16
Schegolev, V.A., 8, 64
Schimpf, E., 15, 16
Schmidt, K.H., 202
Schmidt-Ott, W.D., 10, 72
Schott, H.-J., 15, 16
Schrewe, U.J., 10
Schrijnemakers, P., 156
Semenov, N.N., 67
Seward, N.K., 15
Shalaevski, M.R., 126, 193
Shalaevskii, M.R., 6, 26, 78, 203
Shalayevsky, M.R., 8, 64
Shannon's, R.D., 140
Shchegolev, V.A., 26
Sherer, U.W., 13
Shilov, B.V., 5
Sneh, O., 151
Souza, S.D., 150
Soverna, S., 15, 16
Stallons, J.M., 147, 148, 169, 176
Steele, W.A., 48
Steffen, A., 19, 128
Stender, E., 79
Strellis, D.A., 15
Strom, D.J., 199
Sudowe, R., 15, 16
Suglobov, D.N., 23

T
Taut, S., 205
Taylor, G.I., 95
Thörle, P., 15, 16
Timokhin, S.N., 9, 15, 16, 20, 21, 29, 185, 205
Trautmann, N., 15, 16, 29, 205, 206
Travnikov, S.S., 23
Trubert, D., 195
Tunitskii, N.N., 97
Türler, A., 11, 13, 15, 16, 205

V
Vahle, A., 15, 16, 72, 182, 205
Van Der Voort, E., 156
Vedeneev, M.B., 185
Vermeelen, D., 11, 13
Vitiello, M., 152
Von Dincklage, R.D., 10

W
Wadsak, M., 158
Weber, A., 11, 13
Wirth, G., 15, 16

Y
Yakushev, A.B., 9, 15, 16, 185, 205

V
Vansant, F., 156

Z
Zhuikov, B.L., 29, 74, 99, 100
Zhuravlev, L.T., 148, 149, 151
Zielinski, P.M., 15, 16
Zvara, I., 4–6, 8, 9, 21, 24, 26, 27, 29, 63, 64, 68, 77, 78, 89, 101, 102, 104, 106–108, 126, 127, 137, 185, 193, 203, 205, 208, 211
Zvarova, T.S., 4, 26, 27, 63, 78

Subject Index

A

Actinoids (definition), xxiii
Adsorption. *See* Adsorption enthalpy; Adsorption entropy; Adsorption thermodynamics; Chemisorption; Physical adsorption
   localized, 114, 122–124, 126, 132, 133, 135, 141, 162, 164–166, 174, 180
   intermediate, 133, 162, 164, 173
   mobile, 112, 116, 122–124, 127, 130, 133, 135, 136, 138, 141, 162, 164, 165, 173, 174
Adsorption enthalpy (experimental) by Second Law
   from retention times in IC, 124
   from retention times in temperature programmed chromatography, 125
   from survival yield of short-lived nuclides in IC, 124
   from thermochromatograms at different run duration, 125, 126
   sample measurements, 126–128
Adsorption enthalpy (experimental) by Third Law: ideal surface, mobile adsorption
   calculation formulae for IC, 135
   calculation formulae for TC, 135–137
   correlation of the values with sublimation enthalpies, 71, 138, 139, 178
   proximity of the values to sublimation enthalpies, 138, 139, 178
   rationale lacking, 128, 140, 172, 174, 177
   real surfaces require revision of the values. *See* Desorption energy data by Third Law
Adsorption enthalpy from experimental data, on heterogeneous surface. *See* Desorption energy from experimental data

Adsorption entropy. *See also* Partition functions
   entropy of adsorbate on homogeneous surface from statistical mechanics, 131–134
      mobile model, 131, 132
      localized model, 132–134
      accounting for surface diffusion, 163–165
   on heterogeneous surface, 169–171
   on homogeneous surface
      localized adsorption, 134
      localized adsorption with surface diffusion, 163–165
      mobile adsorption, 131, 132
      uncertainty due to postulating unchanged internal entropy, 162, 163
   quality of experimental values, 127, 128
Adsorption isobar, 89, 100, 126, 127, 209
Adsorption sites, 122, 132, 159, 164–166, 179, 181
   active, 60, 192
   blocking by reagents, 60
   number concentration of, 122, 133
   possibly overestimated, 174
Adsorption sojourn time. *See* Physical adsorption;
Adsorption thermodynamics
   adsorption reference states, 162
   adsorption standard states, xxii, 127, 131, 133, 134, 181
   fractional surface coverage, 123
   molar area, 122, 123, 131, 133
   molar volume, 122, 123, 127, 131
   distribution coefficient (dimensional), 121
   from experiments in uniform isothermal column, 121

219

Adsorption thermodynamics (cont.)
  equilibrium constants (dimensionless), 121
    for ideal mobile adsorption model, 122, 123
    for ideal localized adsorption model, 122, 123
    for real surfaces, effective, 167, 169, 171, 175, 177, 178
Aerosols
  coagulation rate, Smoluchowski equation, 81
  diffusion coefficient, 44, 45
    Cunningham slip factor, 45
  diffusional deposition of. *See* Diffusional deposition in channels
  generators (production), 10, 11, 79, 80
  gravitational settling, 80, 84
  materials of, 10–12, 72
Aerosol flow transportation, 9–12, 79–82
  deposition of particulates by impact, 12, 79
  optimal size of particulates, 80, 81
    necessary lower limit of concentration, 81
  reclustering at IC column exit, 11, 12, 14, 82
  peculiarities compared with molecular transportation
  efficiency for short-lived nuclides, 84
  spike profile change with distance, 83

B

Bayesian statistics, 197, 202, 203, 209. *See also* Poor-statistics data, Bayesian treatment
  Bayesian (confidence) intervals, BI, 197
  BIs for difference of Poisson-distributed quantities (table), 200
  BIs for ratio of Poisson-distributed quantities, (table), 200, 201
  compared with frequentist statistics, 197
  likelihood function, 197, 198, 202, 203, 209
  posterior distribution of parameter, 197, 198
  prior distribution of parameter, 197, 198
    complete ignorance of, 198, 199
  statistical inference, 197
Bimolecular reactions, 67, 186
  rate of, 37–39
Bohrium (Bh, element 107), 12
  longest-lived isotopes, 55
  volatile oxychloride, 12
Boltzmann factor, 42, 100, 136, 160, 161
  integrals containing the factor, 42, 43
  Brominating agents. *See* Synthesis of volatile compounds on-line.

C

Carrier gas (definition), xxii
  hold-up time of. *See* Gas hold-up time
Chemical identification of TAEs (definition), xviii
Chemisorption, 119, 120, 153, 172, 181
Chemical volatilization, xxi, 75
Chlorinating agents. *See* Synthesis of volatile compounds on-line.
Chlorination of adsorbed tracers, 70–72
  conditions for fast kinetics of, 71
  Zr with $TiCl_4$ → $ZrCl_4$ on silica surface, 70–72
Chlorination of gaseous tracers. *See also* Synthesis of volatile compounds on-line; Scavenging impurities in carrier gas
  bimolecular steps involving radicals, 65
  activation energy versus enthalpy change, 67
  conditions for fast kinetics of, 66, 67, 71, 72
  mechanism of Zr with $TiCl_4$ → $ZrCl_4$, 65
  thermochemistry and kinetics, 65
  thermochemistry of all possible reaction paths
    Zr with $TiCl_4$ → $ZrCl_4$, 67–69
    Zr with $SOCl_2$ → $ZrCl_4$, 68
    W(Mo) with $SOCl_2$ → $W(Mo)OCl_4$, 69, 70
Chlorination on hot aerosol filters, 72
Chromathermography, 97, 112
Chromatographic peak shape, 93–100
  statistical moments and cumulants, 93, 94
Chromatographic peaks in IC
  approximate profile formula, 97
  computer simulations. *See* Monte Carlo simulations
  dispersion due to
    laminar flow patterns, 95
    longitudinal diffusion, 95
    migration slower than flow velocity, 95
Chromatographic peaks in TC
  approximate formulae for slow flow, 99, 100
  compression by temperature gradient, 97, 98
  computer simulations. *See* Monte Carlo simulations
  dispersion at very low flow rates, 97–100
  fitting by exponentially modified Gaussian, 108–110
Collisions of molecules. *See* Molecular kinetics
Cunningham slip correction, 44

Subject Index

D

De Broglie wave length, 129
Desorption energy (definition) 165, 166
Desorption energy data by Third Law: heterogeneous surface, localized adsorption
  exceeds sublimation energy, 140, 141, 177, 178. *Cf.* Adsorption enthalpy (experimental) by Third Law
  possible factors enhancing high values of, 175
    adsorption pockets, 173, 174
    incomplete modification of surface, 176, 177
    localized rather than mobile adsorption, 173, 174
    losses of internal entropy in adsorption, 174
    uncertainty of some required quantities, 174, 175
Desorption energy, heterogeneous surface
  fundamentals, 167–169. *See also* Adsorption enthalpy
  spectra of, 167
  spectra of, assumed for discussion, 168, 169
    effective mean value of energy, 168
    Second Law treatment of effective energies, 168, 169
  spectra calculated by molecular dynamics, 176
Desorption entropy, heterogeneous surface. *See also* Adsorption entropy
  accounting for surface diffusion, 169–171
Detection of rare decay events of heavy elements
  ionization chamber for fission events, 17
  semiconductor detectors of α particles and fission fragments, 12, 15
  solid state track detectors of fission fragments, 6
Diffusion. *See* Aerosols, diffusion coefficient; Diffusional deposition in channels; Gaseous diffusion; Knudsen diffusion; Surface diffusion
Diffusional (irreversible) deposition in channels – deposit density and penetration
  analytical solutions for diffusionally developing laminar flow
    for circular channels, 46, 47
    for rectangular channels, 47, 48
    engineering approach, 48
    for developed turbulent flow, 50
    for diffusionally and hydrodynamically developing, laminar flow, 49, 50
Dubnium (Db, element 105), 12, 13, 73
  bromides of, 13
  chlorides of, 192
  longest-lived isotopes, 55

E

Ekahafnium. 7, 202. *See* Rutherfordium
Element 112 (Ekamercury)
  adsorption on gold, 17
  longest-lived isotopes, 55
  volatility in atomic state, 16
Engeworth-Cramer asymptotic expansion, 94
Entropy. *See* Adsorption entropy; Partition functions
Exponentially modified Gaussian, 94, 95
  fitted by Gram–Charlier series, 95
  fitting Monte Carlo simulations by, 107–110
Elution curve, 63, 64, 82, 83, 87, 88, 93, 96, 124

F

Fluorinating agents, 22
Free random
  displacements in VTC column, 114, 116
  flights in gas, 101, 102
  jumps in surface diffusion, 161
Future research needs
  advanced peak profile simulations, 112
  conditioning of open columns, 179
  formulae for thermochromatographic peaks, 98, 100
  more of precise comparative data for known elements, 177, 178, 180

G

Gas hold-up time, 20, 38, 53, 62, 63, 70, 75, 84, 91–93, 101, 202
Gaseous diffusion, 40
  as a result of random flights, 41
  coefficient of mutual diffusion, 40, 41, 45, 77, 96
  for two-dimensional gas, 173
  Gilliland equation for the coefficient, 40
Gas-solid chromatography method and experimental techniques. *See* Chromathermography; Isothermal chromatography (IC); Temperature programmed chromatography; Thermochromatography (TC)
  non-trivial chromatographic mechanisms. *See* Reaction chromatography

Gas-solid chromatography method and experimental techniques (cont.)
  realization of, on-line with accelerator beams
    advantages and disadvantages of TC and IC for transactinoid studies, 13, 14
    first on-line experiments with Hf and Rf, 5
    simulation of, using fission products, 4
Gram–Charlier series, 94

H

Hassium (Hs, element 108), 14–16
  longest-lived isotopes, 55
  volatile tetroxide of, 14–16, 178, 209
Heterogeneous surface. *See* Desorption energy; Desorption entropy; Surface of fused silica; Surface of metals

I

Internal chromatograms, 87, 88, 90
  in isothermal chromatography, 87
  in thermochromatography, 87, 88, 90, 105
Isothermal chromatography (IC). *See also* Reaction chromatography
  characteristic of the method, 87, 88
  theory of ideal, 89–91
    gas hold-up time, 90
    migration distance, 90
    net retention time, 90, 103

K

Knudsen diffusion (regime) in evacuated channels, 112
  description by effective flow, 112
  effective diffusion coefficient, 114, 115
  Monte Carlo simulation by random flights, 116, 117

L

Lanthanoids (definition), xxiii
Lateral diffusion (migration) of adsorbate. *See* Surface diffusion
Localized adsorption model. See Adsorption entropy
Loschmidt number, 36

M

Mobile adsorption model. *See* Adsorption entropy
Molecular kinetics, 36–43
  collisions of gaseous molecules, 38, 39
  collision diameter, 39, 40
  rate of chemical interaction, 37–39
  reduced mass of colliding particles, 38
  collisions of gaseous molecules with walls, 37
  number of, when passing a volume, 38
  concentration of gaseous molecules, 36, 37
  mean speed of gaseous molecules, 37
Monte Carlo simulations of experimental data on few atoms. *See* Poor-statistics data, Bayesian treatment.
Monte Carlo simulations of likelihood function. *See* Poor-statistics data, Bayesian treatment.
Monte Carlo simulations of molecular migration histories and chromatograms
  assumptions and approximations, 101–104, 110, 111
  individual paths in time and distance, 104
  microscopic picture of migrations, 100, 101
  migration distance as sum of long jumps, 102
  effective long jumps (exponential pdf), 103
  jumps of zero length, number of, 102, 103
  simplified pdf of displacements, 101–103
  retention time as sum of multiple sojourns at jump endpoints, 103–105
  pdf of the sum, 103
  simulation flowchart, 106
  graph of simulated individual paths, 104
  simulations of internal chromatograms, examples
    elution TC, long-lived nuclide, 109
    elution TC, short-lived nuclide, 109
    frontal TC, long-lived nuclide, 109
    fits of peaks with exponentially modified Gaussian, 109, 110
    statistical characteristics of simulated and fitting peaks, 110, 111
  variables affecting peak shapes, 110–111

N

Net retention time. *See* isothermal chromatography, theory; Thermochromatography, theory

P

Partition functions, molecular, molar, 128, 129
  rotational, 130
  translational, 129, 130
    for two-dimensional gas, 129
  vibrational, 130

Subject Index

Peclet number (Pe), 96
Physical adsorption,
  adsorption sojourn time, 42, 88, 89, 101, 108, 172, 173, 180
  elementary adsorption–desorption event, xix, 42, 90, 102, 111, 120, 165, 180
  London dispersion forces, 120
  vibrations of adsorbent lattice, 42, 161, 180
Physisorption. See Physical adsorption
Poisson distribution, computer simulation, 207
Poor-statistics data, Bayesian treatment. See also Bayesian statistics
  adsorption enthalpy from corrupted thermochromatogram, 209–211
    persisting ambiguities, 210, 211
  adsorption enthalpy from IC data, 204–208
    evaluation of survival rates, 205, 206
    formulae for survival yield, 204
    likelihood function by Monte Carlo, 207
    uncertainty of final data, 208
  adsorption enthalpy from TC experiment
    basic formulae, 208, 209
  half-life from incomplete decay curve, 202, 203
    likelihood function by Monte Carlo, 203
    sketch of flowchart, 203
Production of transactinoids, 54, 55
  actinoid targets, 54
  effective production cross section, 54, 55
  evaporation residues
    recoil energy and range in target material, 56
    straggling of recoil range, 56
  heavy ion beams (C to Ca), 54
    available intensities, 54, 55, 57
    optimal energy, 55
  simultaneous production of chemical homologs, 57

R
Random flights, 40, 100, 112, 114–116
Reaction chromatography, 180–181
  associative adsorption — dissociative desorption, 183
  atomic silver – silver chloride, 183
  dissociative adsorption — associative desorption, 181–183
  (Ce, Pu, Bk)Cl$_4$ – (Ce, Pu, Bk)Cl$_3$, 181, 182
  complexes with Al$_2$Cl$_6$, 182, 183
  Mo and W oxide-hydroxides, 182
  physical adsorption — substitutive desorption, 184–186

W and Sg oxychlorides, 184–186
  substitutive adsorption — substitutive desorption, 183–184
  (Zr, Hf, Rf)Cl$_4$ – (Zr, Hf, Rf)Cl$_4$, 183
Reference states for mobile and localized adsorption, 162
Retention time, 90, 91, 103, 105, 124, 136, 181
  in vacuum thermochromatography, 112
  measurement of, 5, 10, 12, 28, 62–64
Reynolds number (Re), 48–50
Roughness of surfaces, 141, 142
  indices of, 142
  of fused silica, experimental data, 142–146
  of metals, 158
  reduction of, by chemical etching, 158,
Rutherfordium (Rf, element 104), 6–8
  longest-lived isotopes, 55
  oxychloride and tetrachloride of, 12, 183

S
Scavenging impurities in carrier gas, 73
  deposition of nonvolatile and aerosol species. See also Diffusional deposition
    calculated graphs of deposit density and penetration (laminar flow), 75–77
    turbulent flow, formulae and data, 77, 78
  removing interfering radionuclides by hot CaO and SiO$_2$ filters, 74, 75
  removing water with SOCl$_2$ or BBr$_3$, thermodynamics and mechanism, 73, 74
Schmidt number (Sc), 48–50
Seaborgium (Sg, element 106), 8, 9,
  longest-lived isotopes, 55
  oxide hydroxides of, 182
  oxychlorides of, 9, 69, 70, 184, 204, 209
Separations of groups of related elements,
  elements of groups 7 to 10, 27, 28
  homologs of elements 112 to 117, 27–29
  fission products, 28–30
  lanthanoids and actinoids, 24, 26, 27
Sherwood number (Sh), 48–50
Single atom chemistry, validity of, 191–196.
  See also Poor-statistics data, treatment
  fluctuation of a system property with number of entities, 195
  kinetic limits for exchange of ligands, 194
  probability equivalent to law of mass action, 194, 195
  supported by Monte Carlo simulations of TC experiments, 192, 193
  verified by coprecipitation of Po from solutions, 192

Sizes of ions (atoms) in compounds, 140
  based on additive crystal radii, 140
  visualization of relative, 139, 141, 150, 152
Standard states, xxii, 133. See also Adsorption thermodynamics
Stirling's series and approximation, 195, 196
Superheavy nuclides / elements (definition), xxiii
  atomic electronic ground state, xxiii
  chemical character, xxiv
Surface diffusion on homogeneous surface, 159–162
  as two-dimensional Brownian motion, 161
  diffusion (migration) barrier, 159–162,
  distribution and mean of stochastic jumps, 160, 161
  effective diffusion coefficient, 161
  history of the problem, 159
  observation of atomic jumps, 161
  random migration picture of, 160
Surface of fused silica, bare. See also Surface of fused silica, hydroxylated; Surface of fused silica, modified
  calculated energy potential, 146–148
  calculated adsorption potential for N_2, 148
  heterogeneity (at atomic level)
    distortion of SiO_4 network, holes between the tetrahedra, 146, 147
    strained two- and three membered rings, 151, 152
Surface of fused silica, hydroxylated
  dehydratation, 150, 154
  dehydroxylation, 149–151, 153, 154, 157
  hydratation, 154
  hydroxylation, 148–150
  rehydroxylation, 153–155
  silanols, 148, 149, 151–157, 172, 176, 177, 179
    geminal, isolated, vicinal, 149
    position in nanoscale structures, 151, 152
  siloxanes, 148, 151, 154, 155, 157, 172, 177
Surface of fused silica, modified, 155
  microscopic picture of, 171, 172
  by various agents, 156, 157,
  by $SOCl_2$, 156
  by $TiCl_4$, 155, 156
Surface of metals, 157–159,
  modification by reagents, 158, 159
    nickel modified with bromine, 159
  morphology of bare, 157
    kinks, steps, terraces, 157
  roughness reduction, 158
    by ion bombardment plus annealing, 158
    by polishing, 158

Synthesis of volatile compounds on-line. See also Chlorination of gaseous tracers
  brominating agents, 21, 73, 155
  chlorinating agents, 3–6, 20, 21, 60–73, 155, 182, 183, 186
  experimental evidence for fast synthesis
    of $ZrCl_4$ from fission product Zr, 61
    of $HfCl_4$ from heavy ion produced Hf, 62–64
  in-situ volatilization, xviii, 4, 5, 16, 54, 72–74

T
Temperature-programmed chromatography,
  of fission product chlorides, 30
  of lanthanoid complexes with Al_2Cl_6, 26
  of oxides, 19
Thermal diffusivity, 78, 79
Thermalizing nuclei recoiling from target carrier gas under heavy ion beam
  concentration of ions in gas, 58, 59
  energy absorption rate, cm^3 s^{-1}, 58
  LET and range of heavy ions, 57, 58
  optimal size of target chamber, 56, 57
  range of recoiling evaporation residues, 57
Thermochromatographic columns, 15, 78, 142, 143
  equal temperature of gas and wall, 78
  temperature profile, 3, 88, 97, 111, 116, 137
  measurement of true, 78
Thermochromatography (TC), xvii, xxii. See also Reaction chromatography
  characteristics of method, 87–89
  internal chromatograms in, 87, 88, 90, 105
  net retention and gas hold-up times, 91–93, 136
  at constant column temperature gradient, 92
  at exponential temperature profile, 92
  theory of ideal, 91–93, 208
Thermodynamic and thermochemical properties of (oxy)halides of present interest
  compounds of B, 73, 175
  compounds of Ti, 69–69, 71, 74, 175
  compounds of W, 69, 70
  compounds of Zr, 68, 71, 72, 174, 175
  of $SOCl_2$ and its decomposition products, 68
Tracer (definition), xxi
Transactinoid elements. See also Production of transactinoids
  definition, xxiii

Subject Index

electronic structure of atomic ground state, xxiii
longest-lived isotopes, 55
names and symbols for, xxiv
Trouton's rule, 2, 138

U

Unknown compounds, general, xviii, 187
  (oxy)fluorides of Np to Es, 22
Uranium impurities in detectors, 6, 7

V

Vacuum thermochromatography VCT, 112–117
  non-rigorous definition, 112
  description by random flights, 114
    equivalent diffusion coefficient, 114, 115
    mean lengths and dispersion of flights, 114
  isothermal separation impossible, 112
  Knudsen regime, 112–115
    cosine law, 116
    correct computer simulation, 116
    retention time versus adsorption enthalpy, 113
    using vacuum conductance as effective convective flow, 113
    considering linear diffusion with decreasing coefficient, 115, 116
    simulation of, by Monte Carlo, 116, 117
Van't Hoff equation, 124, 169
Volatile compounds of heavy elements
  early use in radiochemistry, xvii, xviii
  complexes with Al_2Cl_6, 25–27
  halides and oxyhalides, 21, 22
  metals, 18, 24
  oxides and oxide hydroxides, 18
  sulfides, 20
  structural reasons for enhanced volatility, 2, 20, 23, 25

Z

Zone profile. *See* Chromatographic peaks, shapes